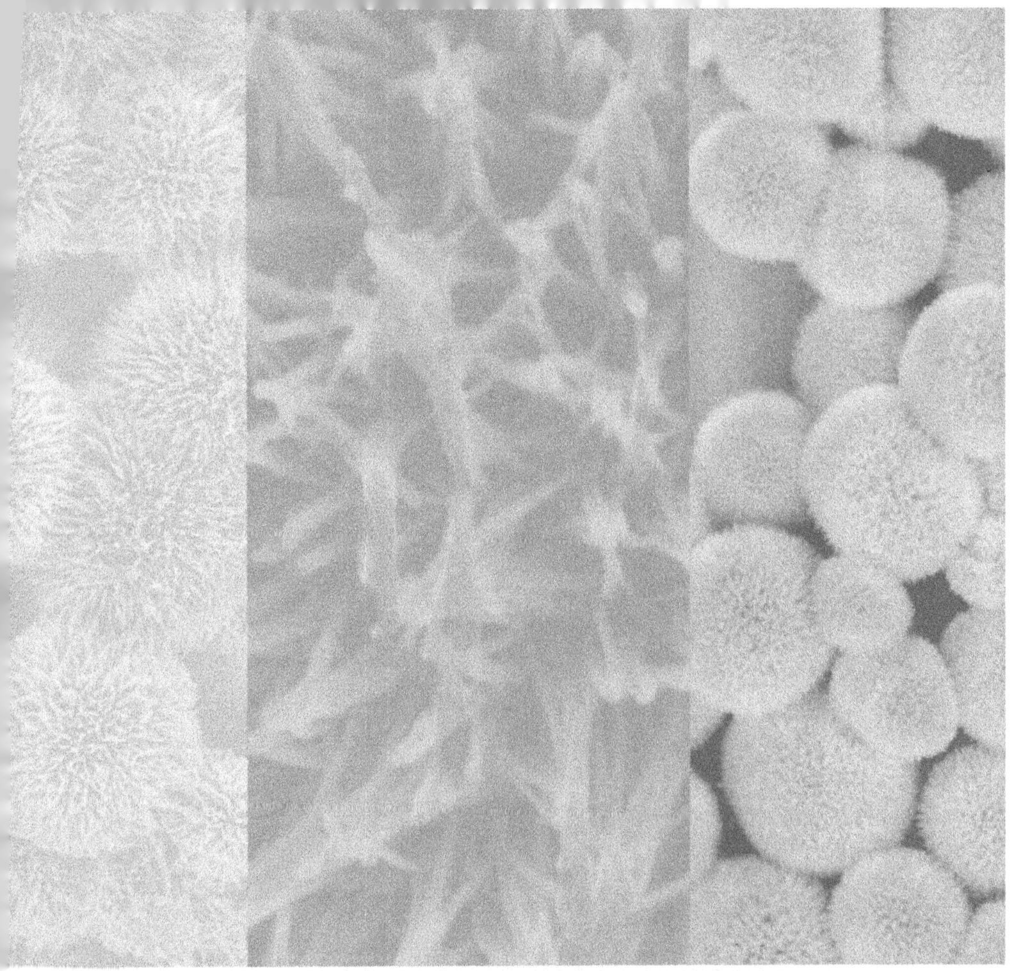

Nanomaterials for Lithium-Ion Batteries

edited by
Rachid Yazami

Nanomaterials for Lithium-Ion Batteries

Fundamentals and Applications

PAN STANFORD PUBLISHING

Published by

Pan Stanford Publishing Pte. Ltd.

Penthouse Level, Suntec Tower 3
8 Temasek Boulevard
Singapore 038988

Email: editorial@panstanford.com
Web: www.panstanford.com

British Library Cataloguing-in-Publication Data

A catalogue record for this book is available from the British Library.

Nanomaterials for Lithium-Ion Batteries: Fundamentals and Applications

ISBN 978-981-4316-40-8 (Hardcover)

ISBN 978-981-4364-23-2 (eBook)

Contents

Preface

Lithium-ion batteries were commercialized in the early 1990s by Sony Corp., Japan. Since then their impact in our mobile society has been tremendous owing to their exceptionally high energy storage performances compared with other batteries operating in ambient conditions. Mobile electronics have become a daily commodity principally because lithium-ion batteries became smaller and lighter while providing longer operation time.

The first-generation Sony lithium-ion batteries provided an energy density of about 90 Wh/kg. Today the energy figure has been more than doubled owing to optimized electrode materials, separators, current collectors, and design engineering.

Yet additional improvements are needed to meet the fast-growing electrical mobility and energy storage application market demand. It is expected that the new applications of lithium-ion batteries will boost the market value to nearly USD 50 billion in 2018, four times larger than the current value.

With this perspective, electrode materials will play a major role in the next-generation battery design. Electric cars manufacturers and clean energy storage facilities require an increase in both energy density and power density of batteries. This is where nanomaterials are expected to provide a sizeable increment in performance owing to their unique properties.

In fact, nanomaterials are potentially able to store more charge than conventional micron-sized ones because they offer additional lithium storage sites, particularly at the grain boundaries. Moreover, nanosized particles can be filled with lithium in much shorter time than micron-sized analogues, making quick charge and discharge possible, thus providing the tremendous power density necessary to accelerate a car and to store energy under peak conditions of wind and solar energy generation.

Both academia and industry have been creative in the formulation, synthesis, processing, and integration of nanostructured anode and cathode materials. Performance optimization involves smart electrode engineering, which allows a large fraction of the active nanoparticles to be electrically interconnected and to adhere well to the current collector, thus ensuring battery reliability and long cycle life.

With 11 chapters contributed by world-class scientists and engineers, this book covers most recent developments in the science and technology of nanomaterials for lithium-ion battery applications. It addresses both academia and industry research and engineering in this fast-developing field.

Rachid Yazami

California Institute of Technology, USA
Centre National de la Recherche Scientifique, France
Nanyang Technological University, Singapore

Autumn 2013

Chapter 1

Silicon Nanowire Electrodes for Lithium-Ion Battery Negative Electrodes

Candace K. Chan,[a] Matthew T. McDowell,[b] and Yi Cui[b]
[a]*Materials Science and Engineering, School for Engineering of Matter, Transport and Energy, Arizona State University, 501 E. Tyler Mall, ECG 301, Tempe, Arizona, USA*
[b]*Department of Materials Science and Engineering, Stanford University, 476 Lomita Mall, Stanford, California, USA*
yicui@stanford.edu

Silicon is an attractive material to replace graphite as the negative electrode in Li-ion batteries due to its 10 times higher theoretical specific capacity. However, the lithiation of silicon is also characterized by a large inherent increase in volume, which can lead to pulverization of the material and subsequent loss of electronic contact with the current collector. The end result is a large irreversible capacity loss in the first cycle and poor capacity retention in following cycles. Recently, there has been much interest in using nanostructured materials in Li-ion battery electrodes due to the advantages imparted by decreasing the Li-ion insertion distance, including improving the ionic and electronic transport properties of the material. Here we show that nanostructured silicon in the form of nanowires (NWs)

Nanomaterials for Lithium-Ion Batteries: Fundamentals and Applications
Edited by Rachid Yazami

ISBN 978-981-4316-40-8 (Hardcover), 978-981-4364-23-2 (eBook)
www.panstanford.com

can exploit additional advantages of nanostructured materials: facile phase transformations, accommodation of lithiation-induced strain, and the ability to undergo large volume changes without pulverization. High specific capacities >3500 mAh/g have been achieved with good capacity retention. Detailed electrochemical and structural characterizations have been performed to understand the good performance of these silicon NW-based electrodes.

1.1 Introduction

Due to the increasing power demands from portable electronics and recent push for electric cars, there has been a great deal of interest in improving the performance of rechargeable batteries. Lithium-ion (Li-ion) batteries have the highest gravimetric and volumetric energy densities and are the battery of choice for these applications.[1] However, the energy densities of current batteries are limited by the low charge-storage capacities of the materials used to comprise the electrodes.

A great deal of research has focused on finding higher specific capacity materials for both the cathode and anode of the battery. Work on improving the graphite anode, which has a theoretical capacity of 372 mAh/g, has revealed several potential candidates. Silicon is a particularly attractive material since it has a theoretical capacity of about 4000 mAh/g, more than 10 times that of graphite. The relatively low potential range at which charge and discharge of Li_xSi occurs (below 0.5 V *vs.* Li/Li^+) is attractive since this is low enough to maintain a high cell voltage while high enough to avoid Li-plating reactions during charging. Thus there has been great interest in trying to use Si as a more energy dense replacement anode for graphite. Additionally, Si is the second most abundant element on earth and already has a mature industrial infrastructure in existence.

There have been a large number of studies on the electrochemical reaction of Li with Si and subsequent cycling behavior. While at high temperatures (415°C) the equilibrium crystalline Li–Si phases are formed during lithiation of crystalline Si,[2] at room temperature a solid-state amorphization occurs to form amorphous lithium silicide (a-Li_xSi) alloys.[3–7] It was later observed that when the potential is driven <50 mV *vs.* Li/Li^+, the a-Li_xSi structure becomes crystalline again, with a composition of $Li_{15}Si_4$ ($Li_{3.75}Si$) instead of the expected $Li_{21}Si_5$.[5] It has further been found that the formation of this transient crystalline phase occurs in thin film electrodes above a critical thickness of 2.5 μm and may be absent or more difficult to detect in nanostructured electrodes.[8] Discharge (delithiation) causes the alloy to become amorphous again, forming amorphous silicon (a-Si) when fully delithiated.

An important feature observed in all of these silicon-based materials is the large inherent change in specific volume (~300%) during the insertion and extraction of large amounts of Li. This volume change causes decrepitation (or pulverization) of the material, which can cause it to delaminate from the electrode or form particles that are disconnected from the current collector (Fig. 1.1). The resulting decrease in electronically connected active material appears as a reduction in the effective capacity obtained during cycling, with as much as 85% capacity loss observed after only 5 cycles in electrodes made with 10 μm Si particles.[9]

To solve the decrepitation problem, several electrode morphologies and architectures have been explored to accommodate the strain and allow the volume changes to occur without subsequent pulverization. Electrodes composed of micron-[9] and nano-sized particles,[10,11] thin films,[12,13] submicrometer pillars,[14–16] 3-D porous structures,[17] and various composite anodes[18] have all been explored. It was soon realized that thin films of silicon could accommodate the volume change and be cycled reversibly with little capacity fade, for example, 50 nm thick Si displayed capacities of 3600 mAh/g for 200 cycles.[12] However, these structures fail when the film thickness increases, and also do not have enough active material for a viable battery. These thin film electrodes also critically rely on roughened substrates upon which the vacuum deposited Si can form a quasi-three-dimensional film with columnar-like structures,[13] which can more easily accommodate the lithiation-induced strain and allow the volume expansion compared with a truly planar film.

For electrodes made with small particles or even nanoparticles, maintaining good electronic contact to all of the active material can become rather difficult when considering the large volume changes. Upon discharge, it has been observed that the contraction of the particles caused them to lose contact with the conducting carbon.[9] Using a polymer binder such as poly(vinylidine fluoride) or sodium carboxymethylcellulose (CMC) can help to hold the particles together but may not be sufficient to accommodate the large volume change.[18] Experiments coupled with modeling found that to accommodate a volume change of 270%, a composite of only 20 vol% Si mixed with binder and carbon is needed to minimize particle movement and avoid extrusion of the material from the electrode.[20] However, such conductive additives and binders add weight to the electrode, thus lowering the overall gravimetric and volumetric capacities of the battery and effectively reducing the gains obtained by using the high specific capacity of Si.

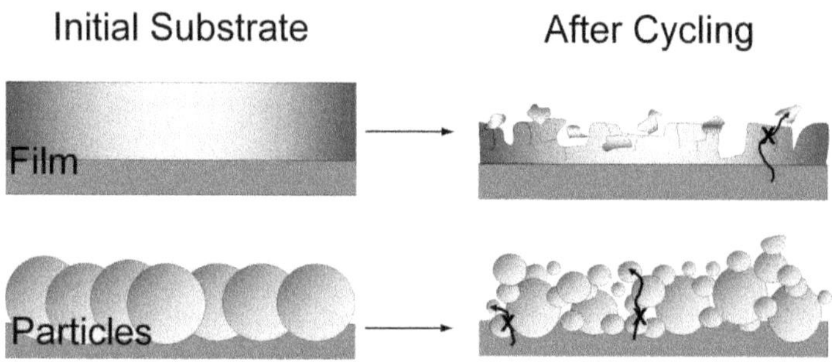

Figure 1.1 Cartoon depiction of morphological changes that occur in Si during electrochemical cycling. The large volume change of ~300% has made it difficult for Si anodes to withstand many cycles. An initial Si film pulverizes due to mechanical failure from the large volume expansion during cycling. Much of the material has lost contact with the current collector, giving poor transport of electrons as indicated by the arrow. A sample of Si particles behaves similarly. Micron-sized particles break apart into smaller particles after cycling, and there is loss of good electronic contact throughout the particle composite. Reprinted with permission from Ref. 19. Copyright 2008, Nature Publishing Group.

Observations on metal hydride materials that undergo similar large volume changes have shown that the pulverization process does not continue indefinitely.[21] Instead, there is a characteristic terminal particle size for each particular material, and particles below this size do not fracture further. Experiments on Li alloy electrodes have shown that electrochemical cycling is significantly improved if the initial particle size is very small,[22] consistent with the terminal particle size phenomenon. With all of these problems in mind, a Si nanowire (SiNW)-based electrode was envisioned. While NW and nanorod materials have been used in Li-battery applications, the advantages imparted have mostly been due to increased interfacial contact with the electrolyte and shorter Li^+ insertion distances, which can result in higher capacities and rate capabilities in insertion compounds such as $LiFePO_4$,[23,24] $LiCoO_2$,[25] $LiMn_2O_4$,[26] and V_2O_5,[27,28] as well as conversion compounds Co_3O_4[29,30] and SnO_2.[31] In the Si case here, the NW geometry is used to exploit the nanometer size in the diameter to avoid decrepitation while simultaneously ensuring good electronic contact with the current collector.

A schematic of the SiNW anode configuration is shown in Fig. 1.2, where the nanowires (NWs) are grown directly on the metallic current collector. This geometry has several advantages. First, the small NW diameter can allow for better accommodation of the large volume changes without the initiation of fracture. Second, each NW is electrically connected to the current collector such that all of the active material (all NWs) can participate in the charge storage and contribute to the capacity. Third, the SiNWs have direct 1-D electronic pathways allowing for efficient charge transport. In electrode microstructures based on particles, electronic charge carriers must move through small interparticle contact areas. In addition, as every NW is connected to the current collector, the need for binders or conducting additives is eliminated.

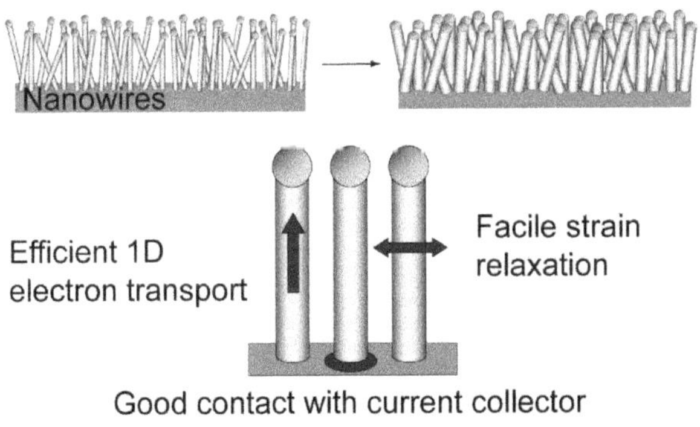

Figure 1.2 NWs grown directly on the current collector have many advantages over films and particles. After cycling, the NWs do not pulverize or break into smaller particles. Rather, facile strain relaxation in the NWs allows them to increase in volume without breaking. This NW anode design has each NW connecting the current collector, allowing for efficient 1-D electron transport down the length of every NW. Reprinted with permission from Ref. 19. Copyright 2008, Nature Publishing Group.

1.2 Fabrication of Silicon Nanowire Electrodes

SiNWs were first synthesized and described (then known as "whiskers") by Wagner and Ellis in 1964[32] and later re-popularized by Lieber *et al.*[33–35] in the late 1990s–early 2000s due to immense interest in the fields of nanomaterials and nanoelectronics. The growth method, coined vapor–liquid–solid (VLS), is a chemical vapor deposition (CVD)-based process performed in a tube furnace (Fig. 1.3a) that exploits the eutectic region in a binary phase diagram (Fig. 1.3b). One of the elements functions as a nanoparticle catalyst for nucleating a NW while the other element is supplied in a vapor form and makes up the bulk of the NW. After the growth, each NW has a catalyst particle on its tip. For Si, a common catalyst is gold (Au). The Au catalyst is heated (II in Fig. 1.3b) above the eutectic temperature (363°C) and SiH_4 is introduced and allowed to decompose into Si. The Si dissolves in the Au particle and forms a eutectic droplet (III in Fig. 1.3b). Upon reaching supersaturation (IV in Fig. 1.3b), nucleation of a single crystalline SiNW occurs. Continuous supply of Si vapor results in increasing growth along the length of the NW (V in Fig. 1.3b). The name VLS comes about because the *solid* crystal grows from the *vapor* source out of a *liquid* eutectic droplet.

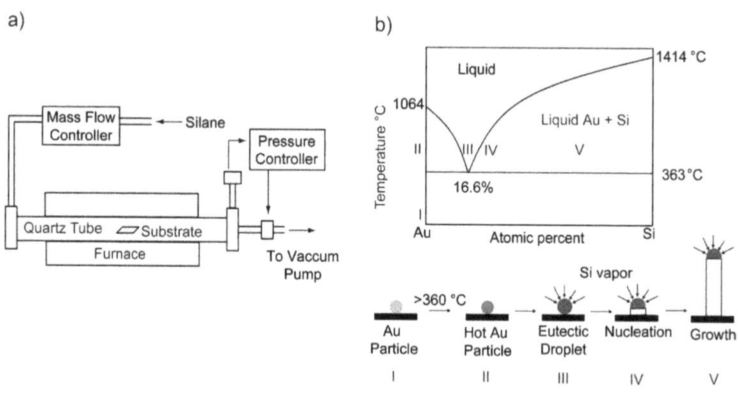

Figure 1.3 (a) Apparatus for synthesizing SiNWs. (b) SiNW growth process and corresponding regions on the Si-Au binary phase diagram. Reprinted with permission from Ref. 36. Copyright 2009, Elsevier.

Other catalysts[37] have been demonstrated for VLS-based growth of SiNWs, including Al,[38] Ni,[39] Cu,[40,41] Pd,[42] Sn,[43] and In.[44] Diameter control can be obtained from the diameter distribution of the catalyst particles.[45] Epitaxial growth on single crystalline Si substrates can also allow for orientation-directed growth and vertical alignment.[40,45] Thus far, SiNWs have found applications as chemical and biosensors,[46] electronic devices such as field-effect transistors,[46] diodes,[46] bipolar transistors,[46] inverters[46] and logic gates,[47] waveguides,[48] solar cell absorbers,[49–52] antireflection materials,[53] photoelectrochemical materials,[54–56] and thermoelectric materials.[57,58]

For electrochemical applications, it is desirable that the NW substrate be metallic in order to efficiently transport electrons with minimum ohmic losses. The substrate should also be chemically and electrochemically inert to Li. For these reasons, using Si wafers as the substrate for SiNW growth is not desirable since the Si wafer could also participate in the electrochemical reactions.[59] However, there have been some recent improvements in developing processes to isolate or detach the SiNWs from the wafer.[60,61] In Li-ion batteries, the negative electrode is typically coated onto a copper foil current collector. Several attempts at growing SiNWs on Cu foil using VLS were largely unsuccessful due to (1) the high mobility of the Au particles inside the Cu, and (2) the high reactivity of SiH_4 with the Cu to form copper silicides. Attempts at using Ni foil as a growth substrate yielded similarly poor results, again most likely due to silicide growth. However, stainless steel (SS) was observed to be a suitable substrate for SiNW growth, perhaps because the native oxide layer made the iron less reactive to the SiH_4, although some α-$FeSi_2$ was also formed [as confirmed by X-ray diffraction (XRD), Fig. 1.13a] and made up about 25% of the total mass deposited during the growth. Other groups have also successfully grown SiNWs on SS substrates coated with a conductive diffusion barrier such as TiN[62] or Ta_2N.[49] Subsequent studies have shown that the amount of silicide formed during SiNW growth on SS 304 can be decreased to as low as ~0.6 wt% by tuning the reaction conditions, in particular by utilizing a low growth temperature, which can improve the specific capacity as well as rate capabilities of the electrode.[63]

Typical scanning electron microscope (SEM) images of VLS-grown SiNWs on SS are shown in Fig. 1.4a,b. The SiNWs had an average diameter of about 100 nm and lengths up to several tens of microns. The cross-section SEM image (Fig. 1.4b) showed that the SiNWs grew vertically off of the substrate, although with a random orientation. While highly oriented arrays of SiNWs can be easily grown,[55] these methods rely on epitaxial growth of the NWs on Si wafers.

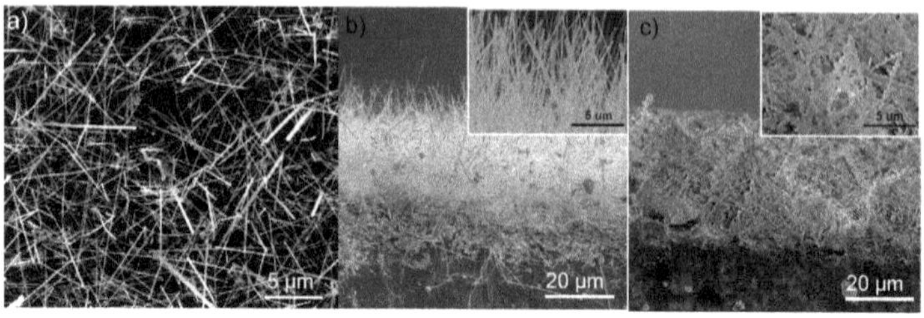

Figure 1.4 SEM images of SiNWs grown on SS, (a) top-down view, (b) side-view as grown, (c) SEM image of side-view of SiNW electrode after assembly into coffee bag cell. Reprinted with permission from Ref. 36 (Copyright 2009, Elsevier) and Ref. 64 (Copyright 2009, American Chemical Society).

The SiNW electrode is in mechanical contact with the separator soaked with the electrolyte (Fig. 1.5a). When these electrodes are assembled into "coffee bag" pouch cells (Fig. 1.5b), a slight amount of pressure is applied to cell, which causes the NWs to lie down mostly parallel to the substrate. As shown in Fig. 1.4c, which was obtained by disassembling the cell after a measurement, the electrode appears as a dense mass of wires packed together between the separator and the substrate.

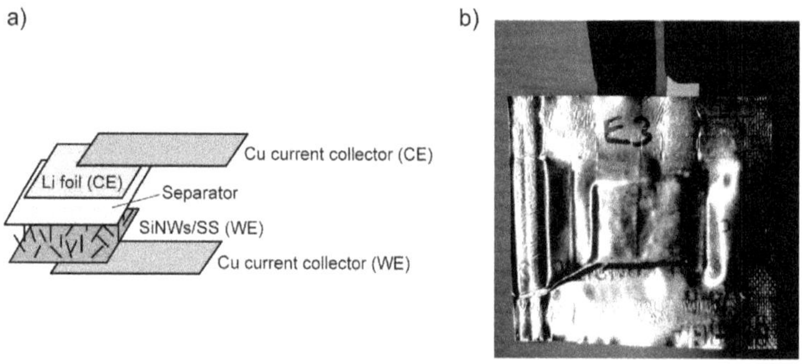

Figure 1.5 (a) Schematic of layers that comprise a two-electrode coffee bag cell. (b) Photograph of coffee bag cell (Ref. 65).

1.3 Electrochemical Measurements

1.3.1 Cyclic Voltammetry

A cyclic voltammogram (CV) of a SiNW electrode is shown in Fig. 1.6. The charge current associated with the formation of the Li–Si alloy started at ~330 mV and became quite large below 100 mV. Upon discharge, current peaks appeared at about 370 and 620 mV. The current–potential characteristics are consistent with previous experiments on microstructured Si anodes.[14] The magnitude of these current peaks increases with cycling due to the reaction of more Si with each scan. Because the scan rate used in the CV measurement is quite fast (a half cycle is acquired in 2000 seconds or 33 minutes, equivalent to a 1.8C rate), this suggests that there are kinetic limitations to the lithiation of the SiNWs. By using electrochemical potential spectroscopy (EPS) to obtain the current–voltage curve (or more accurately, the differential charge *vs.* voltage) at quasi-equilibrium conditions, one can avoid this effect (see Section 1.3.2). The small Li-extraction peak observed at 150–180 mV in the CV curve was due to reaction of the Li with the Au[66] catalyst particles at the ends of the NWs. Control experiments showed that the capacity contribution from Au was negligible in our SiNW electrodes.[65]

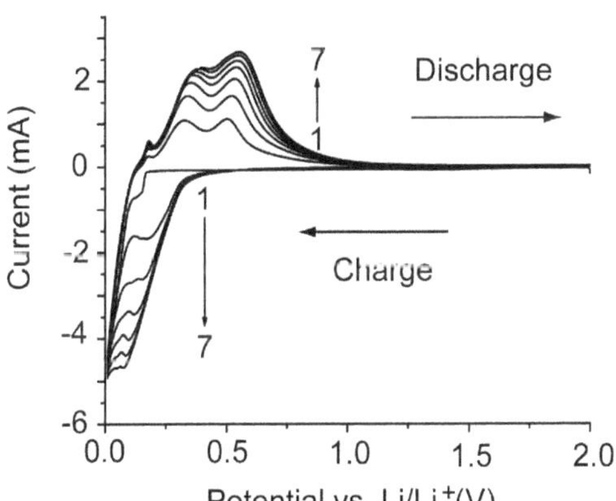

Figure 1.6 CV for SiNWs from 2.0 V to 0.01 V *vs.* Li/Li^+ at 1 mV/s scan rate. The first seven cycles are shown. Reprinted with permission from Ref. 19. Copyright 2008, Nature Publishing Group.

1.3.2 Electrochemical Potential Spectroscopy

EPS[67] was performed to study the differential capacity (charge) as a function of potential in the SiNWs and to obtain information about the structural transformations during lithiation and delithation.[36] The dC/dV *vs.* voltage and corresponding voltage *vs.* capacity curves are shown in Fig. 1.7. During the charge, a small peak is observed at 0.62 V and a larger peak at 0.125 V *vs.* Li/Li^+ (A and B, respectively, in Fig. 1.7b). The peak at 0.62 V has been reported to be due to solid electrolyte interphase (SEI) formation on the Si surface[11] whereas the peak at 0.125 V is a two-phase region where crystalline Si (c-Si) reacts with Li to form an a-Li_xSi.[6] This two-phase region is seen as a long sloping plateau indicated by B in Fig. 1.7a. The corresponding plateau for the SEI formation in the voltage profile (A in Fig. 1.7a) is quite small, although surface studies conducted on the SiNWs (see Section 1.5) found that SEI formation also occurs at potentials lower than that found at 0.62 V (point A), in parallel with lithiation of the Si. In this scenario, EPS would be unable to separate out the current from SEI formation from that associated with lithiation of the Si.

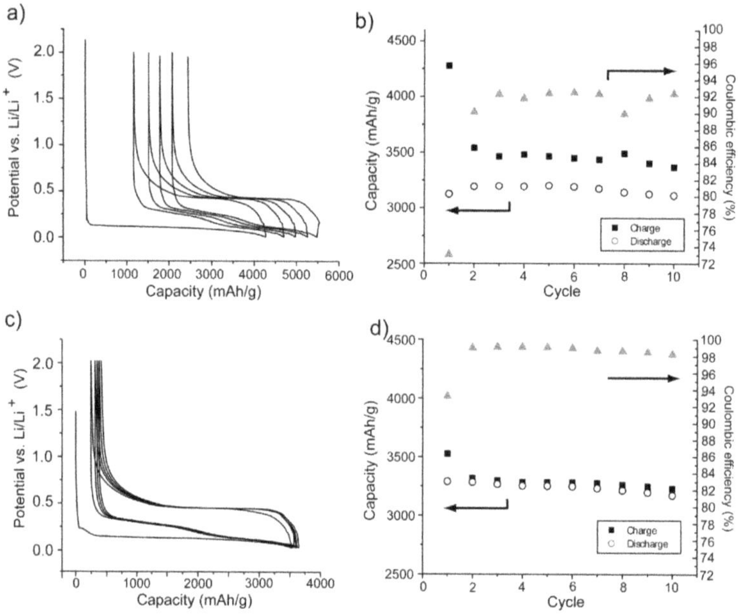

Figure 1.7 (a) Voltage profile and (b) differential capacity curve of the first cycle, starting from crystalline SiNWs and charging to 10 mV. After the first cycle, the SiNWs turned amorphous, resulting in a different (c) voltage profile and (d) differential capacity curve. Reprinted with permission from Ref. 36. Copyright 2009, Elsevier.

It has been reported[5] that charging Si to <50 mV results in the formation of a crystalline $Li_{15}Si_4$ phase, which is not predicted by the equilibrium phase diagram. This can be observed as a small peak in the differential capacity curve <50 mV. During discharging, Li is removed from this crystalline phase to form a-Si in a two-phase region. This can be observed as a plateau in the voltage profile and a peak in the differential capacity curve. The EPS data of the SiNWs clearly show a transformation that leads to a two-phase region. During the discharging, two peaks (D and E) are observed in the differential capacity curve (Fig. 1.7b), which correspond to plateaus in Fig. 1.7a. The two-phase region at E looks similar to the reported[6] delithiation of crystalline $Li_{15}Si_4$ to form a-Si. Later in this chapter, *in situ* structural characterization will be presented that confirms the formation of $Li_{15}Si_4$ at low potentials. Fig. 1.7c,d show the voltage and dC/dV curves after the crystalline Si has been converted to a-Si following the first cycle. During the charge, lithiation into the a-Si occurs in two sloping single-phase regions, as indicated by B′ and B″. The lithiation of a-Si in two different voltage regions has been reported previously.[6] The first plateau was found to be characteristic of a local environment where the inserted Li is surrounded by Si nearest neighbors, whereas in the second plateau, the Li is added to sites where the nearest neighbors are mostly other Li atoms.[68] The transition between these two environments was found to occur at $Li_{2.3}Si$, which is in agreement with the data obtained in the SiNWs.

The features indicated by D were due to alloying of the Au catalyst with the Li. Based on the EPS data, the SiNWs appeared to undergo similar phase transformations as observed in previous studies on micron-sized crystalline Si powders.[6,7] EPS can give valuable information about phase changes, but it is also important to support the data with actual structural studies, which will be discussed in Section 1.4.

1.3.3 Galvanostatic Cycling

1.3.3.1 Capacity

Galvanostatic cycling with potential limitation (10 mV and 2 V *vs.* Li/Li^+ in the charge and discharge, respectively) was performed on the SiNW electrodes. The first cycling data was acquired using a beaker cell flooded with an excess (10 mL or more) of electrolyte. These results,[19] conducted using a C/20 rate in 1 M $LiPF_6$/EC:DEC electrolyte, displayed a coulombic efficiency of 73% in the first cycle and about 92% in subsequent cycles, with reversible capacities of

~3200 mAh/g (Fig. 1.8a,b). Electrochemical cycling of SiNWs inside a coffee bag cell, where the amount of electrolyte was about one drop (about 50 μL), showed much lower irreversible capacities in the first cycle (93%) as well as in subsequent cycles (98%), although the reversible capacities were not affected (Fig. 1.8c,d). Later studies found that the irreversible capacity loss in the first cycle, as well as the poor coulombic efficiency between charge and discharge, were improved at faster rates (Section 1.3.3.2). This improvement may be due to less SEI formation because less time is spent at the low potentials where these reactions occur.

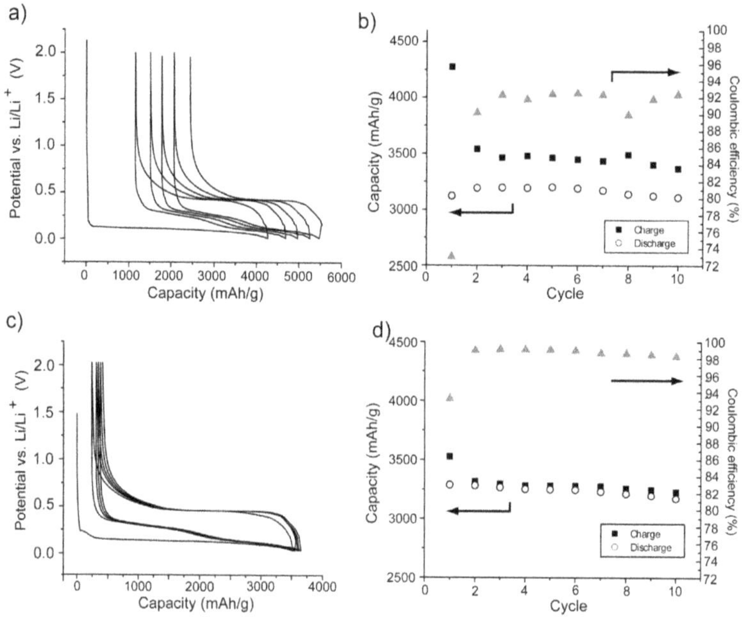

Figure 1.8 (a) Voltage profile and (b) galvanostatic cycling results for SiNWs in beaker cell; (c) voltage profile and (d) galvanostatic cycling results for SiNWs in coffee bag cell using C/20 rate in 1 M $LiPF_6$/EC:DEC electrolyte (Ref. 65).

These observations suggest that minimizing side reactions with the electrolyte are very important for good electrochemical performance of the SiNWs, perhaps even more so because the high surface area of the NWs can increase the reactivity with the electrolyte. Because coulombic efficiency is very important for determining long-term capacity retention, it is necessary to avoid large irreversible capacity losses. Further studies on finding an optimum electrolyte that minimizes side reactions or the use of surface passivation treatments, such that as much of the charge as possible is utilized for the charge storage reactions and not other processes, may be beneficial. A recent report[69] indicates that silane chemistry to modify the surface of SiNWs has promise for achieving surface passivation.

Electrochemical measurements on the SiNW anode showed that not only was this electrode architecture robust over many cycles, but much higher reversible charge storage capacities were observed than for other types of Si electrodes. Figure 1.9 shows the performance of a VLS SiNW anode compared with a planar SiNW anode [a composite film of SiNWs mixed with multi-walled carbon nanotube (MWNT) additive and binder, to be discussed further in Section 1.7.2.1],[70] a film of 12 nm Si nanocrystals (SiNCs, data from Ref. 11), a film of 10 μm Si particles (Bulk Si, from Ref. 9), and the theoretical capacities for both Si and graphite. The capacity for the three-dimensional SiNW anode almost reaches the theoretical capacity for Si and is 10 times higher than that for graphite. It is also clear that the electrode architecture is very effective for obtaining this capacity, as it allows for good electronic contact to all of the active material, enough space for the volume change to occur, and good interfacial contact area with the electrolyte.

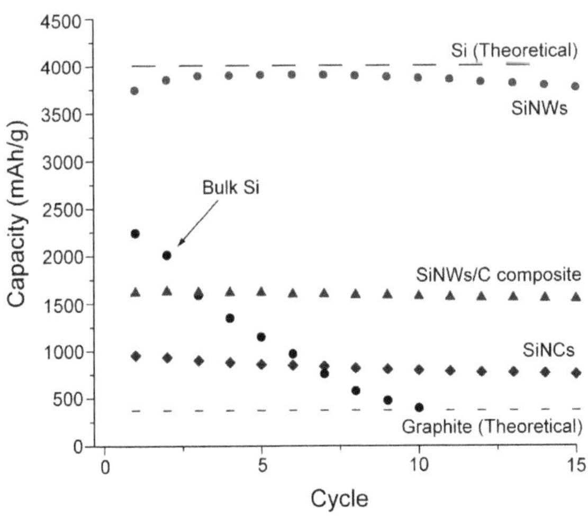

Figure 1.9 Summary of cycling performance for Si electrodes with different morphologies (Ref. 65).

Galvanostatic measurements with capacity limitation were also performed. This technique is useful for determining how long the electrode can be cycled at a certain capacity before failing. The SiNWs were tested using a 2.5 hour charge at the C/5 rate (corresponding to a capacity of 2100 mAh/g) while the discharge was performed using potential limitation at 0.7 V *vs.* Li/Li^+. As shown in Fig. 1.10, the cycling performance was stable until about 40 cycles, where the coulombic efficiency drops and degradation starts to occur. Limiting the charge capacity to 1500 mAh/g and 1050 mAh/g showed much better performance, with no degradation until about 320 cycles for the lower capacity. This shows that better cyclability is observed when the SiNWs are not fully lithiated and also when the charge potential limits remain higher. This may be due to considerations related to the structural integrity of the SiNWs and SEI layer thickness, which will be discussed in later sections.

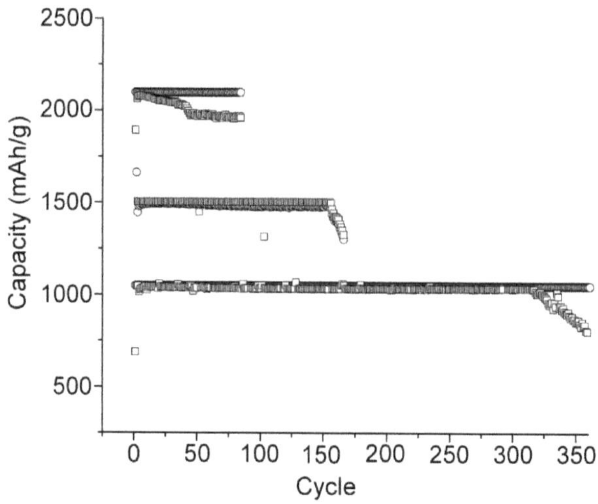

Figure 1.10 Cycling data of SiNWs performed by limiting the charge to 2100, 1500, or 1050 mAh/g and limiting the discharge to 0.7 V *vs.* Li/Li^+ (Ref. 65).

1.3.3.2 Power

Si anodes have the advantage of having higher specific capacities than graphite. However, the Li diffusion coefficients for Li_xSi (10^{-11}–10^{-14} cm^2/sec)[71–73] are lower than those for Li_xC (10^{-9}–10^{-11} $cm^2/$

sec),[74] indicating that Si anodes may not perform as well at higher currents. Nanostructured electrodes are ideal for addressing this problem, since the distances through which Li has to diffuse are in the nanometer range and there is a large interfacial contact area with the electrolyte for easy access to Li-ions.

The capacities observed in the SiNWs when using the C/20 rate and C/5 rate on different samples were similar (the same rate was used for all cycles starting from the first cycle). The SiNWs were also evaluated at increasingly faster rates, that is, first using the C/20 rate, then increasing the current for subsequent cycles (using the same sample). Figure 1.11 shows the cycling data obtained by following this protocol from C/20 to 1C using a beaker cell. As indicated previously, the coulombic efficiency was observed to increase with the higher currents, likely due to less time spent at potentials where the SEI can form. The capacity was observed to drop with increasing current, suggesting kinetic limitations to the Li diffusion in Si. However, even at the 1C rate, the capacities remained >2100 mAh/g. As a comparison, electrodes made of amorphous macroporous Si in an "inverse opal" structure[75] had initial capacities of 2500 mAh/g using the C/10 rate, which faded to about 1500 mAh/g for C/5, 500 mAh/g for C/2, and no capacities for rates 1C and higher. The authors attributed the poor cycling performance at higher currents to the poor electronic conductivity of a-Si. This illustrates the superior electronic transport properties of the SiNW electrode, since each NW is still connected to the current collector even after becoming amorphous.

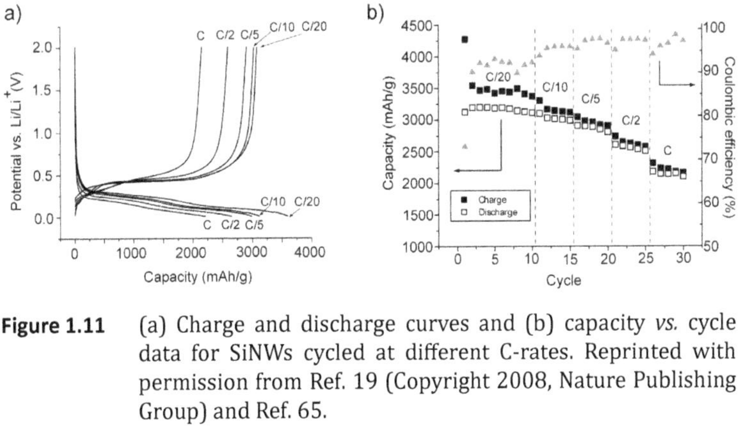

Figure 1.11 (a) Charge and discharge curves and (b) capacity *vs.* cycle data for SiNWs cycled at different C-rates. Reprinted with permission from Ref. 19 (Copyright 2008, Nature Publishing Group) and Ref. 65.

1.4 Structural Characterization

1.4.1 *Ex situ* Studies on Structural Evolution during Cycling

The as-grown SiNWs have an average diameter of 89 nm with a standard deviation of 45 nm, as obtained from SEM observation (Fig. 1.12c). After fully lithiating by charging to 10 mV *vs.* Li/Li^+, the SiNWs had roughly textured sidewalls (Fig. 1.12b) and the average diameter increased to ~141 nm (standard deviation 64 nm) (Fig. 1.12c). An important observation is that despite the large volume change, the SiNWs remained intact and did not pulverize into smaller particles. They also appeared to remain in contact with the current collector so that capacity fade due to electrically disconnected material during galvanostatic cycling is supposed to be minimal.

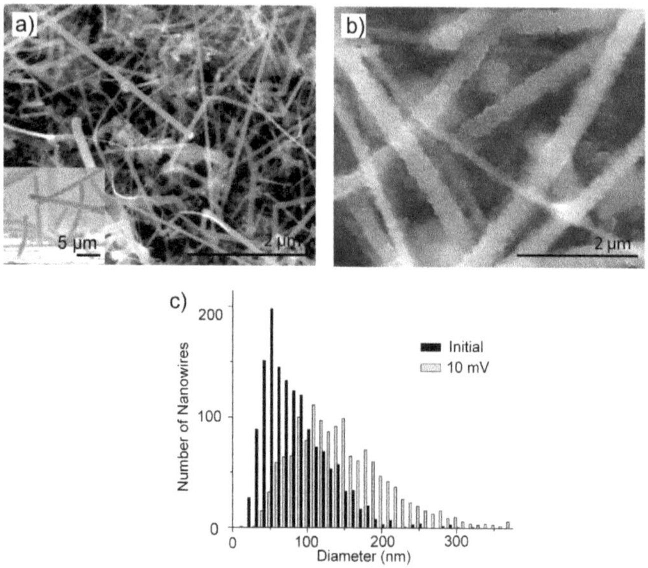

Figure 1.12 (a) SEM image of pristine SiNWs before electrochemical cycling. The inset is a cross-section image showing the NWs are directly contacting the SS current collector. (b) SEM images of SiNWs after Li cycling. The magnification is the same as in (a), showing the increase in size. (c) Size distribution of NWs before and after charge to 10 mV (bin size 10 nm). The average diameter of the NWs increased from 89 nm to 141 nm after lithiation. Reprinted with permission from Ref. 19. Copyright 2008, Nature Publishing Group.

To understand the structural evolution of the SiNWs during lithiation, the NW electrodes were characterized with XRD at different charge potentials. The XRD series is shown in Fig. 1.13a with scans for the initial pristine SiNWs, SiNWs charged to 150 mV, 100 mV, 50 mV, and 10 mV, as well as a scan performed after five charge–discharge cycles. Great care was taken to limit exposure to air in order to preserve the compositional state of the NWs and avoid oxidation. XRD of the as-grown SiNWs had diffracting peaks for Si, α-$FeSi_2$, Au (from the catalyst), and SS. The α-$FeSi_2$ forms at the interface between the SS substrate and the Si wires during the SiNW growth process (see Section 1.2). As the SiNWs are charged to lower potentials, the Si peaks decrease in intensity and eventually disappear, consistent with the disappearance of the crystalline Si and formation of the a-Li_xSi. Notably, at 10 mV, there are no peaks corresponding to the crystalline $Li_{15}Si_4$ phase. An XRD scan taken of a sample discharged to 0.2 V also clearly showed the absence of any crystalline phase (Fig. 1.13b). Using these *ex situ* techniques, the crystalline $Li_{15}Si_4$ phase was not detected; later, we will describe *in situ* experiments that give different results.

The SiNWs were also characterized with *ex situ* transmission electron microscopy (TEM) and selected-area electron diffraction (SAED) at different charge potentials to investigate the local structural features of the SiNWs during the first Li insertion. Figure 1.14a shows an example of a typical as-grown SiNW. Figure 1.14b shows a SiNW that has been charged to 100 mV. Both crystalline and amorphous phases are clearly seen. The distribution of the two phases was observed both across the diameter (a crystalline core and amorphous shell) and along the length. The high-resolution TEM image (HRTEM, bottom) enlarges the interface between the two phases. The SAED shows the spot pattern for the crystalline phase, but weak diffuse rings from the amorphous phase can also be observed. The crystalline phase is still elemental Si while the amorphous phase is a Li_xSi alloy. Li ions must diffuse radially into the NW from the electrolyte, resulting in the core-shell phase distribution. At 50 mV, the SiNW has become mostly amorphous but there are still some crystalline regions imbedded inside the core (Fig. 1.14c). The dark field image shows the crystalline grains as bright regions inside the amorphous wire. The SAED shows spotty rings representative of a polycrystalline sample and diffuse rings for the amorphous phase. The HRTEM image shows several isolated pristine Si grains imbedded in the amorphous phase. At 10 mV

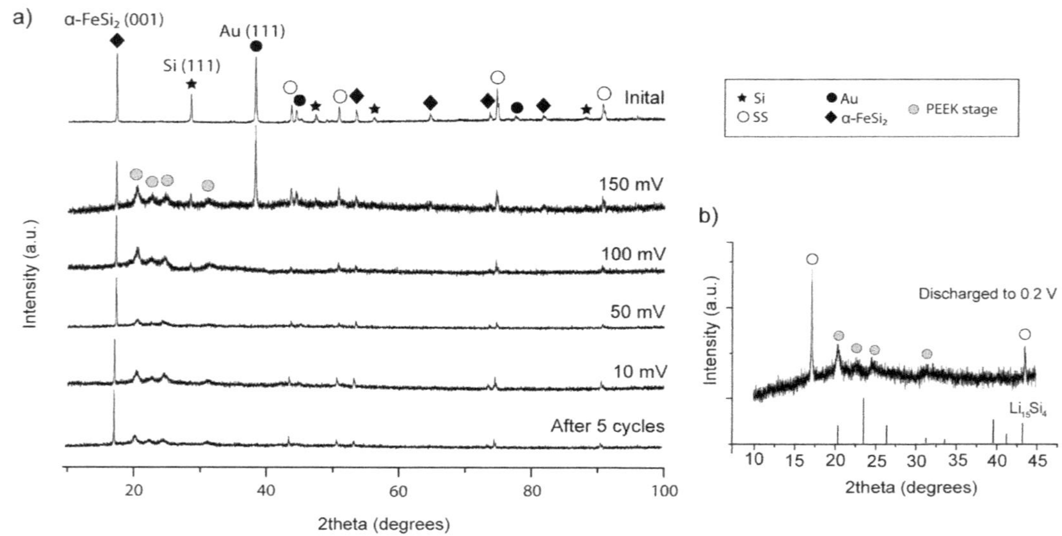

Figure 1.13 (a) XRD of SiNWs before electrochemical cycling (initial), at different potentials (150, 100, 50, and 10 mV) during the first charge and after 5 cycles. The peaks indicated by grey circles are due to the polyaryletheretherketone (PEEK) comprising the enclosure on the sample stage used to prevent oxygen contamination. The crystalline peaks for Si and Au disappear during alloying with Li beginning at 150 mV and 100 mV, respectively, indicating lithiation. The scan taken after five cycles shows that the NWs remain amorphous. Reproduced with permission from Ref. 19. Copyright 2008, Nature Publishing Group. (b) XRD data of SiNWs discharged to 2.0 V. The $Li_{15}Si_4$ peaks, which are overlaid as reference, do not appear in the scans. Reproduced with permission from Ref. 36. Copyright 2009, Elsevier.

(Fig. 1.14d), this *ex situ* experiment indicates that the entire NW has transformed to the amorphous lithiated phase, as indicated by the diffuse rings in the SAED and the lack of contrast in the NW in the TEM image. These TEM observations are consistent with the *ex situ* XRD results (Fig. 1.13) and EPS measurements (Section 1.3.2).

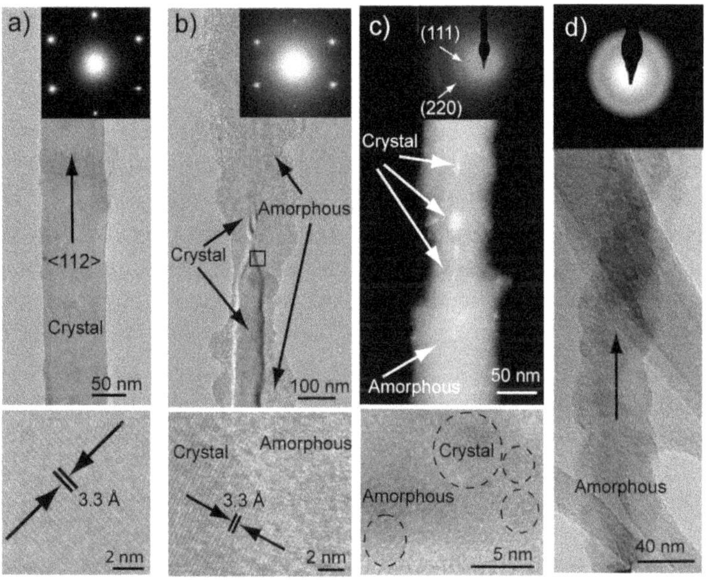

Figure 1.14 *Ex situ* TEM data for SiNWs at different stages of the first charge. (a) A single crystalline, pristine SiNW before electrochemical cycling. The spots in the SAED and the lattice fringes in the HRTEM are from the 1/3(224) planes. (b) A NW charged to 100 mV showing a Si crystalline core and the beginning of a-Li_xSi amorphous shell being formed. The spots in the SAED and lattice fringes in the HRTEM correspond to the 1/3(224) planes in the crystalline core. (c) A dark field image of a NW charged to 50 mV showing an a-Li_xSi NW with crystalline Si grains in the core. The crystalline material appears as bright regions. The spotty rings in the SAED can be indexed to crystalline Si. The HRTEM shows some Si crystal grains imbedded in the amorphous NW. (d) A NW charged to 10 mV showing a completely a-Li_xSi NW. The SAED shows diffuse rings characteristic of an amorphous material. Reprinted with permission from Ref. 19. Copyright 2008, Nature Publishing Group.

According to this *ex situ* XRD and TEM data, the fully lithiated SiNWs always appear amorphous. This contrasts, however, with other studies on Si electrodes,[6,8] which have reported the formation of the crystalline $Li_{15}Si_4$ phase at potentials between 30 mV and 60 mV *vs.* Li. In the next sections, we will show that this phase does form in highly lithiated SiNWs, but it can be detected more readily by *in situ* techniques.

1.4.2 *In situ* Studies on Structural Evolution during Cycling

The aforementioned structural characterization was performed using *ex situ* experiments, with the active material being removed from cells in an Ar atmosphere and then transferred to the designated instrument for analysis. However, because of the reactivity of the lithiated Si phase and the dynamic nature of the electrochemical reactions, *in situ* experiments might provide more detailed observations of the structural changes during lithiation/delithiation. Recently, *in situ* XRD cells were used to monitor the phase transformations in SiNWs during lithiation/delithiation with synchrotron X-ray radiation.[76] The *in situ* XRD electrochemical cells were fabricated by sandwiching SS mesh on which NWs had been grown, polymer separators, and metallic Li foil in transparent polyester pouches. X-ray measurements were performed at the Stanford Synchrotron Radiation Lightsource (SSRL), and the data were gathered in transmission mode (the X-ray beam was transmitted through the entire electrochemical cell). The SS mesh acted as a current collector and it also allowed for sufficient X-ray transmission. The *in situ* studies revealed interesting details about the reaction process. Similarly to the *ex situ* experiments, the SiNWs were observed to amorphize during the lithiation process as the crystalline Si disappeared. However, late in the lithiation process at potentials lower than ~20 mV *vs.* Li/Li^+, the $Li_{15}Si_4$ phase was observed to nucleate. Figure 1.15 shows the peak intensity for the Si (111) and the $Li_{15}Si_4$ (233) peaks during the first lithiation and delithiation. The (111) Si peak gradually decreases in intensity during lithiation, and the $Li_{15}Si_4$ peak appears rather suddenly at the end of lithiation. The $Li_{15}Si_4$ peak intensity is lower than that for the Si peak. Upon delithiation, the $Li_{15}Si_4$ phase disappears soon after reversing the current, leaving only the a-Si phase after full delithiation. This type of phase evolution is consistent with studies on Si anodes with other morphologies.[8]

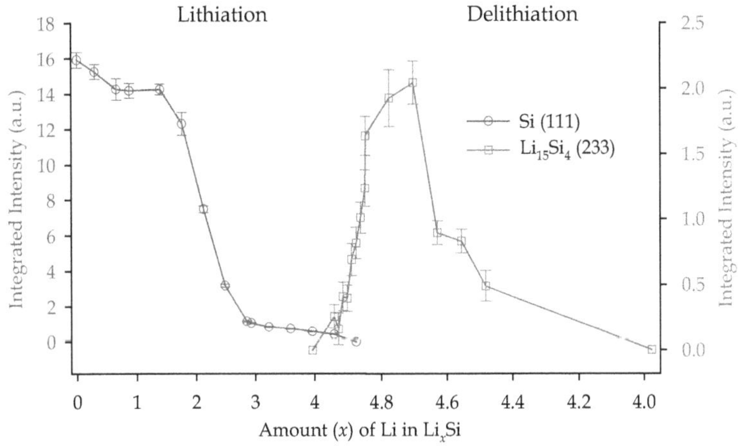

Figure 1.15 *In situ* XRD data showing the peak intensity for the Si (111) peak and the $Li_{15}Si_4$ (233) peak during lithiation and delithiation. The Si (111) peak intensity decreases during lithiation, and at the very end of lithiation, the $Li_{15}Si_4$ peak appears. Upon delithiation, the $Li_{15}Si_4$ peak disappears quickly (the *x*-axis scale is different during lithiation and delithiation). Reprinted with permission from Ref. 76. Copyright 2012, American Chemical Society.

These *in situ* XRD results illustrate the importance of monitoring the electrochemical reaction during battery operation. The differences between the *ex situ* and *in situ* experiments are probably a result of the reactivity of the lithiated Si phase and the different X-ray sources used (high-intensity synchrotron radiation for the *in situ* test *vs.* lower-intensity Cu-kα radiation for the *ex situ* test). In the *ex situ* experiments, it is possible that slight oxidation of the lithiated Si might occur during transfer, but care was taken to minimize air exposure time. Instead, it has been shown that the $Li_{15}Si_4$ phase is reactive toward the electrolyte,[77] which suggests that the $Li_{15}Si_4$ phase could disappear during a side reaction with the electrolyte when the current is turned off. This effect might be mitigated by employing protective coatings, which is an important area for future study.

In addition to *in situ* XRD, powerful *in situ* TEM techniques for monitoring electrochemical reactions in real time have recently been developed.[78,79] These experiments involve the use of a specialized dual-probe electrical biasing TEM holder on which a nanoscale electrochemical cell is fabricated, as shown in the schematic in

Fig. 1.16. In this schematic, a SiNW is connected to the Au probe at the bottom of the image, which acts as the current collector for the working electrode. $LiCoO_2$ powder is attached the other probe and serves as a counter electrode. An ionic liquid electrolyte comprised of 10 wt% lithium bis(trifluoromethylsulfonyl) imide (LiTFSI) in a solvent of 1-butyl-1-methylpyrrolidinium bis(trifluoromethylsulfonyl) imide (P_{14}TFSI) is coated onto the counter electrode. The ionic liquid has a negligible vapor pressure and does not evaporate in the high vacuum of the TEM column, thus enabling these experiments. After the sample is within the TEM, the NW electrode can be positioned so that the tip of the NW is submerged in the ionic liquid. By applying a bias of approximately −4 V to the NW electrode, the cell operates in charging mode, where $LiCoO_2$ is oxidized and Li is reduced at the working electrode, causing Li to diffuse into the NW.[80] The lithiation/delithiation reactions can be observed in the section of the NW that is outside of the electrolyte, since the ionic liquid is not transparent in the electron beam. Because the TEM column is under vacuum, environmental oxidation of the lithiated NW is not a concern. There have been a few *in situ* TEM studies that have focused on the lithiation of SiNWs.[79–81] Note that in these *in situ* TEM experiments, Li-ions enter the SiNW at the tip and must diffuse down the length of

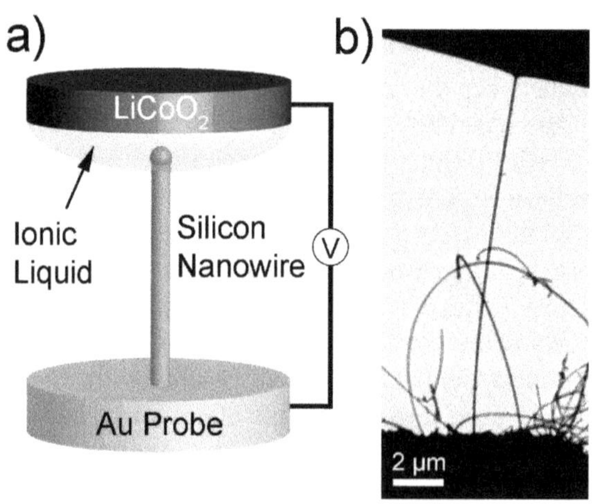

Figure 1.16 (a) Schematic of an *in situ* TEM electrochemical cell. See text for details. (b) Low-magnification TEM image of a device. A NW spans the gap between the Au probe at the bottom of the image and the ionic liquid at the top of the image. Reproduced with permission from Ref. 79. Copyright 2012, Elsevier.

the NW, whereas in a conventional battery cell, the NW is completely immersed in electrolyte and Li-ions can diffuse both axially and radially into the NW. It is still not clear how (or if) this difference in diffusion affects the correlation between the *in situ* TEM studies and understanding macroscopic electrochemical properties in conventional cells. Although many of the characteristic features of Si lithiation are present in both *in situ* and *ex situ* experiments, the actual volume change process is different because the entire NW is not surrounded by a Li-ion containing electrolyte. However, these *in situ* studies do provide a valuable tool for understanding fundamental structural transformations that *ex situ* studies may not be able to observe, with the formation of the $Li_{15}Si_4$ phase as a notable example.

Here, we focus on the structural changes that have been observed; in subsequent sections, additional results will be discussed. Figure 1.17a shows an image of a typical pristine SiNW before reaction in an electrochemical cell; the associated diffraction pattern indicates that there is a twinning defect running along the axis of the NW, which is commonly seen in VLS-grown SiNWs. During lithiation, NWs in these experiments typically first transform into a-Li_xSi and then crystallize into the $Li_{15}Si_4$ phase. Bright-field and dark-field images of fully lithiated $Li_{15}Si_4$ NWs and an associated SAED pattern are shown in Fig. 1.17b. The rings in the diffraction pattern indicate that the $Li_{15}Si_4$ is polycrystalline. The bright spots in the dark-field image correspond to single $Li_{15}Si_4$ crystals (all the crystals are not visible because of the imaging conditions used). Figure 1.17c shows the same NWs after delithiation; the NWs are now a-Si, and the diameter has decreased significantly. Although the electrochemical conditions are different, these observed structural changes are consistent with the *in situ* XRD experiments, indicating that the metastable $Li_{15}Si_4$ phase does form at high state of charge in SiNWs.

To gain a physical understanding of the volume change that occurs during the lithiation of SiNWs, Fig. 1.18 shows a time series of TEM images of a single <112>-axially oriented NW during lithiation. Figure 1.18a(i) is an image of the pristine ~85 nm NW with a ~20 nm Cu coating on one side. The Cu coating improves the electronic conductivity, which results in faster and more repeatable lithiation in these experiments.[80] The Cu does not react or alloy with Li. The NW is inserted into the ionic liquid above the image frames, so Li will diffuse down the length of the NW. After nine seconds [Fig. 1.18a(ii)], the upper part of the NW begins to expand as the Li–Si reaction front moves into the frame; the reaction front is marked by the arrow on the right side of the NW. Here, the movement of the reaction front involves the lithiation of Si near the surface. After 17 seconds [Fig. 1.18a(iii)], the reaction front has advanced further down the NW. Thirty-six seconds into lithiation [Fig. 1.18a(iv)], the initial reaction front has passed, but the core of the NW is still crystalline Si, and lithiation is now occurring radially. Finally, after ~393 seconds of lithiation, the NW has expanded to a diameter of ~150 nm. A corresponding series of schematics illustrating the lithiation process is shown in Fig. 1.18b.

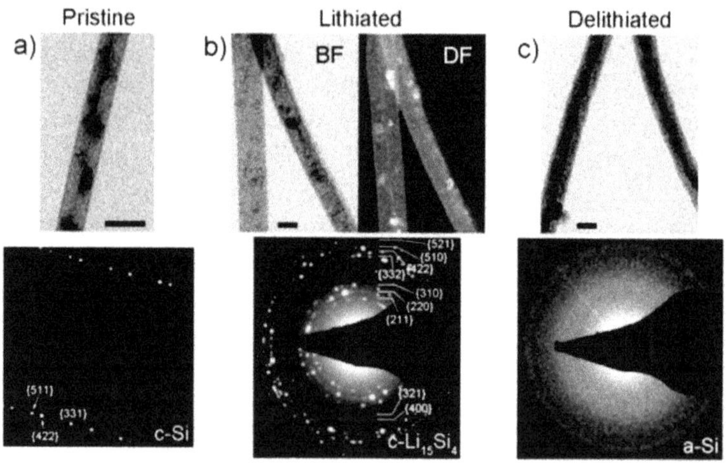

Figure 1.17 (a) TEM image and associated SAED pattern of a pristine crystalline SiNW. (b) A bright-field (BF) image, a dark-field (DF) image, and a SAED pattern of SiNWs after *in situ* lithiation. The indexed diffraction pattern indicates that the NWs have converted to polycrystalline $Li_{15}Si_4$; a few of the many crystals in the NWs are visible as light areas in the dark-field image. (c) The same NWs after delithiation. The diameter of the NWs has shrunk, and the SAED pattern indicates that the NWs have converted to a-Si. All scale bars are 100 nm. Reproduced with permission from Ref. 79. Copyright 2012, Elsevier.

This series of images shows that volume expansion of this <112>-oriented NW primarily occurs in a radial fashion. This NW/Cu structure does not bend after lithiation; bending would be expected if the SiNW changed in length while the Cu layer did not change. Instead, radial diffusion of Li from the surface through the Li_xSi phase to react with the crystalline Si is the dominant feature throughout most of the lithiation process. The expansion anisotropy observed here has to do with the crystallographic orientation of the NW sidewalls, and it will be further discussed in the next section.

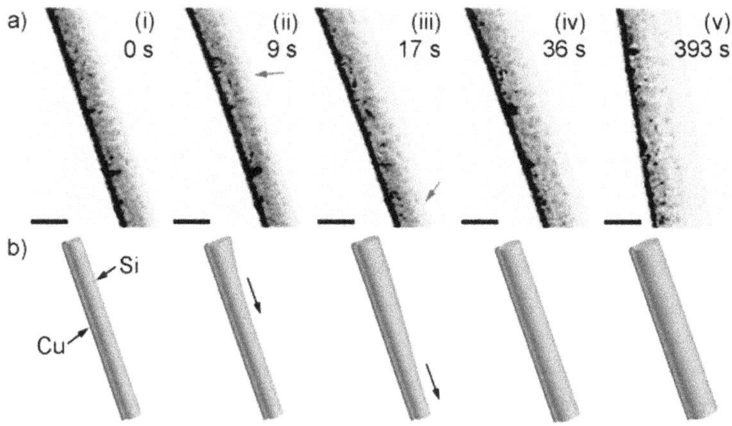

Figure 1.18 (a) Time series of *in situ* TEM images of the lithiation of a single SiNW with a ~20 nm thick Cu coating on the left sidewall. The arrows indicate the position of the initial reaction front that progresses down the axis of the NW. (b) Illustrations showing the progression of the initial reaction front and subsequent radial expansion. All scale bars are 100 nm. Reproduced with permission from Ref. 79. Copyright 2012, Elsevier.

1.4.3 Anisotropic Expansion during the Lithiation of Crystalline Si

In previous sections, it was demonstrated that the lithiation of crystalline Si proceeds via a two-phase reaction in which lithiated Si (Li_xSi) is interfaced with un-lithiated crystalline Si. In SiNWs, this is manifested as a crystalline Si core that shrinks as the lithiated shell grows. Interestingly, a few different experimental studies have shown that the direction of lithiation and volume expansion of crystalline Si nanostructures is strongly influenced by the crystallography of the shrinking crystalline Si core.[81–83] In one set of experiments, Si nanopillars with three different axial orientations were tested.[82] In order to test the effect of lithiation on crystallographic orientation, a top-down approach was used to fabricate the Si nanopillars rather than using the bottom-up VLS growth method, which can give SiNWs with various crystallographic growth directions. The nanopillars were fabricated by first depositing spherical silica particles onto Si wafers as etch masks, then dry etching the Si, and finally removing the particles to create nanopillars that are a few microns tall and between 100 nm and 500 nm in diameter. The pillar size is precisely controlled by the size of the nanoparticle etch masks and the etch time. These pillars are similar in structure and morphology to NWs, but by etching wafers with different crystallographic orientations, the crystallography of the pillars can be controlled. Another notable difference between the VLS-grown SiNWs and the etched Si nanopillars is that the diameters of the nanopillars tend to be larger, and the nanopillars also remain connected to the Si wafer, which acts as the substrate and current collector. As a result, long-term cycling tests were not performed on the Si nanopillars, but rather they were used as a tool for understanding the fundamental mechanisms behind lithiation of Si.

After fabrication, *ex situ* SEM observation before and after lithiation allows for visualization of volume changes. Figure 1.19a–c show top-down SEM images of pristine pillars with <100>, <110>, and <111> axial crystallographic orientations, respectively. The pillars have circular cross-sections, and the small circle visible on the top of each pillar is a protruding piece of Si that is a remnant of the fabrication process. This area is where the silica sphere contacts the top of the pillar during etching, which causes this region to resist etching and to therefore slightly protrude. These pillars were lithiated in electrochemical cells by sweeping to and holding at a low potential (10 mV *vs.* Li/Li^+) for 10 hours. Figure 1.19d–f show similar images of the three types of pillars after lithiation. From these images, it is evident that lithiation and volume expansion occur anisotropically, with the three different pillars expanding into different shapes. Comparing the final cross-sectional shapes to the original crystallography of each pillar, as shown in Fig. 1.19g, reveals that preferential expansion occurs at lateral {110} planes of the crystalline Si. This results in the <100> pillar expanding into a cross shape, the <110> pillar expanding into an ellipse, and the <111> pillar expanding more evenly into a circular shape because of the six lateral {110} planes.

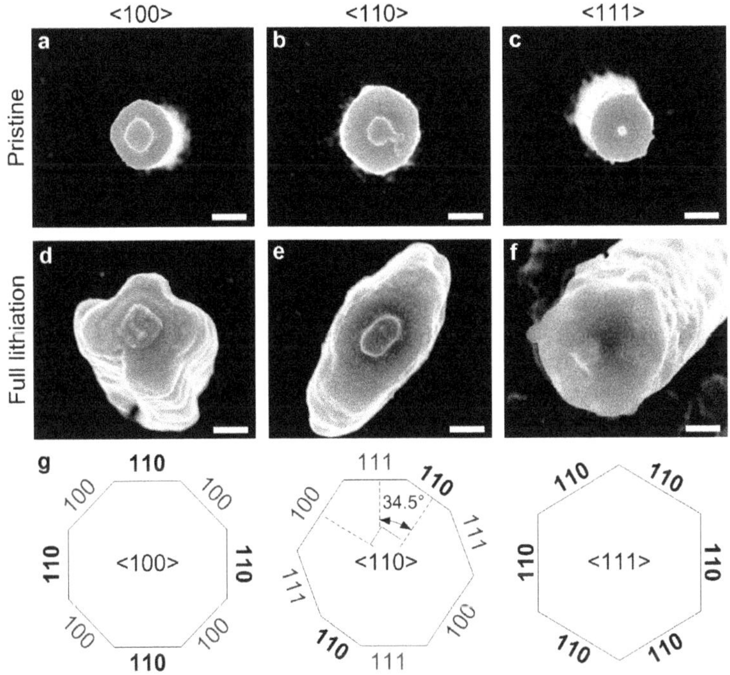

Figure 1.19 (a–c) Top-down SEM images of pristine Si nanopillars with <100>, <110>, and <111> axial crystallographic orientations. (d–f) Top-down SEM images of nanopillars with each axial orientation after lithiation in a conventional pouch-type electrochemical cell. Anisotropic expansion is evident, with preferential expansion occurring at {110} sidewalls, as shown in the crystallographic schematics in (g). All scale bars are 200 nm. Reproduced with permission from Ref. 82. Copyright 2011, American Chemical Society.

The preferential lithiation and expansion at {110} crystallographic planes has been observed in many types of Si nanostructures, including nanoparticles, nanopillars, NWs, and lithographically defined bars.[81–84] Since these original observations, it has been suggested that the expansion anisotropy is a result of differing reaction kinetics at different crystallographic planes, with {110} planes reacting at the fastest rate.[85–87] Modeling of the two-phase lithiation process, along with concurrent volume expansion and mechanical deformation, has shown that the cross-sectional shapes experimentally observed in Fig. 1.19 arise when different planes exhibit different reaction kinetics.[85,86] These results suggest that the rate-limiting step during the lithiation of crystalline Si nanostructures is actually the reaction of crystalline Si to form Li_xSi at the Li_xSi/Si interface, not diffusion of Li through the Li_xSi shell. Above some critical size, however, the diffusive transport of Li atoms through the shell is expected to become the rate-limiting step.

Finally, we note that the preceding discussion on expansion anisotropy has focused on the lithiation of crystalline Si nanostructures. The lithiation of a-Si is a different process altogether; it is a single-phase reaction in which the concentration of Li is governed by diffusion. The volume expansion during lithiation of amorphous structures has been shown to be more isotropic.[79,88] This has important consequences for the stress state and fracture properties of amorphous *vs.* crystalline Si nanostructures, as will be examined later.

1.4.4 Fracture of One-Dimensional Crystalline Nanostructures during Lithiation

As discussed earlier in this chapter, the mechanical degradation and fracture of Si structures during cycling is a major source of capacity decay in Si electrodes. Therefore, it is important to understand the conditions that result in fracture of individual nanostructures, particularly the critical diameters needed for fracture to occur. *Ex situ* SEM experiments of crystalline Si nanopillars with diameters ranging from 140 nm to 390 nm have revealed that fracture occurs at the surface of the pillars during lithiation for pillars with initial diameters between 240 nm and 360 nm.[89] Example of SEM images of fractured pillars with three different axial orientations are shown in Fig. 1.20.

These images show that axially oriented cracks form at the surface of the larger diameter pillars during lithiation. Other studies have also revealed similar fracture behavior, where crack formation was observed in a SiNW ~250 nm in diameter[81] and Si nanoparticles greater than 150 nm in diameter.[84] To explain these observations, mechanical stress modeling has shown that the two-phase lithiation of spherical or cylindrical structures results in tensile hoop stress at the surface, which can be large enough to cause fracture to occur.[84,85] This is different than in the lithiation of a-Si, where only compressive hoop stress is predicted to exist at the surface.[90] This again illustrates the importance of studying electrochemical processes in both crystalline and a-Si.

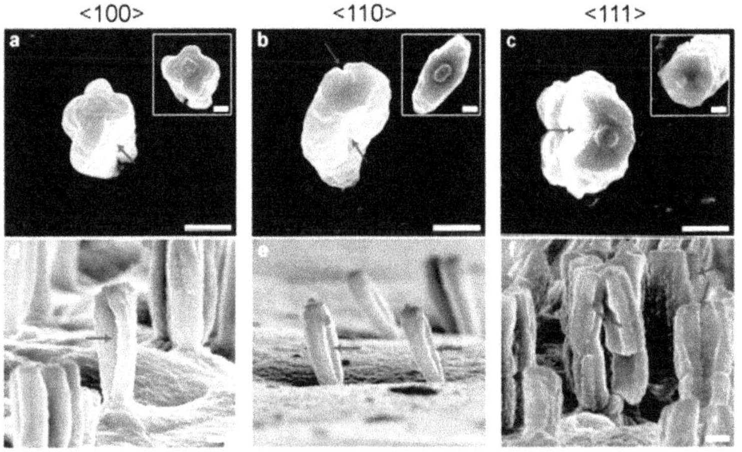

Figure 1.20 (a–c) Top-down SEM images of lithiated Si nanopillars with <100>, <110>, and <111> axial orientations, respectively, showing the presence of surface cracks that run along the axis. The inset images in (a–c) show smaller lithiated pillars that do not fracture. (d–f) Side-view SEM images of example pillars with each axial orientation showing the morphology of fractured pillars. The scale bars in the main images are 500 nm, and those in the insets are 200 nm. Reproduced with permission from Ref. 89. Copyright 2012, Proceedings of the National Academy of Sciences.

By analyzing fracture in many Si nanopillars, the effect of crystallographic orientation was elucidated.[89] Figure 1.21 shows statistics of the fracture location for many pillars of each type; the fracture location is measured as an angle from the horizontal. For all three axial orientations, cracks were primarily observed to form at surfaces between lateral {110} planes (the planes of preferential lithiation and expansion). These data indicate that anisotropic expansion plays a role in generating intensified tensile stress at these fracture locations. So, anisotropic expansion has a direct influence on both the stress state and fracture properties during lithiation of one-dimensional Si nanostructures. In a similar study, it was shown that a <112> axially oriented SiNW fractured and almost split in two during lithiation observed with *in situ* TEM.[81] We would like to point out that the SiNWs in this particular study, although grown

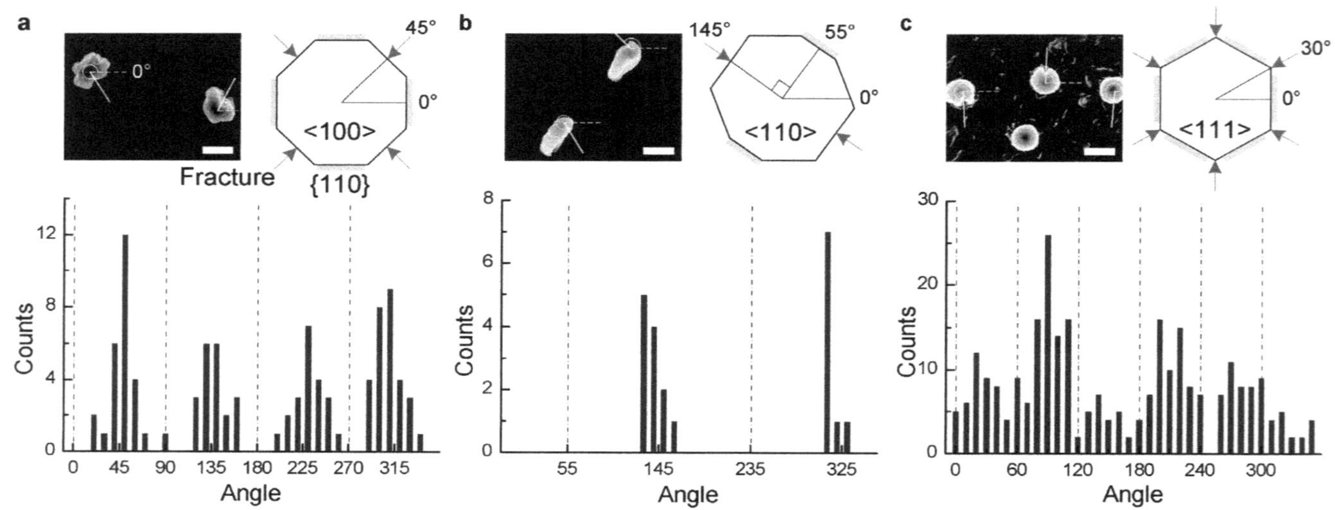

Figure 1.21 Statistical study of the fracture locations at the surface of Si nanopillars with different axial orientations during lithiation. Each panel (a–c) shows data from pillars with a certain axial orientation, including an example SEM image, a schematic of the pillar cross-section, and a compilation of fracture location data from many pillars presented as a column chart. The fracture location is measured as an angle from the horizontal, as shown in the schematics. The column charts show that each crystalline orientation shows favored fracture locations; these locations are indicated by the arrows in the schematics and are aligned between the preferential expansion directions ({110} sidewalls). This suggests that anisotropic expansion causes intensified tensile hoop stress at these locations. All scale bars are 1 μm. Reprinted with permission from Ref. 89. Copyright 2012, Proceedings of the National Academy of Sciences.

with VLS, displayed diameters of 150–200 nm; moreover, lithiation was performed by contacting the SiNW with a solid-state electrolyte rather than an ionic liquid. Thus, it is not clear if these results are representative of the behavior of smaller diameter VLS SiNWs cycled in a conventional cell. However, the anisotropic expansion observed is consistent with the nanopillar data presented here.

Finally, based on statistical analysis, it was found that the critical diameter for fracture of Si nanopillars upon lithiation was between ~240 nm and ~360 nm. Below this size, very few pillars fractured, while 90–95% of pillars larger than 360 nm fractured. This can be rationalized since fracture depends on both the mechanical stresses present and a characteristic length scale; for smaller pillars, the elastic energy present during deformation is not large enough to drive crack extension. This is important because most NWs used in battery electrodes have diameters that are less than 200 nm, resulting in resistance to fracture and decrepitation.

1.4.5 Structural Evolution during Cycling

In Sections 1.4.1–1.4.4, the structural and morphological changes during lithiation of SiNWs and nanopillars were discussed. After the first lithiation/delithiation cycle, crystalline Si never reappears; instead, the Si remains amorphous after delithiation. Because of this, the volume expansion and contraction are different in the first cycle compared to subsequent cycles. This is illustrated in Fig. 1.22, which shows an *in situ* TEM image of a NW with Cu on one side after delithiation. As previously discussed, this Cu coating does not react with Li but instead acts as a passive constraint to the expansion and contraction of Si. In the lithiated state (Fig. 1.18), the NW/Cu structure is straight; this is because during lithiation, expansion occurs primarily along the radial direction due to the anisotropic effects discussed in Section 1.4.3. Most NWs of this type have a <112> axial orientation, which means there are two {110} sidewalls where initial expansion preferentially occurs. After delithiation (Fig. 1.22), the composite Si/Cu structure bended and buckled with the Cu layer on the outside of the bent structure. This indicates that the SiNW contracted in diameter and length upon delithiation while the Cu layer did not change; the SiNW length contraction caused the bending. This simple experiment illustrates that the lithiation/delithiation of a-Si is more isotropic than the lithiation of crystalline Si.

Figure 1.22 TEM image of a delithiated SiNW with a ~20 nm Cu coating on one sidewall inside an *in situ* electrochemical cell. The NW bends and buckles due to a length decrease in the SiNW while the Cu remains the same length. The tip of the NW can be seen submerged in the ionic liquid electrolyte on the right side of the image. The scale bar is 500 nm. Reproduced with permission from Ref. 79. Copyright 2012, Elsevier.

As discussed in Section 1.4.1, the SiNWs underwent the expected change in volume during lithiation. Based on these observations, a decrease in the surface-to-volume ratio was expected. However, Brunauer–Emmett–Teller (BET) surface area measurements on the SiNWs after one cycle showed that the opposite occurred—a slight increase in the surface area (Fig. 1.23).[75] Continued monitoring of the surface area over several cycles showed it continued to increase until 5 cycles, then decreased at 10 cycles. Such a phenomenon is possible if, as SEM has confirmed, there is dramatic surface texturing and roughening (Fig. 1.23a). The pore volume as determined from the Barret–Joyner–Halenda (BJH) calculation scheme also revealed the existence of pores in the delithiated SiNWs (Fig. 1.23b), presumably formed due to tensile stresses when the Li is de-alloyed from the Si. The pore volume increases with cycling, even when the surface area decreases, which suggests that an Ostwald ripening effect may be occurring in the pores. Indeed, the observed pore size distribution revealed an increase in larger pores (>10 nm in size) and decrease in small pores as the number of cycles increased (Fig. 1.23c). Further analysis also showed that the pore formation is potential dependent, and that more voids form when charging the Si anodes to 0.05 V than 0.17 V *vs.* Li/Li^+ (Fig. 1.34).[91]

1.5 Surface Chemistry and Solid Electrolyte Interphase of SiNWs

In general, because the negative electrode operates at low potentials close to metallic Li, the electrolyte is outside its thermodynamic

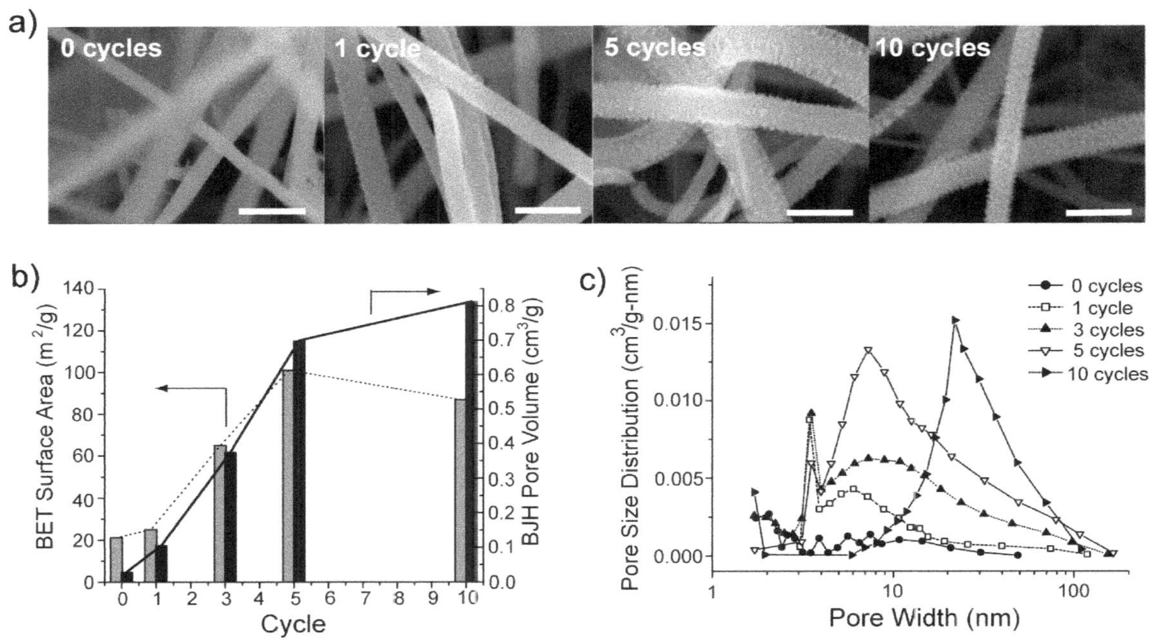

Figure 1.23 (a) SEM images showing increased surface texturing in the SiNWs after several cycles, scale bar is 200 nm. (b) BET surface area and BJH pore volume measurements for the SiNWs. (c) Volumetric pore size distributions as a function of pore width over several cycles. Reproduced with permission from Ref. 92. Copyright 2010, American Chemical Society.

stability region. Electrochemical reduction of both the solvent and salt of the electrolyte can cause the formation of a film on the surface of the anode. This film is composed of insoluble reduction products of the electrolyte. The presence of this film, the SEI, has been known to be an important feature of graphite anodes that allows for reversible cycling and long-term stability due to surface passivation.[93–96] The components of the SEI on graphite electrodes have been well studied and have shown that the decomposition products of the ethylene carbonate solvent, namely Li alkyl carbonates and Li carbonate, dominate the SEI layer.[94, 97–99]

However, little attention has been directed toward understanding the surface chemistry and SEI formation in silicon anodes. The first charge profile of Si when it is reacted with Li is characterized by a flat plateau at about 0.1 V *vs.* Li/Li^+ (Fig. 1.8a). Between 1.0 V and 0.2 V *vs.* Li/Li^+, the potential region where the SEI formation in graphite is known to occur, there is often very little capacity. This has caused some to report that there is no SEI formation on the Si electrode.[9] However, because the cell is operating at the potential where the organic electrolyte is thermodynamically unstable,[100] we can still expect the decomposition of the electrolyte to form some kind of an SEI layer. Often times, a plateau is observed at about 0.5–0.6 V[11] but does not account for the entire observed irreversible capacity loss after the first cycle. This capacity loss can be as high as 30% of the first charge capacity, whereas the capacity of the plateau can account for <10%. The rest of the irreversible capacity loss has been attributed to loss of contact of the active material after the large volume change in the first charge.[9]

Because analysis of the SiNW electrodes showed that the NWs did not pulverize during cycling, irreversible capacity loss due to lost or disconnected material seems unlikely for low cycle numbers, but it may rather be due to side reactions and SEI formation. For high–surface area materials such as NWs, there may be increased reactivity of the electrode with the electrolyte, which may have implications for long-term cycling, depending on which species are being formed.

To further understand the SEI formation on SiNWs, several SiNW samples were cycled to different states of charge or discharge, washed to remove excess electrolyte, and then characterized for changes in SEI composition and morphology using X-ray photoelectron spectroscopy (XPS) and SEM.[101]

1.5.1 SEI Composition

The molecular composition of the SEI layer was determined by analyzing the XPS high-resolution scans. The peaks were assigned according to assignments made by other SEI studies on graphite electrodes[99,102,103] (Fig. 1.24a). The distribution of the elements varies for each sample, suggesting that the SEI layer is dynamic over the voltage range. The signal from the Si disappears at 0.1 V *vs.* Li/Li^+ in the first charge, indicating that the SEI layer has grown thicker than the XPS penetration depth (~10 nm). This confirms that SEI formation is not limited to the region at 0.6 V as previously thought, but rather occurs in parallel with Li-ion insertion at lower potentials. From our cycling data, this does not appear to affect the insertion of Li into the SiNWs. This is different from graphite electrodes, where Li-ion insertion can be obstructed if good, passivating surface layers are not formed at potentials higher than Li intercalation.[95,96]

During the first charge, the SiNWs are slowly polarized to low potentials. As in graphite electrodes,[94] this results in a gradual, voltage-dependent formation of the SEI layer. Figure 1.24b shows a summary of the composition of the surface of each sample as determined by identifying the compounds based on peaks from the high-resolution spectra. As the voltage is initially lowered to 0.5 V, the SEI layer is composed almost equally of hydrocarbons, PEO-type oligomers, LiF, Li_xPF_y, and $Li_xPF_yO_z$ products, with smaller amounts of Li alkyl carbonates and ethers. These species are consistent with previous studies and are the main reduction products observed in this electrolyte.[98,100,102,104] At 0.1 V, the SEI surface consisted predominately of hydrocarbons and PEO-oligomers, indicating that more electrolyte was being reduced. Li_2CO_3 was also observed but no Li alkyl carbonates or ethers were observed, suggesting that these less stable products transformed into Li_2CO_3. Sputtering the SEI layer formed at 0.1 V revealed that the hydrocarbons and PEO are part of an organic layer on the surface. $LiPF_6$ species were also found on the outer layer, with more LiF on the surface of the electrode. This is consistent with previous observations on Li and Li_xC electrodes, where the SEI is actually a mosaic consisting of an inorganic layer closer to the surface of the electrode with an organic, polymer-containing layer at the interface with the electrolyte.[99] At the full charge to 10 mV, the SEI surface consisted mainly of Li_2CO_3 and LiF. The Li_2CO_3 is formed as a product of electrochemical

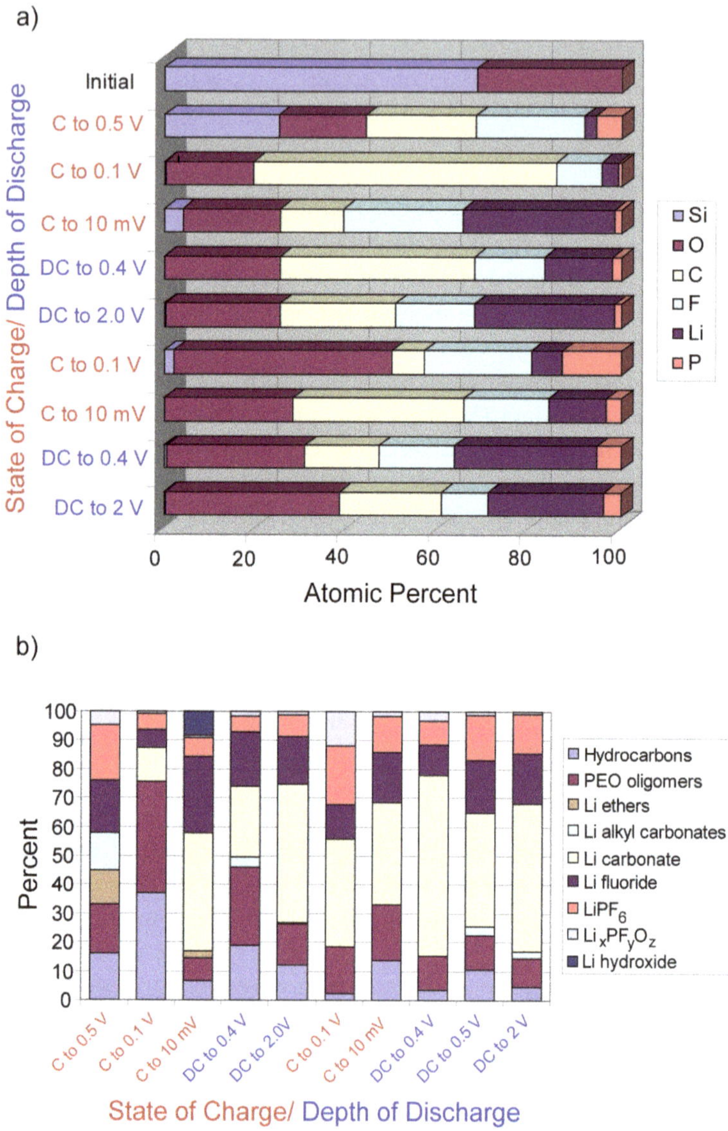

Figure 1.24 (a) Surface composition of the SiNW samples as derived from the XPS survey scan. The atomic percentages are shown for different potentials during the charge and discharge. (b) Molecular species found on the surface of the SiNW samples at different potentials as determined from the XPS high-resolution scans. Reprinted with permission from Ref. 64. Copyright 2009, American Chemical Society.

reduction of the electrolyte, but the LiF is formed as a decomposition product rather than an electrochemical one. LiF can be formed from various mechanisms as discussed in the literature,[102] which include reactions between $LiPF_6$, Li_2CO_3, HF, and/or water.

During the discharge of the first cycle, the SEI was found to be composed predominately of Li_2CO_3, with other major species being LiF and $LiPF_6$. In the second cycle, the composition remained the same both during the charge and discharge. This suggests that the compositional formation of the SEI is an important feature during the first cycle and afterward stays mostly the same, with Li_2CO_3 as the major species. This is in contrast to the work done on a-Si thin films,[105] where LiF was found to be the main component of the SEI. While the reason for this difference is currently not clear, it is possible that the higher surface area of the SiNWs may play a role.

1.5.2 SEI Morphology

Previous SEM analysis of the SEI formed on a-Si thin-film electrodes[105,106] at different states of charging and discharging revealed several key points. First, the morphology of the surface film changes drastically during cycling. The initial flat layer becomes rougher until it appears to be covered in a film composed of nanometer to micron-sized particles. At the full state of charge as well as during the discharge, cracks can be observed in the films due to volume expansion of the Si layer beneath it. The SEM characterization is complicated by the fact that it is difficult to visually distinguish the a-Si thin films from the SEI layer. In our case, the characteristic shape of the NWs facilitates the differentiation between the morphology changes of the Si and of the SEI layer itself.

SEM observation of the SiNW samples at different points of charge and discharge confirmed that the surface layer changes as a function of the potential. The pristine SiNWs are characterized by clean surfaces and sharp boundaries (Figs. 1.4a and 1.14a). In contrast, after the SiNWs had been charged to 0.1 V, the surfaces of the NWs were no longer smooth but rather rough in appearance due to what appears to be a conformal coating (Fig. 1.25a, inset). On the surface of the NW network, there also appeared to be regions covered with deposits of SEI products. From the sputter profile, it was estimated that the layer covering the NWs was about 40 nm thick.

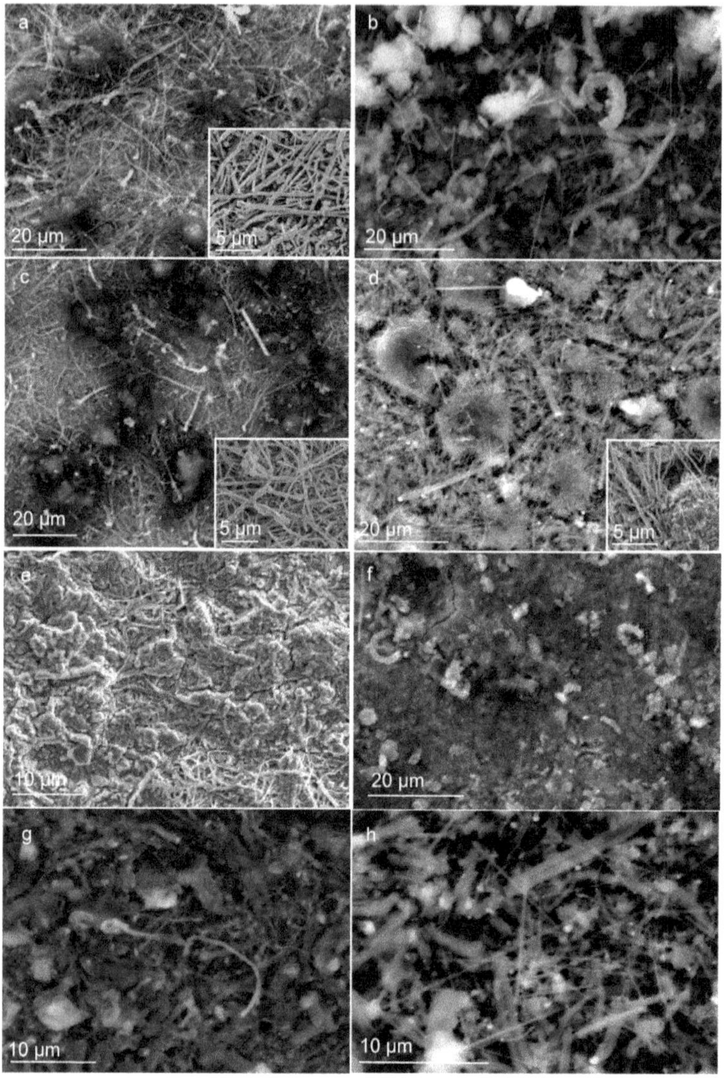

Figure 1.25 SEM images of SiNWs and SEI layer at different potentials *vs.* Li/Li^+: the first charge to (a) 0.1 V and (b) 10 mV; the first discharge to (c) 0.4 V and (d) 2.0 V; the second charge to (e) 0.1 V and (f) 10 mV; the second discharge to (g) 0.4 V and (h) 2.0 V. Reprinted with permission from Ref. 101. Copyright 2009, Elsevier.

Further charging to 10 mV (Fig. 1.25b) revealed a very different type of film that consisted of a composite of different kinds of particles and deposits. Many of the particles are covering the SiNWs, but some are found adjacent to the NWs. The location of these deposits suggests that there is deposition on or close to the substrate and growth outward. Because the NW layer can be considered a highly porous film rather than a planar solid, it is possible that SEI products can form on the substrate itself between adjacent NWs. This may be the case particularly for electrochemical reduction products, since the metallic current collector is more conducting than the SiNWs, making it a preferential nucleation site for the formation of these species.

During the discharge, the large particles were no longer present, suggesting partial dissolution of the SEI layer (Fig. 1.25c). A similar dissolution of the SEI during the discharge has been observed in graphite electrodes using XPS and atomic force microscopy (AFM) imaging.[102] At the discharge to 2 V, the morphology of the film looks very different (Fig. 1.25d). Instead of appearing like a cohesive film consisting of both particles and NWs, the film appears to be broken up, with the particles residing as islands separated from the NWs (Fig. 1.25d, inset). Additionally, the NWs look more separated from each other and appear to have less of a surface coating as seen from the higher Si signal in the survey scans at 2 V than at 0.4 V in the discharge (Fig. 1.24a). These observations suggest that the SEI film may have further dissolved during the discharge to 2 V. It should be emphasized that the NWs themselves do not appear to have pulverized. Although cracks are observed in the film as well, it is clear that this is due to the SEI layer and not the SiNW active material.

During the second charge, the SEI film has reformed on the surface of the SiNWs at 0.1 V (Fig. 1.25e). The film appears thicker than in the first cycle, indicating that the SEI growth still plays a large role after the first cycle. At further charging to 10 mV, the SEI has almost completely covered the NWs (Fig. 1.25f). During the discharge, more of the NWs are visible at 0.4 V and 2 V (Fig. 1.25g and h, respectively), indicating that the SEI film is indeed dissolving.

1.5.3 Silicon Surface Structure and Bonding

High-resolution XPS is often used to characterize the local structure and bonding environments and can be used to better understand the nature of the interface between the SEI and Si surface. The

XPS spectrum of the as-grown SiNWs is shown in Fig. 1.26. It is characterized by the Si–Si peak at 99 eV and the Si–O peak at about 103 eV due to SiO_2. The presence of the native oxide is due to the SiNWs being exposed to air after the growth, which results in the

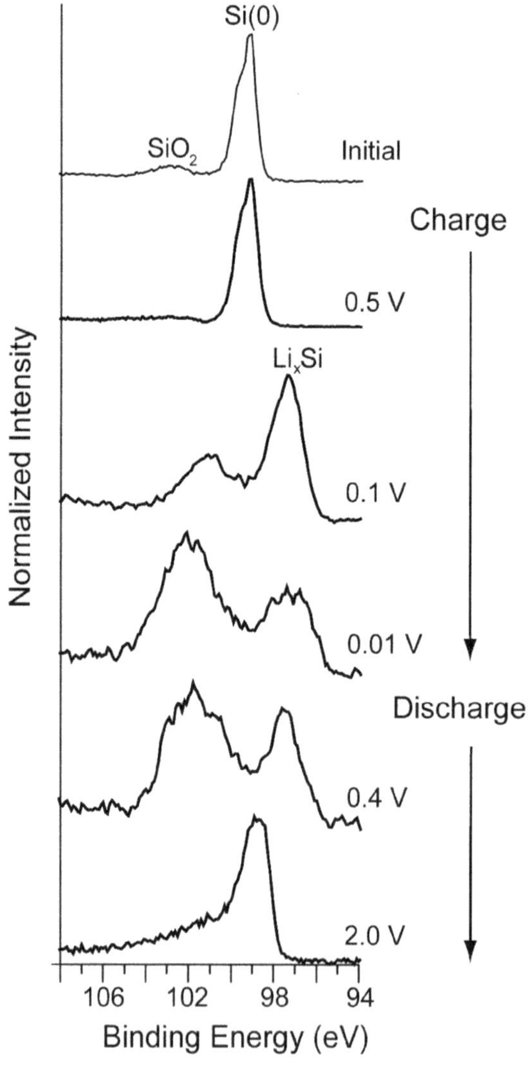

Figure 1.26 XPS high-resolution spectra of Si 2p peak of SiNWs at different potentials during the first cycle. Reprinted with permission from Ref. 101. Copyright 2009, Elsevier.

formation of 1–3 nm SiO_2 on the surface.[107] After charging, analysis of the Si 2p peak showed that the native oxide was removed, either electrochemically through the reaction $4Li + SiO_2 \rightarrow 2Li_2O + Si$, or chemically by hydrofluoric acid in the electrolyte, which is present either as impurities or formed *in situ* due to the decomposition of $LiPF_6$. Further charging showed that the Si 2p peak shifted to lower binding energies, indicating it is in a reduced state due to alloying with Li.[106] The peak observed at around 101 eV may be due to Si bonding to carbonaceous species in the SEI layer. At the full charge to 10 mV, the high resolution scan of the Si 2p surface showed two different environments for the Si. The peak due to Si bonding to the SEI became broader and shifted to higher energies, suggesting that there is Si bonded to many species, including carbon, oxygen, and possibly fluorine.[106,108] The peak at 98 eV, representing the Si alloyed with Li was also present, but the peak area was smaller than that for Si bonded to the SEI. Based on our previous SEM and TEM observations, at the end of the charge, the NWs have a heavily textured surface and have an increase in porosity. This results in the exposure of more Si on the surface to bond with SEI and electrolyte. Due to the high surface-to-volume ratio of the NWs, it appears that surface bonding plays a large role in the bonding of the Si compared to Li alloy formation in the bulk of the NW.

Analysis of the Si 2p peak at 0.4 V in the discharge showed a similar set of environments as observed at the fully charged state. Because only about a quarter of the Li has been removed at 0.4 V, the Si is still highly lithiated. At the fully discharged state at 2 V, the Si 2p main peak was observed at 99 eV, indicating that the Si was no longer in the reduced state alloyed with Li, but was instead close to the the original Si(0) state. The shoulder of the peak toward higher energies suggests that there were still some surface species bonded to the Si.

1.5.4 Nanostructured Electrode Designs for SEI Management

For reliable and efficient use of Si anodes in Li-ion batteries, it is vitally important to form a passivating SEI of limited thickness on the surface of the anode structures. However, the large volume changes inherent to the cycling of Si can result in unstable SEI formation. This is because the SEI formed on the surface during lithiation can rupture during delithiation (due to contraction of the Si), resulting in the exposure of fresh Si surface to the electrolyte for the formation of additional SEI (Fig. 1.27a). This can cause the growth of very thick SEI films, which (1) limits electrode kinetics due to increased ionic resistance, (2) causes active material to become electrically isolated due to the electrically insulating nature of the SEI, and (3) consumes Li and leads to low coulombic efficiency.

To combat this problem, tailored nanostructures have been developed that allow for stable SEI growth while still accommodating the lithiation-induced strain without fracture. The underlying design principle is the formation of a static interface with the electrolyte, so that the SEI that forms does not break due to dimensional changes of the surface of the nanostructure. One way to implement this principle is with hollow structures with constraining surface layers. The mechanically strong surface layer prevents outward expansion of Si during lithiation; instead, the Si expands to fill the empty space within the hollow structure, leaving the surface dimensions unchanged. An example of this type of nanostructure is the double-walled Si nanotube, where Si tubes with wall thickness of ~30 nm and diameters of ~400 nm are coated with a thin layer of SiO_x that is ~10 nm[109] (Fig. 1.27a). These nanotube structures were synthesized by electrospinning polymers to form nanofibers, followed by carbonization, a-Si deposition using CVD, and then removal of the carbon core to make tubes through an oxidation process that also leaves a SiO_x layer on the surface of the silicon. This SiO_x layer acts as a mechanical clamping layer, such that the Si only expands inward and hence does not disrupt the SEI layer. This is in contrast to SiNW and Si nanotube electrodes without mechanical clamping layers, where the SEI layer is disrupted during repeated volume changes and can form on fresh surfaces exposed during cycling, resulting in the formation of very thick SEI layers (Figs. 1.25 and 1.27c). However, Si nanotubes with the SiO_x coating showed an SEI layer thickness of only 110 nm. Moreover, impedance analysis of the Si nanotube electrodes showed very little change in SEI layer growth after many cycles, in contrast to unmodified SiNWs (that only have a native oxide 1–2 nm thick, as described in Section 1.5.3), as will be discussed in the next section.

Due to the thin and passivating SEI formed on these nanostructures, stable cycling of double-walled Si nanotube electrodes was possible for thousands of cycles. This is illustrated in Fig. 1.28, which shows the specific capacity and the galvanostatic voltage curves of a double-walled Si nanotube electrode tested in a half cell at a rate of 12C over

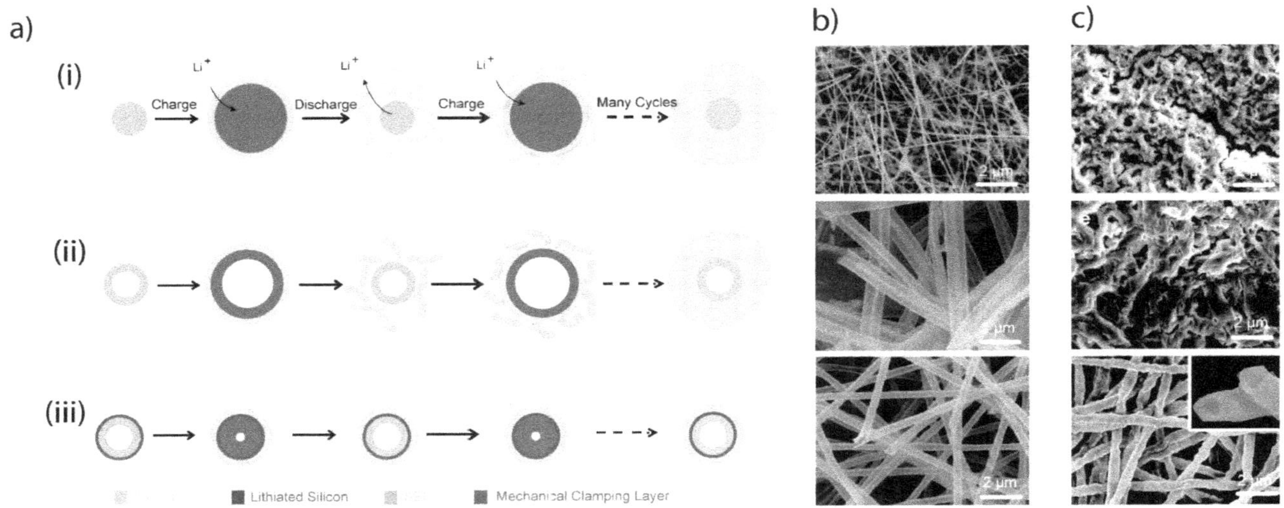

Figure 1.27 SEI formation on (i) Solid SiNWs, (ii) silicon nanotubes without a mechanical constraining layer, and (iii) Si nanotubes with a SiO_x mechanical constraining layer. (a) Schematic of SEI formation during the lithiation and delithiation on various Si nanostructures. On SiNWs, the SEI layer is ruptured during discharge, and new SEI can form on the newly exposed silicon surface during subsequent charges, resulting in a thick SEI layer. On silicon nanotubes without SiO_x, the SEI formation is similar as the SiNW case since the surface has a varying and unstable interface with the electrolyte. Silicon nanotubes with SiO_x are prevented from expanding toward the electrolyte during lithiation. As a result, a thin and stable SEI layer is formed. (b) SEM image of electrodes before cycling and (c) after 200 cycles. Reprinted with permission from Ref. 109. Copyright 2012, Nature Publishing Group.

6000 cycles. Since the Si nanotubes with SiO_x layer demonstrated stable cycling whereas Si nanotubes without the SiO_x layer only retained about 50% of the initial capacity after 400 cycles,[109] this indicates that good long-term cycling is a direct result of the stable SEI that forms on these structures.

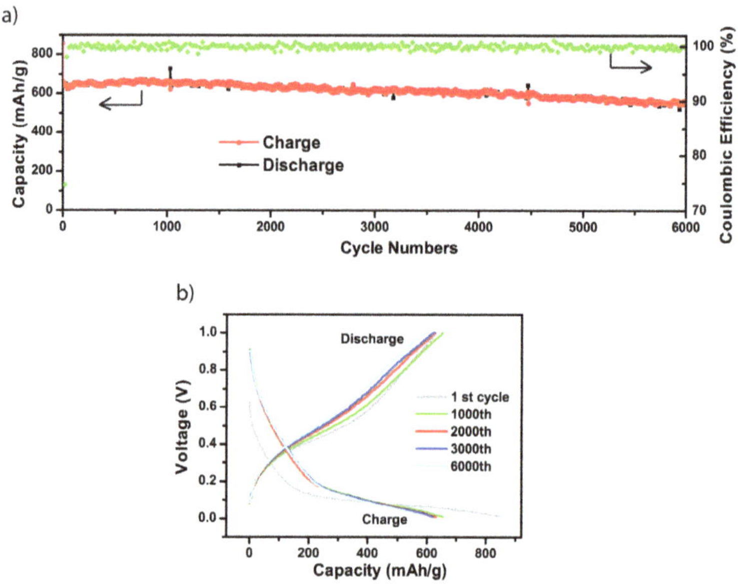

Figure 1.28 (a) Specific capacity with cycling for a double-walled Si nanotube electrode tested in a half cell. (b) Galvanostatic voltage curves from various cycles. Reprinted with permission from Ref. 109. Copyright 2012, Nature Publishing Group.

1.6 Electrochemical Impedance Spectroscopy

Electrochemical impedance spectroscopy (EIS) was used to determine how the potential-dependent morphological and compositional changes of the SEI affect the performance of the SiNW cycling[64] (Fig. 1.29). A three-electrode configuration was utilized for the EIS measurements, modeled after the cell designed by Dolle *et. al.*[110] Galvanostatic pulses (0.21 A/g) were used to change the state of charge or depth of discharge in the SiNW electrodes and EIS measurements were obtained at the end of each galvanostatic step after waiting for the electrode to reach equilibrium.

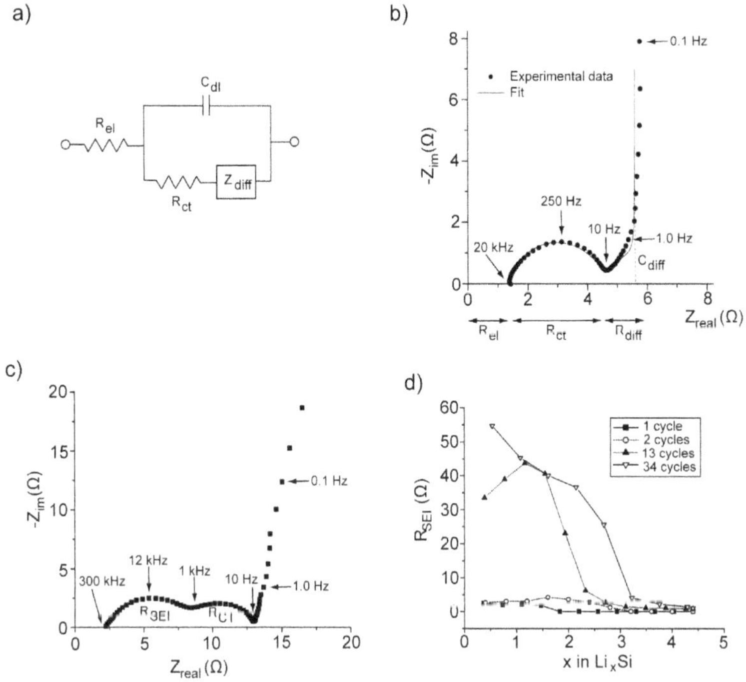

Figure 1.29 (a) Equivalent circuit used to fit impedance data. R_{el} is the series resistance from the electrolyte, R_{ct} is the charge-transfer resistance, C_{dl} is the double-layer capacitance, and Z_{diff} is the diffusion element. (b) Typical Nyquist plot observed at intermediate and high SOC, with contributions from the circuit elements labeled. (c) Nyquist plot observed after cycling >10 cycles, with new R_{SEI} contribution observed. (d) Increase in R_{SEI} as a function of cycle number at different SOC. Reproduced with permission from Ref. 64 (Copyright 2009, American Chemical Society) and Ref. 65.

At intermediate and high SOC, the Nyquist plots (Fig. 1.29b) showed contributions from the resistance of the electrolyte, charge-transfer resistance at the SiNW–electrolyte interface, and diffusion of Li through the SiNW. The equivalent circuit diagram used to model this system is shown in Fig. 1.29a. After cycling the SiNWs several times, a new semicircle appeared in the Nyquist plots at higher frequencies than the semicircle due to charge transfer (Fig. 1.29c). This appears to be related to a contribution from the resistance of the SEI layer (labeled as R_{SEI} in Fig. 1.29c). This SEI resistance became prominent after 10 cycles (Fig. 1.29d) due to the thickening of the

SEI layer during extended cycling as shown in Figs. 1.25 and 1.27.

In contrast, the Si nanotube structures with SiO_x constraining layer as discussed in Section 1.5.4 showed very different impedance behavior. As shown by the SEM images in Fig. 1.27, the SEI growth on the Si nanotubes was well controlled and only reached thickesses of ~110 nm. EIS analysis on Si nanotube electrodes before cycling and after 1000 and 2000 cycles (Fig. 1.30) showed no obvious impedance increase or R_{SEI} contribution as was observed in the SiNWs.

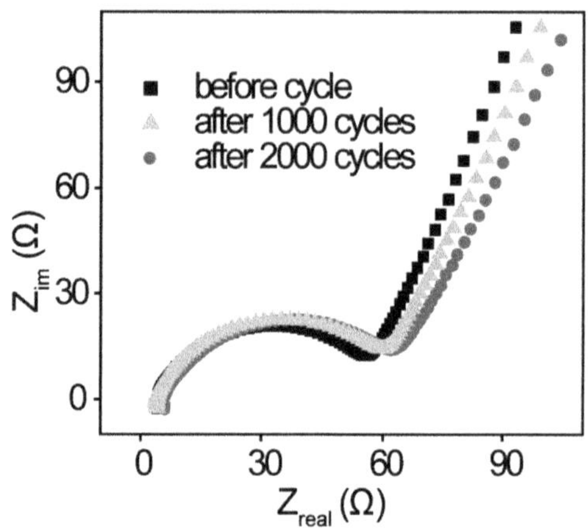

Figure 1.30 Impedance measurements for Si nanotubes with SiO_x constraining layer. Reprinted with permission from Ref. 109. Copyright 2012, Nature Publishing Group.

These results, along with the galvanostatic cycling results shown in Fig. 1.28, highlight the importance of SEI growth and control for achieving Si electrodes with good capacity retention.

1.7 Advanced Electrode Morphologies and Architectures

Based on the characterization studies presented here, it is clear that dramatic structural and morphological changes occur in the SiNWs upon lithiation and delithiation, both within the course of a single cycle and over several cycles. While the cycling performance of the SiNWs is quite promising, further improvements can be possible by modifying the electrode morphology and architecture to improve structural integrity and electronic conductivity throughout the SiNW electrode. Several example strategies using core-shell structures and composites are described in the following sections.

1.7.1 Core-Shell Nanowire Structures

The motivation for using core-shell NW structures is to engineer a stable backbone from which the Si can remain in good electronic contact. This can be achieved in various ways using different materials to comprise the core. To be an effective core, the material typically is inert to the lithiation reactions and will maintain good electronic conductivity throughout charging and discharging.

1.7.1.1 Crystalline silicon–amorphous silicon core-shell

Core-shell NWs consisting of a crystalline Si core and a-Si shell (Fig. 1.31a) were synthesized on SS substrates by varying the growth time and using a higher pressure than what is typically used when synthesizing entirely crystalline SiNWs[111] (Fig. 1.31b). Because a-Si reacts with Li at slightly higher potentials than crystalline Si (Section 1.3.2), it is possible to utilize only the amorphous shell material for Li-ion storage while preserving the crystalline core so that it can serve as both mechanical support and efficient electron transport pathway (Fig. 1.31c). This can be achieved by limiting the charging potential to >150 mV *vs.* Li/Li^+ so that the crystalline Si core is not lithiated. Observation of the NWs after cycling revealed that the core-shell structure was maintained as a robust mechanical and electronic support (Fig. 1.31d). Cycling measurements on these core-shell NWs showed that excellent capacity retention was obtained by using this electrode structure, with a capacity of ~1000 mAh/g and only 0.15% capacity fade per cycle observed.

1.7.1.2 Amorphous silicon–metal core-shell

A more common strategy to achieve core-shell structures is to deposit a film of a-Si onto metallic NWs or nanofibers. C-Si and TiC/C-Si core-shell nanofiber structures were obtained by using CVD of SiH_4 onto carbon nanofiber (CNF) and TiC/C nanofibers.[112,113] In the case of the C–Si core-shell nanofiber structures (Fig. 1.32), the amorphous CNF cores displayed a reversible capacity of ~280 mAh/g, much smaller than the capacity for lithiation of crystalline Si.

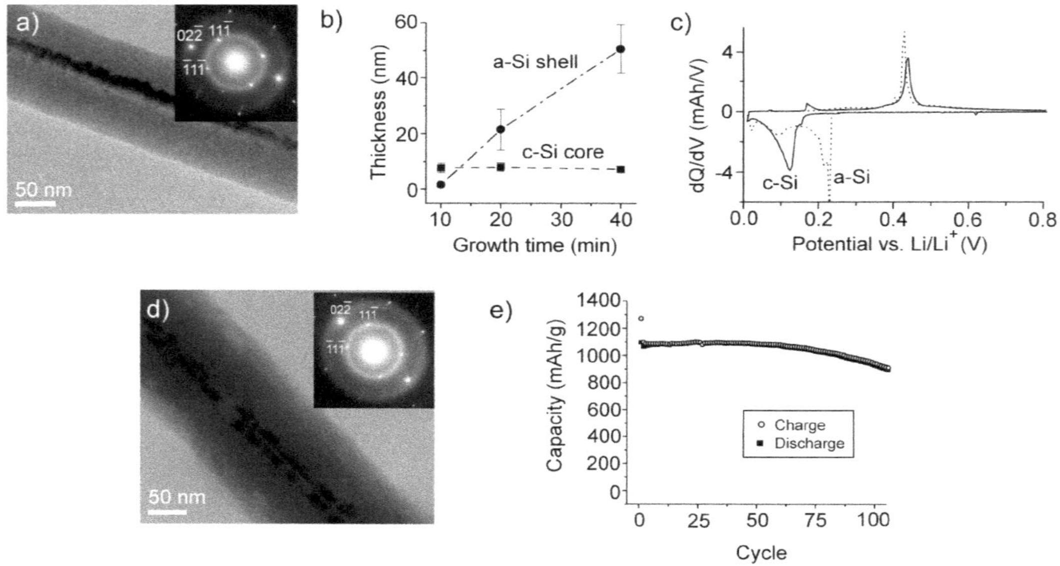

Figure 1.31 (a) TEM image of as-grown core-shell NW, with c-Si core and a-Si shell. SAED of the core shows it consists of c-Si. (b) The thickness of the a-Si shell can be controlled by increasing the growth time. (c) Differential capacity curves of c-SiNWs and c-Si/a-Si core-shell NWs when reacting with Li show that the a-Si shell reacts at higher potentials than the c-Si core. By limiting the potential to >150 mV *vs.* Li/Li^+, the a-Si shell can be used to alloy with Li, whereas the c-Si core remains unlithiated and instead functions as a mechanical backbone. (d) SEM and SAED of the c-Si/a-Si core-shell NWs when cycling with potential limitation to >150 mV confirms that the c-Si core remains unlithiated and crystalline Si. (e) Cycling results for c-Si/a-Si core-shell NWs with potential limitation to >150 mV. Reproduced with permission from Ref. 111. Copyright 2009, American Chemical Society.

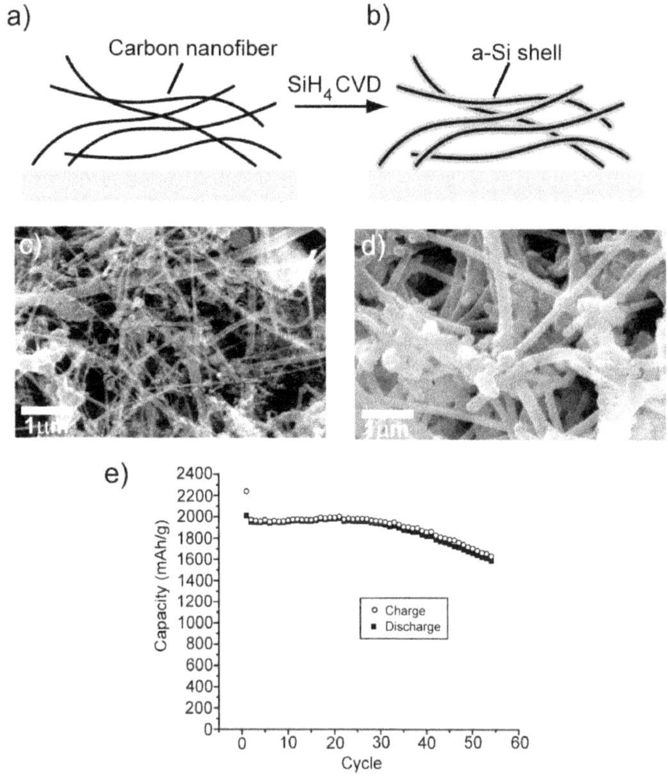

Figure 1.32 Synthesis of CNF–Si core-shell nanostructures (a) CNFs are coated with a-Si using CVD of SiH_4 to form (b) CNF–Si core-shell structures. SEM images of (c) CNFs and (d) a-Si coated CNFs. (e) Galvanostatic cycling performance of CNF–Si core-shell nanostructures. Reproduced with permission from Ref. 112. Copyright 2009, American Chemical Society.

As a result, the volume changes and structural stresses experienced by the carbon cores during lithiation were much lower than in the case of the crystalline SiNW cores. The lighter weight of the carbon also allowed the specific capacities of the C-Si core-shell structures to be higher, with ~2000 mAh/g observed, considering the weight of both carbon and silicon (Fig. 1.32e). Again, limiting the charging potential so that the core was not lithiated (>100 mV *vs.* Li/Li^+ for the amorphous CNFs) resulted in better long-term cycling stability but with a lower specific capacity since the Si was not fully lithiated.

To overcome this voltage restriction, TiC/C-Si core-shell NWs utilize an inert TiC/C scaffold (Fig. 1.33a,b) so that charging can occur down to low potentials, for example, 10 mV *vs.* Li/Li^+, in order to fully lithiate the Si shell. TiC/C nanofibers were synthesized through thermochemical reaction of a titanium alloy (Ti 6Al-4V) foil in acetone vapor at 850°C. Detailed TEM analysis showed that the nanofibers consisted of a 30 nm crystalline TiC core coated with a 20–40 nm amorphous carbon shell (Fig. 1.33c). Cyclic voltammetry on

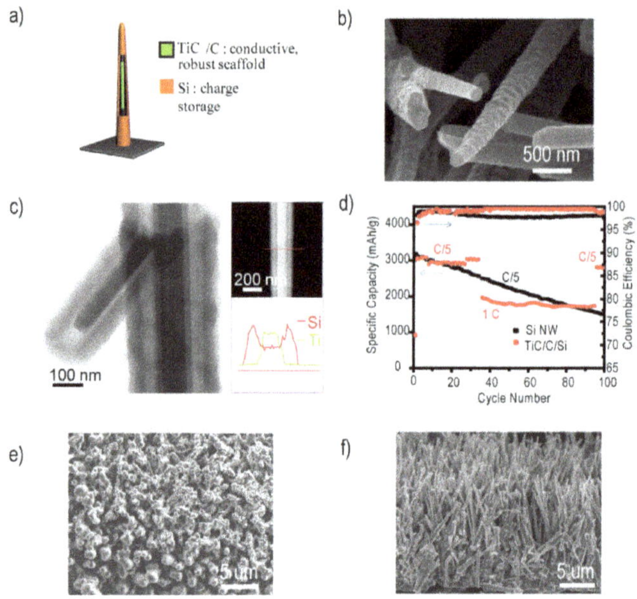

Figure 1.33 (a) Schematic illustration of a TiC/C nanofiber scaffold for Si anode architecture. The TiC/C nanofibers (green/black) provide a conductive, mechanically robust scaffold, and the Si coating (orange) functions as the charge-storage layer. (b) SEM image of TiC/C/Si nanofibers. (c) Bright field scanning TEM (STEM) image revealing the trilayer structure with contrast due to atomic number. Top inset shows the dark field STEM image and EDX line scan profile of Si (red) and Ti (green). (d) Discharge-specific capacity *vs.* cycle number for 100 cycles at a rate of C/5 and 1C in comparison to SiNWs (only considering mass of Si). (e) SEM image of TiC/C/Si electrode after 50 cycles and (e) rinsing in acetonitrile to remove residual electrolyte, (f) SEM image after second rinsing in 1 M HCl to remove SEI. Reproduced with permission from Ref. 113. Copyright 2011, American Chemical Society.

the TiC/C nanofibers showed only double-layer charging, indicating that the TiC/C was inert to lithiation.[113] After the TiC/C nanofiber synthesis, a 60 nm-thick coating of a-Si was deposited using CVD. Electrochemical cycling of the TiC/C SiNWs showed a stable capacity of 2800 mAh/g that degraded only 0.08% per cycle over 100 cycles with better capacity retention than SiNWs without the TiC/C core (Fig. 1.33d). This excellent cycling stability was attributed to the highly conductive nature and mechanical stability of TiC. This was evident by SEM imaging of the electrodes after 50 cycles, as shown in Fig. 1.33e,f, which shows that the TiC/C SiNWs remained rigid after cycling. This is in contrast to the case of the SiNWs without the TiC/C cores, which become pressed down onto the substrate (Fig. 1.4c).

To further improve the volumetric capacity of the Si-based anode, a porous, highly conductive carbon nanotube (CNT) sponge-like structure was investigated as the backbone and substrate for a 30 nm–thick coating of a-Si (Fig. 1.34).[91] These aerogel-like sponges are synthesized using CVD from ferrocene and 1,2-dichlorobenzene and consist of self-assembled, interconnected CNT skeletons with a porosity >99%, high structural flexibility and robustness, and wettability to organic solvents.[114] The highly porous nature allowed for Si deposition to be easily performed on the CNT sponge with high uniformity. These electrodes showed specific capacities of 2800 mAh/g when cycling from 0.05–1.0 V *vs.* Li/Li^+ and 1300 mAh/g when cycling from 0.17–1.0 V *vs.* Li/Li^+. The lower voltage range was found to affect the morphology of the coaxial structure, which resulted in a rapid capacity fade (Fig. 1.34c). The Si coating was observed to become roughly textured and contain 10–20 nm pores, which can cause the Si to debond from the CNT backbone (Fig. 1.34d). Increasing the charge cutoff potential to 0.17 V improved the capacity retention and minimized nanopore formation (Fig. 1.34e).

1.7.1.3 Si nanoparticle–decorated SiNWs

A variation of the core-shell NW structure is the decoration of silicon nanoparticles (SiNPs) onto the surface of the SiNWs to fill the empty space between adjacent NWs[115] (Fig. 1.35). This can increase the mass loading of silicon from ~0.3 mg/cm^2 for the bare SiNWs to 1.5 mg/cm^2, which increases the areal capacities from about 0.6 mAh/cm^2 to 2 mAh/cm^2.

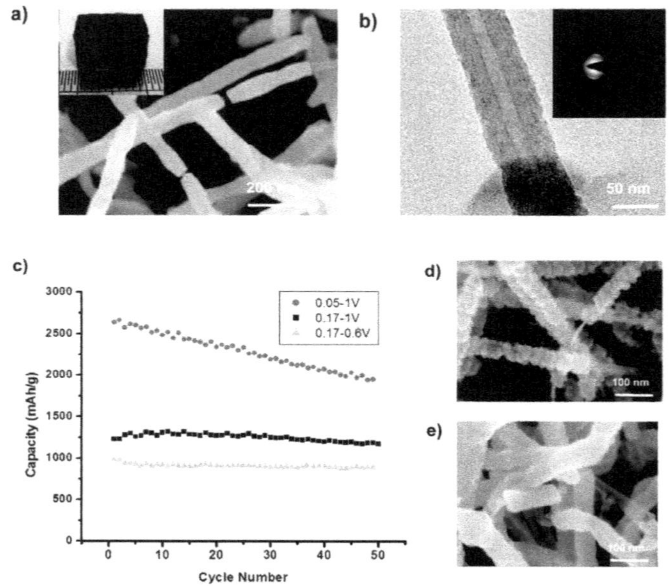

Figure 1.34 (a) SEM image of Si-CNT sponge showing the coaxial structure of a-Si conformally coated around CNTs. The inset shows a picture of a CNT sponge 1 cm × 1 cm × 8 mm in size before Si coating. (b) TEM image of Si-CNT structure showing the thickness of the Si coating is ~30 nm. Inset shows the TEM diffraction pattern for the Si-CNT structure. (c) Cycling performance of Si-CNT coaxial structure at different voltage ranges. (d) Morphology of Si-CNT sponge after cycling with voltage range of 0.17–1.0 V *vs.* Li/Li^+ and (e) 0.17–0.6 V *vs.* Li/Li^+. Reprinted with permission from Ref. 91. Copyright 2011, WILEY-VCH Verlag GmbH & Co. KGaA, Weinheim.

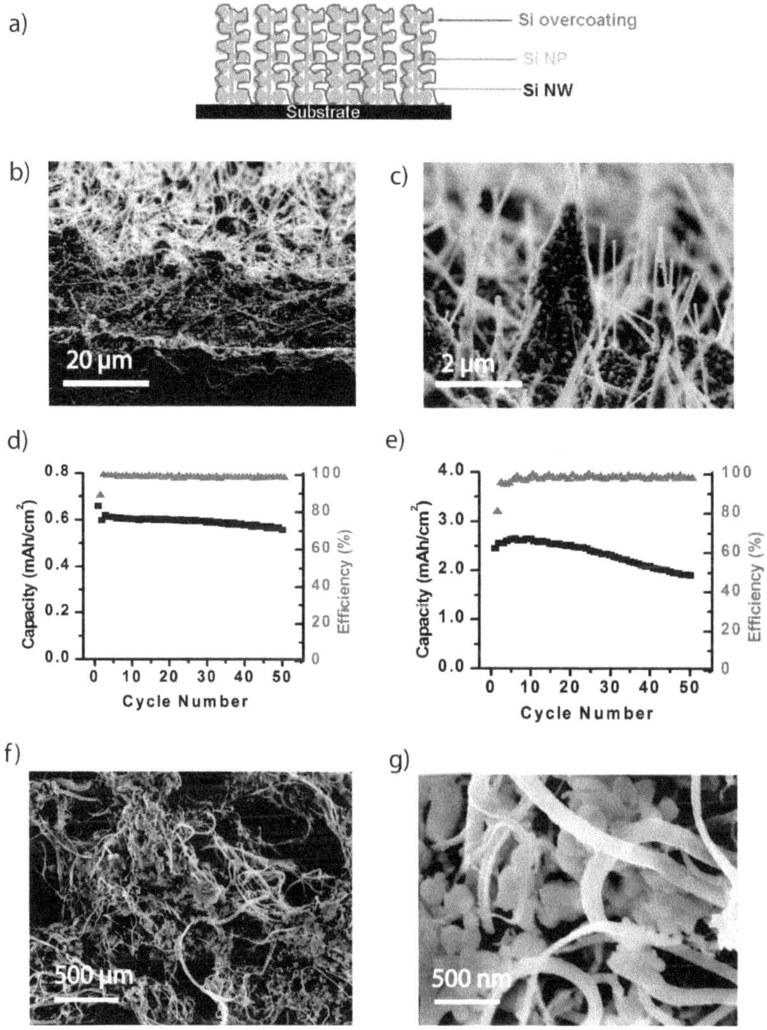

Figure 1.35 (a) Schematic of network comprising SiNP-decorated SiNWs. (b) SEM of SiNWs filled with SiNPs. (c) Cross-section SEM image of SiNWs after being filled with SiNPs. (d) Areal capacity of bare SiNW anode. (e) Areal capacity of SiNP–SiNW network. (f) and (g) Top-down SEM view of SiNP–SiNW network after cycling Reprinted with permission from Ref. 115. Copyright 2010, Royal Society of Chemistry.

Commercial SiNPs with diameters of 50 nm were dispersed into an aqueous, binder-free ink with low viscosity, which allows the

SiNPs to easily penetrate into the SiNW network when deposited. Because Si has a native oxide, which can act as an insulating layer that prevents good electrical transport at the interface between the SiNPs and the SiNWs, a thin film of Si was deposited using CVD to overcoat the entire structure as well as "glue" the SiNPs to the SiNWs (Fig. 1.35a–c). This coating was found to be important, as electrodes without the coating only showed capacities of ~500 mAh/g. This SiNP/SiNW electrode exhibited a discharge capacity of 1600 mAh/g with minimal fading capacity over 50 cycles with higher areal capacities than SiNWs alone (Fig. 1.35d,e). The decorated, porous network structure was also maintained after cycling, as shown by SEM analysis (Fig. 1.35f,g), indicating that this is an effective strategy for improving the mass loading of nanostructured Si anodes.

1.7.2 SiNW Composite Electrodes

Although the SiNW electrode structure has several advantages for obtaining high-capacity SiNW anodes, as described in Section 1.1, it also has several drawbacks when considering commercial application. This electrode architecture cannot be easily integrated into traditional battery electrode–coating processes, where the active material is obtained as a powder and mixed with a conducting additive and binder, suspended into a slurry, and coated onto a current collector as a composite. The advantages of using the slurry method are that the mass loading and thickness can be controlled and that the process is compatible with industrial practices already established for mass production of Li-ion battery electrodes. The VLS method for growing SiNWs is limited in that it is difficult to synthesize large quantities of NW material using this technique (only ~200–250 μg/cm^2 or ~0.75 mg/h. Therefore, different strategies to integrate SiNWs or nanostructured Si into more commercially viable electrode architectures have also been explored and are described in this section.

1.7.2.1 Supercritical fluid–liquid–solid SiNW/carbon composite electrodes

Solution-based analogues of the VLS method have been developed by Korgel's group[116–119] and the large-scale synthesis of SiNWs (>45 mg/h) has been demonstrated using the supercritical fluid–liquid–solid (SFLS) growth technique by decomposing phenylsilane in supercritical toluene with 2 nm dodecanethiol-capped Au nanocrystal seeds[119] as catalysts. SEM analysis showed that the synthesized product consisted of single crystalline SiNWs that had a thin shell, as shown in the TEM image in Fig. 1.36c, as well as some micron-sized spherical particles as by-products (Fig. 1.36a).

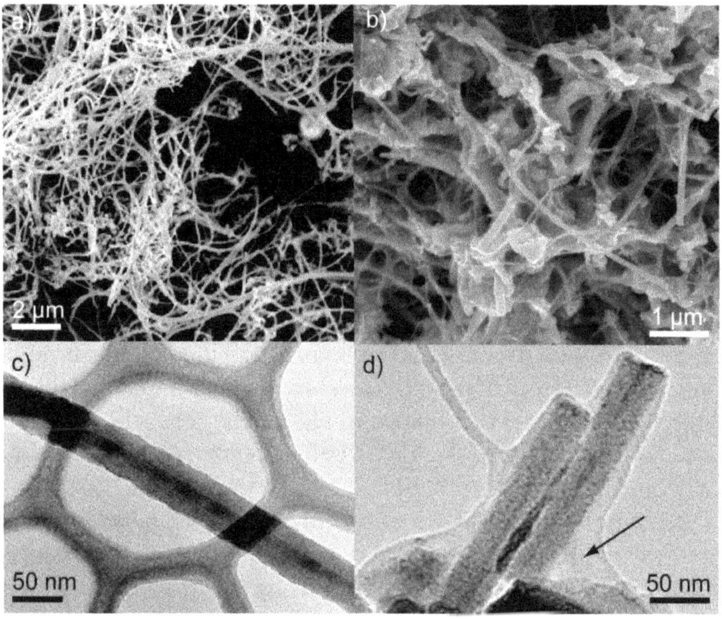

Figure 1.36 SEM image of SFLS SiNWs (a) as-grown, (b) after carbon coating. TEM image of SFLS SiNWs (c) as-grown, (d) after carbon coating, with arrow indicating carbon layer. Reprinted with permission from Ref. 70. Copyright 2010, American Chemical Society.

Thermogravimetric analysis performed on the SFLS SiNWs showed a ~6% weight loss during heating to 700°C, and mass spectrometry identified the major species as water. However, benzene was also observed, which may originate from the decomposition products of the phenylsilane used to synthesize the SFLS SiNWs. These results suggest that the surfaces of the SFLS SiNWs were coated with a residual organic layer that was not removed after the synthesis.

Similar to the example with the SiNP/SiNW electrode, the contact between adjacent NWs may be impeded by the native oxide. Carbon coating[120] of battery materials has been frequently used to improve the electron flux to the surface of the Li insertion material, aid electron transport between particles, and improve rate performance. For Si, carbon coating may also be used to buffer the large volume changes and control the SEI formation, as well as improve the electronic conductivity of the electrode. Carbon coatings based on CVD[121] or high-energy milling with polymers[122] may form SiC on the surface of the Si, which is electrochemically inert to Li.[121] Thus, the SiC plays no role other than to add extra unwanted weight, although some reports have claimed improved contact/conductivity at the interface between Si and carbon.[123] High-energy ball milling will also likely destroy the NW morphology. For these reasons, carbon coating was performed using the carbonization of sucrose, which is known to form amorphous carbon without any SiC phases.[124–127] Figure 1.36b,d shows an SEM and TEM image of the SiNWs after carbon coating. From these images, it is clear that the morphology and crystallinity of the SiNWs were preserved after the carbonization process. The carbon coating did not appear to be conformal around all of the SiNWs; rather, pairs or bundles of SiNWs may have been coated, as shown in Fig. 1.36d. This is likely due to inadequate mixing and penetration of the sucrose molecules into the SiNW bundles prior to pyrolysis.

Because the SFLS SiNWs are synthesized as a powder rather than grown on a substrate, these NWs can be suspended in a slurry and coated onto the current collector using the traditional method. The composite electrodes were made by mixing the SiNWs with sodium CMC binder and commercially available conducting carbon additive in water (Fig. 1.37). CMC was the binder of choice in this study because it has been previously found to be a superior binder for silicon electrodes compared to other polymers.[128,129] However, recent research on binders for Si anodes has shown other promising low-cost candidates, such as conducting polymers[130] and naturally derived polysaccharides.[131] The carbon used was either SuperP® Li-conducting carbon black (Timcal) or MWNTs (CNano). The carbon black (CB) was used as received, but the MWNTs were rather hydrophobic and had to undergo surface oxidation treatments to aid suspension in water.[70]

The galvanostatic cycling data of the SFLS SiNW electrodes with different slurry compositions are shown in Fig. 1.38. The first charge capacity has been omitted for clarity, but the charge and discharge capacities for the first three cycles are shown in the inset. The uncoated SiNWs mixed with carbon black and binder at a weight ratio of 78:12:10 performed poorly, with an irreversible capacity loss of 66% in the first cycle and rapid capacity loss from an initial discharge capacity of 1077 mAh/g fading to 151 mAh/g after 75 cycles. The low discharge capacities suggested that a large amount of the Si was not actively participating in the electrochemical reactions due to being electrically disconnected from the current collector. The large irreversible capacity loss may also be due to NWs losing electronic contact after the initial volume change. An SEM image of the electrode is shown in Fig. 1.37a, with the inset image showing the CB particles alone. Because the size of the carbon particles was on the same order as the size of the SiNWs, the CB did not form a percolating network with the SiNWs. Rather, the particles only made contact to some of the NWs.

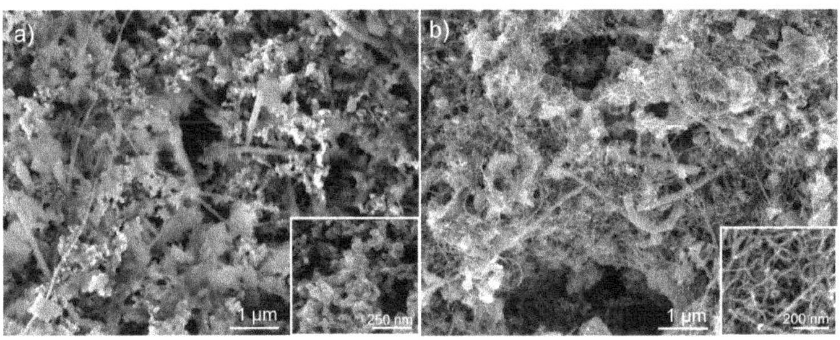

Figure 1.37 SEM images of SFLS SiNWs mixed with (a) carbon black and (b) MWNTs. Reprinted with permission from Ref. 70. Copyright 2010, American Chemical Society.

The carbon-coated SiNW sample mixed with carbon black and CMC at a weight ratio of 65:9:8 (with the carbon coating accounting for 18 wt%) also displayed a high irreversible capacity loss of 55% and fast degradation, although the capacities achieved were higher than without the carbon coating. This illustrates the importance of the coating for improving electronic conductivity in the electrode. With the coating, a high initial charge capacity of 5450 mAh/g was observed in the first cycle. As the theoretical capacity for the full lithiation of Si is only ~4000 mAh/g, all the excess capacity was likely due to SEI formation on the additional carbon from the coating.

A sample containing carbon-coated SiNWs mixed with MWNTs and CMC at a weight ratio of 78:12:8.5 (with the carbon coating accounting for 1.5 wt%) was found to display even higher capacities than the electrodes made with CB as the conducting agent, although still much lower than the theoretical capacity. This may be due to improved electronic conductivity of the MWNT network compared to the CB, which has been previously observed.[132] Also, the MWNTs may improve the resiliency of the electrode and help to buffer the large volume changes.[133,134] Although CNTs have been studied as active materials,[135–138] large irreversible capacity loss in the first cycle due to SEI formation, a large hysteresis between charge and discharge voltages, and insufficient long-term cyclability make CNTs unsuitable replacements for graphite. However, CNTs have been used to replace CB as the conducting additive and have been found to lead to improvements in capacity and cycle life in different anode[139,140] and cathode[141–143] systems. There have been several reports[133,144,145] of the growth of CNTs on the surface of Si particles. However, this can lead to SiC formation. To avoid this, simple mixing (with magnetic stirring) of commercially available MWNTs with the SiNW slurry was used. As shown in Fig. 1.37b, this mixing was adequate to make a film with good contacts between the MWNTs and the SiNWs. For small-diameter SiNWs, there could be multiple points of contact with the MWNTs, with the MWNTs often completely wrapping around the SiNW. This may help to ensure good connectivity throughout the composite. However, for larger diameter SiNWs and particles, this may not be possible. Also, the MWNTs formed a good network with the SiNWs and other MWNTs, ensuring good transport of electrons throughout the electrode.

Control tests on MWNTs mixed with binder only showed a capacity of 506 mAh/g in the first charge, while only 97 mAh/g was obtained in the discharge.[70] This large irreversible capacity loss was likely due to SEI formation on the MWNTs.[98] The extra charge stored in the MWNTs does not account for all of the improvement in the specific capacities observed in the SiNW/MWNT sample. Rather, the MWNTs allow more electron flux to reach the SiNWs, allowing more of the active material to participate in the electrochemical reactions.

All of the observed capacities for the SFLS SiNW electrodes were lower than those achieved with the VLS SiNW electrodes (Fig. 1.9). There may be several reasons for this. First, the quality of the SFLS SiNWs may not equal that of the VLS SiNWs. The organic residue on the surface of the SFLS NWs may act as a barrier to Li-ion insertion into the Si. Second, the micron-sized a-Si particles (by-products in the synthesis) contribute to the weight of the electrode without displaying good electrochemical performance, lowering the performance for the ensemble as a whole. Fine-tuning the synthesis or using post-synthetic separation techniques to acquire a purified NW sample will be beneficial in further understanding this matter. Third, the VLS electrode design may be more effective for transporting charge from the current collector to the SiNWs. This is because the NWs are grown directly on a metallic substrate and do not have to rely on binder and carbon to be electronically connected.

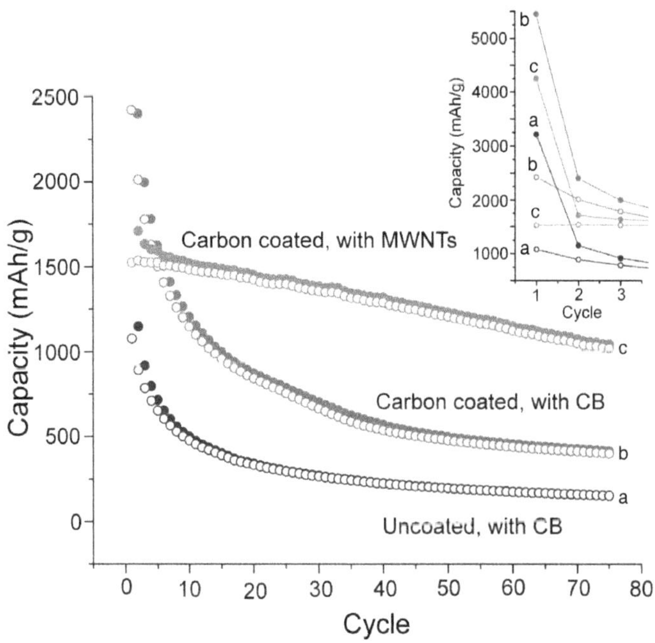

Figure 1.38 Capacity *vs.* cycling data for SFLS SiNWs uncoated and mixed with carbon black, carbon coated and mixed with carbon black, and carbon coated and mixed with MWNTs. The first cycle is omitted for clarity but is shown in the inset. Reprinted with permission from Ref. 70. Copyright 2010, American Chemical Society.

Nonetheless, the SFLS SiNW/MWNT composite electrode still displays better performance than other reports in the literature that used ball-milling[134,146] or ultrasonication[123] of Si and CNTs or direct CVD growth[133,145,147] of CNTs onto Si. These techniques are energy-intensive and require many steps. Our procedure shows that better performance can be achieved using simple mixing. We attribute this improvement to the good conductivity and connectivity of the network made from the SiNWs and MWNTs. Future work will focus on improving the capacities and cycle life through optimization of synthesis, surface treatments to reduce side reactions and irreversible capacity losses, and evaluation of rate performance and different mass loadings.

1.7.2.2 Free-standing silicon–carbon composites

In order to realize higher energy density batteries, most research efforts have been focused on improving the gravimetric capacities of the active electrode materials. However, the copper current collector is relatively heavy and takes up about 10% of the total weight of a battery.[148] To address this, nanocomposites using carbon nanostructures and SiNWs were investigated,[149,150] using the carbon component to also serve as a lighter-weight alternative current collector to the copper or SS foil (Fig. 1.39).

Nanocomposites of CNFs and SiNWs[149] were made by preparing free-standing CNF films by vacuum filtration of a solution of dispersed CNFs onto an alumina membrane followed by removal of the alumina using etching in NaOH. The CNF films were about 20 μm thick and displayed mass loadings of 0.8 mg/cm^2 and sheet resistances of 20–80 Ω/ϒ. Although these are higher sheet resistances than metallic current collector foils, these are not limiting factors for small-scale batteries on the several cm^2 scale. These films were coated with Au catalyst and used as 3-D substrates for VLS SiNW growth, which increased the mass loading of SiNWs to 1.2–2.8 mg/cm^2 compared to 0.3–0.7 mg/cm^2 for a planar substrate. Although some of the SiNWs could be found growing inside the CNF network (due to Au nanoparticle diffusion inside the film), the SiNWs predominately were found on the surface of the CNF film (Fig. 1.39b). High-resolution TEM imaging of the interface between the SiNWs and CNFs showed firm binding as a result of the VLS SiNW growth process (Fig. 1.39c). Electrochemical cycling of the CNF–SiNW nanocomposites were performed with different mass loadings of Si (Fig. 1.39d). The nancomposite films showed comparable capacity retentions to similar SiNW electrodes grown on metallic current collectors despite the higher amount of active material due to the porous 3-D structure of the CNF networks. Replacing the conventional graphite on Cu current collector anode with this SiNW on CNF nanocomposite would lead to a 23% increase in specific energy density given an areal capacity of 4 mAh/cm^2 (a typical commercial value for 18,650 Li-ion cells).

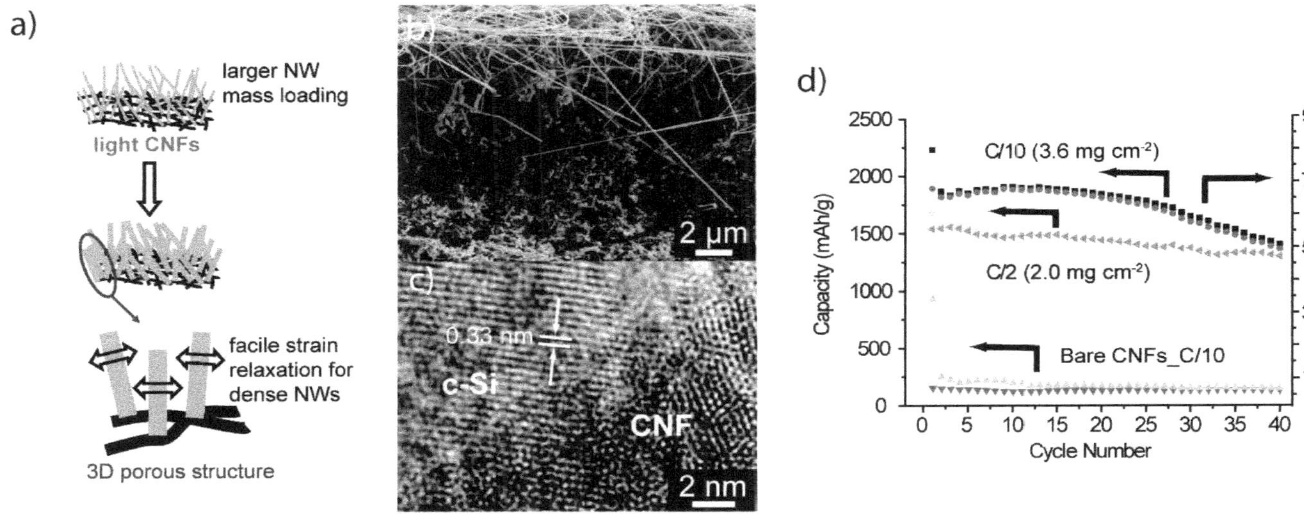

Figure 1.39 (a) Schematic of free-standing nanocomposites consisting of SiNWs grown on CNF nanofiber films. (b) Cross-section of CNF-SiNW nanocomposite. (c) High-resolution TEM image showing the interface between a SiNW and CNF. (d) Capacity retention data for CNF–SiNW nanocomposites with different mass loadings and charging/discharging rates as well as those for a control sample with bare CNFs. Capacities are based on the combined mass of both CNFs and SiNWs. The symbols in squares, diamonds, and upright triangles are for charging and the symbols in circles and side-facing and inverted triangles are for discharging. Reprinted from Ref. 149. Copyright 2010, Elsevier.

1.8 Conclusions and Outlook

Electrodes comprising SiNWs have been found to display excellent Li-ion storage characteristics and cycling performance. Detailed structural and morphological characterization, both *ex situ* and *in situ*, has revealed that the NW shape and morphology is advantageous for obtaining high lithiation capacities without pulverization for small NWs, in addition to the other gains from using nanomaterials in electrical storage electrodes, such as high surface area and contact with the electrolyte and short ionic-diffusion distances. Impedance spectroscopy and surface analysis of the SiNW electrodes revealed that the SEI layer formation and growth may play a large role in good long-term cyclability. Nanostructured designs for SEI management, increasing mass and areal capacities and developing light-weight and scaleable NW-based composite electrodes indicate that there is great promise for the application of SiNWs as high capacity anodes in Li-ion batteries.

Acknowledgments

The authors would like to acknowledge financial support from the Global Climate and Energy Project at Stanford, the U.S. Office of Naval Research, King Abdullah University of Science and Technology, DOE BATT Program, National Science Foundation Graduate Fellowship, Stanford Graduate Fellowship, and the National Defense Science and Engineering Graduate Fellowship. The authors would also like to thank Prof. Robert Huggins, Prof. Riccardo Ruffo, Seung Sae Hong, Prof. Jang Wook Choi, Dr. Lifeng Cui, Prof. Liangbing Hu, Dr. Seok Woo Lee, and Prof. William D. Nix for their contributions and technical discussions.

References

1. G.-A. Nazri, G. Pistoia, *Lithium Batteries: Science and Technology* (Kluwer Academic/Plenum, Boston, 2004).
2. C. J. Wen, R. A. Huggins, *J. Solid State Chem.* **37**, 271 (1981).
3. H. Li, *et al.*, *Solid State Ionics* **135**, 181 (2000).
4. P. Limthongkil, Y.-I. Jang, N. J. Dudney, Y.-M. Chiang, *J. Power Sources* **119–121**, 604 (2003).

5. M. N. Obrovac, L. Christensen, *Electrochem. Solid-State Lett.* **7**, A93 (2004).
6. M. N. Obrovac, L. J. Krause, *J. Electrochem. Soc.* **154**, A103 (2007).
7. J. Li, J. R. Dahn, *J. Electrochem. Soc.* **154**, A156 (2007).
8. T. D. Hatchard, J. R. Dahn, *J. Electrochem. Soc.* **151**, A838 (2004).
9. J. H. Ryu, J. W. Kim, Y.-E. Sung, S. M. Oh, *Electrochem. Solid-State Lett.* **7**, A306 (2004).
10. H. Li, X. Huang, L. Chen, Z. Wu, Y. Liang, *Electrochem. Solid-State Lett.* **2**, 547 (1999).
11. J. Graetz, C. C. Ahn, R. Yazami, B. Fultz, *Electrochem. Solid-State Lett.* **6**, A194 (2003).
12. T. Takamura, S. Ohara, M. Uehara, J. Suzuki, K. Sekine, *J. Power Sources* **129**, 96 (2004).
13. T. Takamura, M. Uehara, J. Suzuki, K. Sekine, K. Tamura, *J. Power Sources* **158**, 1401 (2006).
14. M. Green, E. Fielder, B. Scrosati, M. Wachtler, J. S. Moreno, *Electrochem. Solid-State Lett.* **6**, A75 (2003).
15. Recent progress in batteries and future problems. In *Solid State Ionics for Batteries*, T. Minami, *et al.*, Eds. (Springer, Tokyo, 2005), pp. 5–30.
16. Fujimoto, M. *et al.*, Electrode for use in lithium battery and rechargeable lithium battery. **US 7,195,842 B1,** (2007).
17. H. Kim, B. Han, J. Choo, J. Cho, *Angew. Chem. Int.* **47**, 1 (2008).
18. U. Kasavajjula, C. Wang, A. J. Appleby, *J. Power Sources* **163**, 1003 (2007).
19. C. K. Chan *et al.*, *Nature Nanotech.* **3**, 31 (2008).
20. S. D. Beattie, D. Larcher, M. Morcrette, B. Simon, J.-M. Tarascon, *J. Electrochem. Soc.* **155**, A158 (2008).
21. R. A. Huggins, W. D. Nix, *Ionics* **6**, 57 (2000).
22. J. Yang, M. Winter, J. O. Besenhard, *Solid State Ionics* **90**, 281 (1996).
23. C. R. Sides, F. Croce, V. Y. Young, C. R. Martin, B. Scrosati, *Electrochem. Solid-State Lett.* **8**, A158 (2005).
24. A. Manthiram, A. V. Murugan, A. Sarkar, T. Muraliganth, *Energy Environ. Sci.* **1**, 621 (2008).
25. F. Jiao, K. M. Shaju, P. G. Bruce, *Angew. Chem. Int.* **44**, 6550 (2005).
26. D. K. Kim *et al.*, *Nano Lett.* **8**, 3948 (2008).
27. C. J. Patrissi, C. R. Martin, *J. Electrochem. Soc.* **146**, 3176 (1999).

28. C. J. Patrissi, C. R. Martin, *J. Electrochem. Soc.* **148**, A1247 (2001).
29. K. M. Shaju, F. Jiao, A. Debart, P. G. Bruce, *Phys. Chem. Chem. Phys.* **9**, 1837 (2007).
30. K. T. Nam *et al., Science* **312**, 885 (2006).
31. M. -. Park *et al., Angew. Chem. Int.* **46**, 750 (2007).
32. R. S. Wagner, W. C. Ellis, *Appl. Phys. Lett.* **4**, 89 (1964).
33. A. M. Morales, C. M. Lieber, *Science* **279**, 208 (1998).
34. Y. Cui, L. J. Lauhon, M. S. Gudiksen, J. Wang, C. M. Lieber, *Appl. Phys. Rev.* **89**, 2214 (2001).
35. Y. Wu *et al., Nano Lett.* **4**, 433 (2004).
36. C. K. Chan, R. Ruffo, S. S. Hong, R. A. Huggins, Y. Cui, *J. Power Sources* **189**, 34 (2009).
37. F. Iacopi *et al., Mater. Res. Soc. Symp. Proc.*, 1017 (2007).
38. Y. Wang, V. Schmidt, S. Senz, U. Gosele, *Nature Nanotech.* **1**, 189 (2006).
39. S. D. Hutagalung, A. Ahmad, K. A. Yaacob, Growth of silicon nanostructures by thermal evaporation using nickel catalyst. In *Solid State Science and Technology*, (2008); pp. 100–106.
40. B. M. Kayes *et al., Appl. Phys. Lett.* **91**, 103110 (2007).
41. Y. Yao, S. Fan, *Mater. Lett.* **61**, 177 (2007).
42. S. Hofmann *et al., Nature Mater.* **7**, 372 (2008).
43. M. Jeon, H. Uchiyama, K. Kamisako, *Mater. Lett.* **63**, 246 (2009).
44. Z. W. Wang, Z. Y. Li, *Nano Lett.* **9**, 1467 (2009).
45. A. I. Hochbaum, R. Fan, R. He, P. Yang, *Nano Lett.* **5**, 457 (2005).
46. Y. Cui, Q. Wei, H. Park, C. M. Lieber, *Science* **291**, 851 (2001).
47. Y. Huang, X. Duan, Q. Wei, C. M. Lieber, *Science* **291**, 630 (2001).
48. B. C. Lee *et al., IEEE Photonic Tech L.* **20**, 398 (2008).
49. L. Tsakalakos *et al., Appl. Phys. Lett.* **91**, 233117 (2007).
50. B. Tian *et al., Nature* **449**, 885 (2007).
51. E. C. Garnett, P. Yang, *J. Am. Chem. Soc.* **130**, 9224 (2008).
52. M. D. Kelzenberg *et al., Nano Lett.* **8**, 710 (2008).
53. J. Zhu *et al., Nano Lett.* **9**, 279 (2009).
54. Y. J. Hwang, A. Boukai, P. Yang, *Nano Lett.* **9**, 410 (2009).
55. J. R. Maiolo III *et al., J. Am. Chem. Soc.* **129**, 12346 (2007).
56. A. P. Goodey, S. M. Eichfeld, K. K. Lew, J. M. Redwing, T. E. Mallouk, *J. Am. Chem. Soc.* **129**, 12344 (2007).

57. A. I. Hochbaum *et al.*, *Nature* **451**, 163 (2008).
58. A. I. Boukai *et al.*, *Nature* **451**, 168 (2008).
59. K. Peng, J. Jie, W. Zhang, S. T. Lee, *Appl. Phys. Lett.* **93**, 033105 (2008).
60. E. Quiroga-Gonzalez, E. Ossei-Wusu, J. Carstensen, H. Foll, *J. Electrochem. Soc.* **158**, E119 (2011).
61. H. Foll *et al.*, *J. Electrochem. Soc.* **158**, A580 (2011).
62. B. Laik *et al.*, *Electrochim. Acta* **53**, 5528 (2008).
63. J. H. Cho, X. Li, S. T. Picraux, *J. Power Sources* **205**, 467 (2012).
64. R. Ruffo, S. S. Hong, C. K. Chan, R. A. Huggins, Y. Cui, *J. Phys. Chem. C* **113**, 11390 (2009).
65. C. K. Chan, One-dimensional nanostructured materials for Li-ion battery and supercapacitor electrodes. Ph.D. Dissertation, Stanford University (2009).
66. G. Taillades, N. Benjelloun, J. Sarradin, M. Ribes, *Solid State Ionics* **152–153**, 119 (2002).
67. A. H. Thompson, *J. Electrochem. Soc.* **126**, 608 (1979).
68. J. Li *et al.*, *J. Electrochem. Soc.* **156**, A283 (2009).
69. W. Xu, S. S. S. Vegunta, J. C. Flake, *J. Power Sources* **196**, 8583 (2011).
70. C. K. Chan, R. N. Patel, M. J. O'Connell, B. A. Korgel, Y. Cui, *ACS Nano* **4**, 1443 (2010).
71. T. L. Kulova, Y. V. Pleskov, A. M. Skundin, E. I. Terukov, O. I. Kon'kov, *Russ. J. Electrochem.* **42**, 708 (2006).
72. N. Dimov, S. Kugino, M. Yoshio, *J. Power Sources* **136**, 108 (2004).
73. E. M. Pell, *Phys. Rev.* **119**, 1222 (1960).
74. M. D. Levi, D. Aurbach, *J. Phys. Chem. B* **101**, 4641 (1997).
75. A. Esmanski, G. A. Ozin, *Adv. Funct. Mater.* **19**, 1 (2009).
76. S. Misra *et al.*, *ACS Nano,* Article ASAP (2012).
77. B. Key *et al.*, *J. Am. Chem. Soc.* **131**, 9239 (2009).
78. J. Y. Huang *et al.*, *Science* **330**, 1515 (2010).
79. M. T. McDowell, S.W. Lee, C. Wang, Y. Cui, *Nano Energy* **1**, 401 (2012).
80. X. H. Liu *et al.*, *Nano Lett.* **11**, 2251 (2011).
81. H. L. Liu *et al.*, *Nano Lett.* **11**, 3312 (2011).
82. S. W. Lee, M. T. McDowell, J. W. Choi, Y. Cui, *Nano Lett.* **11**, 3034 (2011).
83. J. L. Goldman, B. R. Long, A. A. Gewirth, R. G. Nuzzo, *Adv. Funct. Mater.* **21**, 2412 (2011).

84. X. H. Liu *et al.*, *ACS Nano* **6**, 1522 (2012).
85. K. Zhao *et al.*, *J. Electrochem. Soc.* **159**, A238 (2012).
86. H. Yang *et al.*, *Nano Lett.* **12**, 1953 (2012).
87. M. K. Y. Chan, B. R. Long, A. A. Gewirth, P. Greeley, *J. Phys. Chem. Lett.* **2**, 3092 (2011).
88. A. Timmons, J. R. Dahn, *J. Electrochem. Soc.* **154**, A444 (2007).
89. S. W. Lee, M. T. McDowell, L. A. Berla, W. D. Nix, Y. Cui, *Proc. Nat. Acad. Sci. USA.* **109**, 4080 (2012).
90. K. J. Zhao, M. Pharr, S. Q. Cai, J. J. Vlassak, Z. G. Suo, *J. Am. Ceram. Soc.* **94**, S226 (2011).
91. L. Hu *et al.*, *Adv. Energy Mater.* **1**, 523 (2011).
92. J. W. Choi *et al.*, *Nano Lett.* **10**, 1409 (2010).
93. E. Peled, *J. Electrochem. Soc.* **126**, 2047 (1979).
94. D. Aurbach *et al.*, *J. Power Sources* **81–82**, 95 (1999).
95. D. Aurbach, B. Markovsky, I. Weissman, E. Levi, Y. Ein-Eli, *Electrochim. Acta* **45**, 67 (1999).
96. J. S. Gnanaraj, R. W. Thompson, J. F. DiCarlo, K. M. Abraham, *J. Electrochem. Soc.* **154**, A185 (2007).
97. A. M. Andersson, A. Henningson, H. Siegbahn, U. Jansson, K. Edstrom, *J. Power Sources* **119–121**, 522 (2003).
98. P. B. Balbuena, Y. Wang, *Lithium-Ion Batteries: Solid-Electrolyte Interphase* (Imperial College Press, London, 2004).
99. K. Edstrom, M. Herstedt, D. P. Abraham, *J. Power Sources* **153**, 380 (2006).
100. D. Aurbach, Y. Gofer, in Nonaqueous Electrochemistry, D. Aurbach, Ed. (Marcel Dekker, New York, 1999), pp. 137–212.
101. C. K. Chan, R. Ruffo, S. S. Hong, R. A. Huggins, Y. Cui, *J. Power Sources* **189**, 1132 (2009).
102. S. Leroy *et al.*, *Surf. Interface Anal.* **37**, 773 (2005).
103. H. Bryngelsson, M. Stjerndahl, T. Gustafsson, K. Edstrom, *J. Power Sources* **174**, 970 (2007).
104. A. Xiao, W. Li, B. L. Lucht, *J. Power Sources* **162**, 1282 (2006).
105. Y. M. Lee, J. Y. Lee, H. T. Shim, J. K. Lee, J. K. Park, *J. Electrochem. Soc.* **154**, A515 (2007).
106. N. S. Choi *et al.*, *J. Power Sources* **161**, 1254 (2006).
107. D. D. D. Ma, C. S. Lee, F. C. K. Au, S. Y. Tong, S. T. Lee, *Science* **299**, 1874 (2003).

108. P. Ram, J. Singh, T. R. Ramamohan, S. Venkatachalam, V. P. Sundarsingh, *J. Mater. Sci.* **32**, 6305 (1997).

109. H. Wu *et al.*, *Nature Nanotech.* **7**, 310 (2012).

110. M. Dolle, F. Orsini, A. S. Gozdz, J.-M. Tarascon, *J. Electrochem. Soc.* **148**, A851 (2001).

111. L.-F. Cui, R. Ruffo, C. K. Chan, H. Peng, Y. Cui, *Nano Lett.* **9**, 491 (2008).

112. L.-F. Cui, Y. Yang, C. M. Hsu, Y. Cui, *Nano Lett.* **9**, 3370 (2009).

113. Y. Yao *et al.*, *ACS Nano* **5**, 8346 (2011).

114. X. Gui *et al.*, *Adv. Mater.* **22**, 617 (2010).

115. L. Hu *et al.*, *Chem. Commun.* **47**, 367 (2011).

116. J. D. Holmes, K. P. Johnston, R. C. Doty, B. A. Korgel, *Science* **287**, 1471 (2000).

117. A. T. Heitsch, D. D. Fanfair, H.-Y. Tuan, B. A. Korgel, *J. Am. Chem. Soc.* **130**, 5436 (2008).

118. T. Hanrath, B. A. Korgel, *Adv. Mater.* **15**, 437 (2003).

119. H. Y. Tuan, B. A. Korgel, *Chem. Mater.* **20**, 1239 (2008).

120. A. V. Murugan, T. Muraliganth, A. Manthiram, *J. Phys. Chem. C* **112**, 14665 (2008).

121. A. M. Wilson, J. R. Dahn, *J. Electrochem. Soc.* **142**, 326 (1995).

122. W. Wang, M. K. Datta, P. N. Kumta, *J. Mater. Chem.* **17**, 3229 (2007).

123. W. Wang, P. N. Kumta, *J. Power Sources* **172**, 650 (2007).

124. G. X. Wang, J. H. Ahn, J. Yao, S. Bewlay, H. K. Liu, *Electrochem. Commun.* **6**, 689 (2004).

125. Z. P. Guo, E. Milin, J. Z. Wang, J. Chen, H. K. Liu, *J. Electrochem. Soc.* **152**, A2211 (2005).

126. S. H. Ng *et al.*, *Angew. Chem. Int. Ed.* **45**, 6896 (2006).

127. T. Zhang *et al.*, *Pure Appl. Chem.* **78**, 1889 (2006).

128. J. Li, R. B. Lewis, J. R. Dahn, *Electrochem. Solid-State Lett.* **10**, A17 (2007).

129. M. Wachtler, M. R. Wagner, M. Schmied, M. Winter, J. O. Besenhard, *J. Electroanal. Chem.* **510**, 12 (2001).

130. G. Liu *et al.*, *Adv. Mater.* **23**, 4679 (2011).

131. I. Kovalenko *et al.*, *Science* **334**, 75 (2011).

132. M. Kaempgen, C. K. Chan, J. Ma, Y. Cui, G. Gruner, *Nano Lett.* **9**, 1872 (2009).

133. T. Kim, Y. H. Mo, K. S. Nahm, S. M. Oh, *J. Power Sources* **162**, 1275 (2006).
134. Y. Zhang *et al.*, *Electrochim. Acta* **51**, 4994 (2006).
135. G. Che, B. B. Lakshmi, E. R. Fisher, C. R. Martin, *Nature* **393**, 346 (1998).
136. E. Frackowiak, S. Gautier, H. Gaucher, S. Bonnamy, F. Beguin, *Carbon* **37**, 61 (1999).
137. B. Gao *et al.*, *Chem. Phys. Lett.* **327**, 69 (2000).
138. V. L. Pushparaj *et al.*, *Proc. Nat. Acad. Sci. USA* **104**, 13574 (2007).
139. T. P. Kumar, R. Ramesh, Y. Y. Lin, G. T. K. Fey, *Electrochem. Commun.* **6**, 520 (2004).
140. S. Yoon, B. H. Ka, C. Lee, M. Park, S. M. Oh, *Electrochem. Solid-State Lett.* **12**, A28 (2009).
141. T. Muraliganth, A. V. Murugan, A. Manthiram, *J. Mater. Chem.* **18**, 5661 (2008).
142. X. Lu, F. Kang, W. A. Shen, *Electrochem. Solid-State Lett.* **9**, A126 (2006).
143. J. S. Sakamoto, B. Dunn, *J. Electrochem. Soc.* **149**, A26 (2002).
144. T. Ishihara, M. Nakasu, M. Yoshio, H. Nishiguchi, Y. Takita, *J. Power Sources* **146**, 161 (2005).
145. Y. Zhang *et al.*, *Int. J. Nanomanufacturing* **2**, 4 (2008).
146. B. Lestriez *et al.*, *Electrochem. Solid-State Lett.* **12**, A76 (2009).
147. J. Shu, H. Li, R. Yang, Y. Shi, X. Huang, *Electrochem. Commun.* **8**, 51 (2006).
148. L. Hu *et al.*, *Proc. Nat. Acad. Sci. U. S. A.* **106**, 21490 (2009).
149. J. W. Choi, L. Hu, L. Cui, J. R. McDonough, Y. Cui, *J. Power Sources* **195**, 8311 (2010).
150. L. F. Cui, L. Hu, J. W. Choi, Y. Cui, *ACS Nano* **4**, 3671 (2010).

Chapter 2

Nanoscale Anodes of Silicon and Germanium for Lithium Batteries

Jason Graetz and Feng Wang

Brookhaven National Laboratory, Sustainable Energy Technologies Department, Upton, NY 11973, USA

graetz@bnl.gov

A key challenge in the development of a new generation of lithium batteries for electric and hybrid electric vehicles is to increase energy and power density while maintaining an overall safe and low-cost system. One method being explored is to replace the conventional carbon anode with a host material that can accommodate a much greater lithium density (e.g., Si or Ge). However, these high-capacity electrodes were generally considered unsuitable for rechargeable lithium batteries since they exhibit extremely low cycling rates at room temperature and undergo large structural changes during cycling. Many of these problems can be mitigated through the use of nanoscale and nanocomposite electrode materials. Rapid lithium diffusion and lattice relaxation in nanoscale electrodes is believed to mitigate microstructural damage that is a common problem in bulk lithium alloy electrodes during cycling. In addition,

Nanomaterials for Lithium-Ion Batteries: Fundamentals and Applications
Edited by Rachid Yazami

ISBN 978-981-4316-40-8 (Hardcover), 978-981-4364-23-2 (eBook)
www.panstanford.com

nanoscale materials tend to circumvent conventional mechanisms of mechanical deterioration. These nanoscale electrodes offer enhanced kinetics, material stability (reversibility), and gravimetric capacity with respect to their bulk counterparts. Lithium alloy anodes such as Li–Si have a real potential to replace the conventional carbon anode, but this development will require a better understanding of how nanostructure affects electrochemical performance and new methods to tailor or tune these materials on the nanoscale. In this chapter, we present an overview of our research on the electrochemical reaction of lithium with nanoscale Si and Ge electrodes, degradation mechanisms, routes to improve performance, along with an update of recent developments in this area.

2.1 Introduction

Widespread commercialization of plug-in hybrid-electric and all-electric vehicles is dependent upon the development of safe electrical energy storage systems with high energy and power densities. The lithium battery appears to be the best candidate to replace nickel–metal hydride batteries in these vehicles, but a number of challenges remain. A key requirement is the development of new, stable, nontoxic electrode materials that can accommodate greater lithium densities. For the anode, the simplest and greatest energy densities are achieved with pure lithium metal. However, issues of safety resulting from internal shorts that can form during cycling have confined lithium anodes to primary and small rechargeable batteries. Larger, rechargeable lithium batteries require a framework material to host the lithium during cycling. The most common lithium anode is graphite, which is stable upon lithium intercalation up to LiC_6 (372 mAh/g). Lithium alloy electrodes (e.g., $M + x\text{Li} \leftrightarrow \text{Li}_xM$) and lithium conversion electrodes (e.g., $MO_x + x\text{Li} \leftrightarrow (x/2)\ \text{Li}_2O + M$) are currently of interest as "next generation" lithium anodes due to their extremely high energy densities (3–10 times greater than that of LiC_6) and low operating voltage *versus* lithium (0–1 V). The average voltage and theoretical capacity of lithium metal, LiC_6, and a number of alternative lithium anodes are shown in Table 2.1.

Table 2.1 Average voltage (*vs.* Li metal) and the theoretical specific capacity of a variety of possible lithium anodes. The theoretical specific capacities are given for the charge (lithiation) reaction except for lithium metal, which is given for the discharge (delithiation) reaction

Anode	Potential *vs.* Li metal (V)	Theoretical capacity (mAh/g)
Li Metal	0.0	3860
Li–Al	0.1–0.6	990
Li–C	0.0–0.5	372
Li–Si	0.1–0.5	4010
Li–Ge	0.2–0.5	1620
Li–Sn	0.1–1.2	790

Due to the low voltage and high theoretical capacities, silicon and germanium are obvious candidates for the lithium anode. The Li–Si and Li–Ge phase diagrams are replete with alloy phases with high lithium content. A few of the possible Li–Si compositions are shown in the reaction below:[1-5]

$$Si + xLi^+ + xe^- \rightarrow Li_xSi \ (x = 1.71, 2.33, 3.25, 4.20) \quad (2.1)$$

Similarly, germanium also exhibits a number of lithiated phases (see Ref. 6 and references therein):

$$Ge + xLi^+ + xe^- \rightarrow Li_xGe \ (x = 1.00, 1.83, 2.25, 3.00, 3.50, 3.75, 4.40) \quad (2.2)$$

In the highest lithiated phase, silicon and germanium accommodate over four lithium atoms per host atom. Figure 2.1 shows the crystal structure of a unit cell of silicon [germanium has the same structure with a slightly larger lattice parameter (4%)] compared with a single unit cell of $Li_{21}Si_5$ and $Li_{22}Ge_5$. The crystalline densities of Si and $Li_{21}Si_5$ are 2.34 g/cm^3 and 1.16 g/cm^3, respectively, which implies a volume expansion of just over 400% between Si and the fully lithiated $Li_{21}Si_5$ phase. Similarly, a roughly 370% volume expansion occurs between Ge (5.32 g/cm^3) and $Li_{22}Ge_5$ (2.04 g/cm^3).

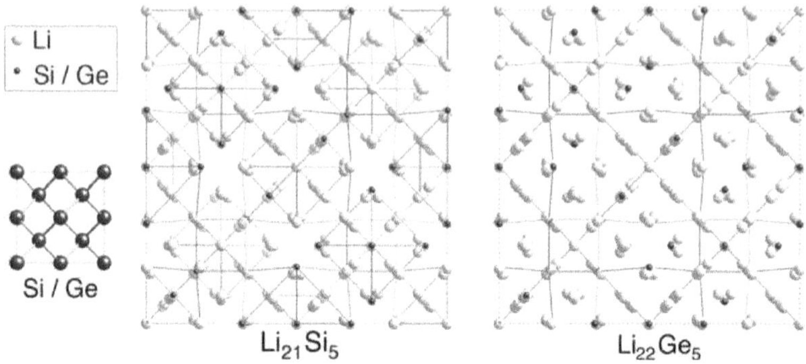

Figure 2.1 Crystallographic structures of a unit cell of Si/Ge, $Li_{21}Si_5$, and $Li_{22}Ge_5$ along the (001) direction.

For complete lithiation, the theoretical specific capacities of silicon and germanium are 4010 mAh/g and 1620 mAh/g, respectively. In recent studies, it has been shown that the maximum capacities achieved by electrochemical lithiation at room temperature are 3579 mAh/g and 1385 mAh/g, which correspond to fully lithiated phases of $Li_{15}Si_4$ and $Li_{15}Ge_4$, respectively.[7,8] However, these capacities are still far greater than the capacity of the commercially available graphitic carbons (372 mAh/g), which are currently the standard anode material in lithium secondary cells. Furthermore, lithium alloys do not suffer from the solvent co-intercalation that can occur in graphitic carbons, and may degrade the storage capacity.[9–13] As a consequence, silicon has received considerable attention in recent years as a leading candidate to replace carbon at the anode. Investigations of the analogous Li–Ge system have received little attention owing to the high cost and lower energy density with respect to the silicon system. Despite these issues, germanium has a number of advantages over silicon, one example of which is its weak affinity to form a native oxide layer. In a high surface area material a thin contamination layer can lead to a large irreversible capacity on the first cycle, a larger cell impedance and an overall lower capacity. Therefore, Li–Ge is a model alloy for investigating the intrinsic properties of the Li–X (X = Si, Ge, Sn) system while avoiding artifacts due to a native oxide layer (such as the displacement reaction involving lithium oxidation and tin reduction in the SnO_2 system).

Although costly, Li–Ge may still be a reasonable candidate anode for some lithium battery applications. Germanium is an abundant element in the earth's crust and the current price is maintained by the lack of demand. The Li–Ge anode also has a high lithium diffusivity, which is approximately 15 times greater than that of lithium in silicon at 360°C (2.14×10^{-7} cm^2/sec for Ge and 1.47×10^{-8} cm^2/sec for Si).[14] At room temperature the diffusion rate of lithium in germanium is approximately 400 times that of lithium in silicon (using the empirical diffusion equation $D = D_0 \exp[-Q/RT]$ with Q_{Si} = 64 kJ/mol and $D_0^{Si} = 2.3 \times 10^{-3}$ cm^2/sec and Q_{Ge} = 49 kJ/mol and $D_0^{Ge} = 2.5 \times 10^{-3}$ cm^2/sec).[14]

The challenge with silicon and germanium electrodes is that, unlike graphite, they do not insert lithium topotactically and undergo substantial structural changes during lithiation. As a result, these anodes are generally plagued by slow kinetics at room temperature. More importantly, a large volume expansion occurs during lithiation (up to 370% for Ge and 400% for Si). This generates enormous mechanical stresses within the brittle material, which is pulverized during the first few charge/discharge cycles as illustrated in Fig. 2.2.[12,13,15–17] Ultimately, the capacity is considerably reduced due to the loss of electrical integrity between the particles and the current collector. It has been demonstrated that bulk silicon loses approximately 90% of its initial capacity after five cycles at ambient temperature.[18,19] In addition, the solid electrolyte interphase (SEI) can also be a problem in these anodes since this layer is reformed on fresh surfaces with each cycle, which ultimately lowers the capacity and increases the impedance.

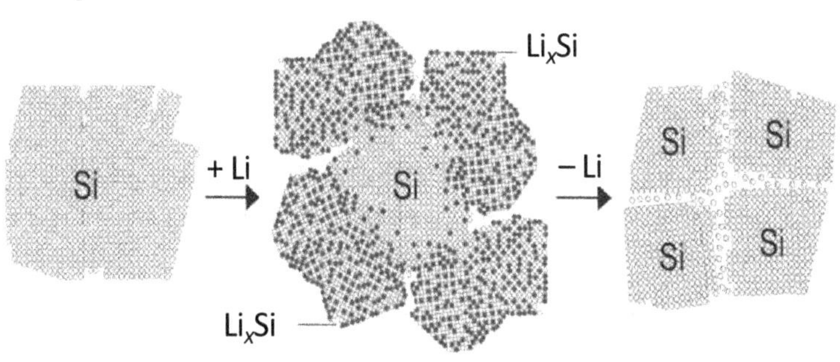

Figure 2.2 Schematic model of decrepitation occurring during electrochemical cycling of lithium in a bulk silicon anode.

2.2 Improved Performance with Nanoscale Electrodes

Many of the problems associated with electrochemical cycling of lithium in alloy and conversion electrodes may be mitigated with nanoscale and nanostructured materials. Their high surface area and small dimensions allow the host to accommodate the lithium quickly with minimal internal strain. There has been considerable effort recently to utilize nanoscale electrode materials to improve the cycle life in electrodes that typically suffer from mechanical deterioration due to large structural changes. This is especially true for the lithium alloy anodes such as aluminum,[20] silicon,[18–21] germanium,[22] tin,[23,24] and other metals.[25,26] It was shown that silicon nanoparticles (~80 nm) could achieve reversible capacities of up to 1700 mAh/g over 10 cycles,[18] which is a large improvement over bulk silicon. Other nanostructured silicon anodes such as silicon nanowires have also demonstrated high capacity with improved cycle life.[21,27] Similar investigations into nanosized intermetallic silicon alloys have shown some reversibility with reduced additives, but with lower gravimetric capacity (1100 mAh/g on the second cycle).[28] Other attempts to suppress particle decrepitation include the replacement of the large metallic host crystals with a nanophase host material such as Sn/SnSb.[12,15] In addition to nanoparticles, thin films (~1.2 μm) of silicon have been prepared by chemical vapor deposition in an attempt to mitigate the capacity loss in the silicon system.[15] Amorphous thin films of silicon and germanium have demonstrated reversible capacities of >1800 mAh/g and >1600 mAh/g, respectively over 50 cycles.[19,22]

2.2.1 Cycle Life: Strain Reduction and Defect Elimination

There are a number of reasons for the enhanced cycle life in nanoscale materials. First, the relatively open structure of nanoscale materials seems capable of accommodating a large volume expansion during lithiation. Secondly, the conventional mechanisms for microstructural damage are often absent in nanoscale materials. Dislocations are usually unstable in crystals with nanoscale dimensions (<20 nm) due to image forces, which attract the dislocation to the surface. For brittle materials such as silicon and germanium, decrepitation occurs through the formation of cracks and their propagation by dislocation emission from the crack tip. For a crack to propagate, however, it must exceed a critical size (a_c), where

$$a_c = (2/\pi)(K_{1c}^2/\sigma^2) \quad (2.3)$$

The fracture toughness, K_{1c}, and yield strength, σ, in polycrystalline silicon are approximately 0.751 MPa $m^{1/2}$ and 1.1 GPa, respectively.[29,30] These values yield a critical size of ~300 nm for silicon and a roughly similar value for germanium, which has comparable material properties. The size dependence of fracture, was recently verified by the direct observation of lithiation of single particles.[31,32]

Gradients in strain can cause defect generation in solids, so gradients of lithium concentration could perhaps cause microstructural damage in bulk silicon and germanium. The diffusion of lithium into the host distorts the lattice parameter, a, to approximately $4^{1/3}a$. An advantage of nanoscale materials is that their relaxation time for diffusion is short, owing to their small dimensions. The strain is not intrinsically harmful to the host, but rather it is the gradients in strain that are responsible for breakup of the electrode during cycling. Therefore, when the diffusion lengths are short, the relaxation time is short and the material is strained quickly and uniformly and is less likely to decrepitate.

2.2.2 Kinetics

The cycling kinetics of a lithium cell is typically limited by the solid-state diffusion of lithium in the host electrodes. Recent studies on nanostructured lithium anodes of carbon,[33] silicon,[19] germanium,[22] and tin[34,35] have demonstrated fast charging rates (less than 1 min) in materials with extremely low room temperature diffusivities. Similarly, rapid cycling rates have also been reported for various nanoscale metal oxides with rates as high as 4000 C (where a rate of 1 C corresponds to a complete charge or discharge in 1 hr) for a nanostructured V_2O_5 cathode.[36] The rapid kinetics of Li transport in nanoscale materials can, in part, be attributed to the large volume fraction of grain boundaries, which provide additional diffusion pathways and also have a lower activation barrier.

A more significant factor in improving the kinetics is the short diffusion distances in these nanoscale materials. The rate of atomic diffusion in a three-dimensional solid is given by $\tau = d^2/4D$, where τ is the time required for an atom to diffuse a distance d, and D is the diffusivity. Therefore, the time required for Li to saturate (fully charge) a 10 nm particle is 10^4 times faster than for a 1 μm particle. This is further illustrated in Fig. 2.3, a plot of the time scales required for Li to diffuse through a few different electrodes. This figure shows that fast cycling rates can be achieved even in materials with low lithium diffusivity (e.g., Si and Ge), provided the diffusion lengths are short. These results indicate that nanoscale (10–100 nm) cathodes of Li(Ni, Co)O_2 should cycle up to and beyond 4000 C (~1 second charge/discharge). Similarly, nanophase lithium alloys should tolerate rates of at least 100 C in germanium and a few C in silicon without a considerable reduction in cell capacity.

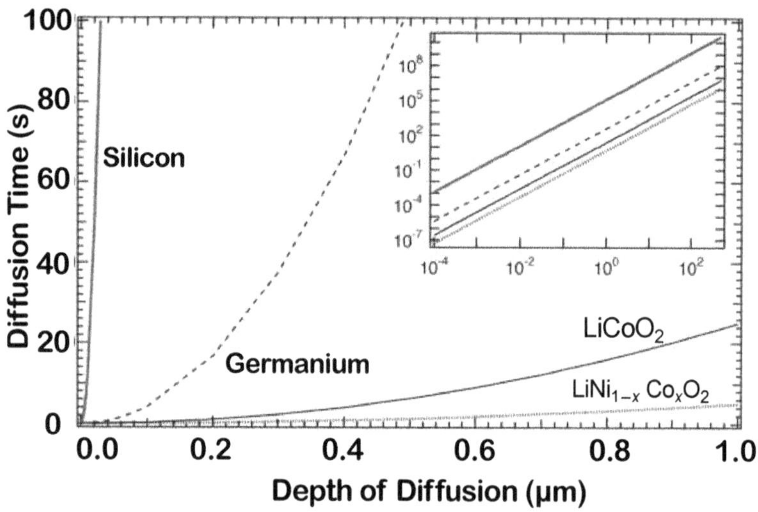

Figure 2.3 Timescales for lithium diffusion in bulk Si, Ge, $LiCoO_2$, and $LiNi_{1-x}Co_xO_2$ at ambient temperature with lithium diffusivities of $D_{Si} = 1.8 \times 10^{-14}$ cm²/s, $D_{Ge} = 7.7 \times 10^{-12}$ cm²/s, $D_{LiCoO_2} = 10^{-10}$ cm²/s, and $D_{LiNiCoO_2} = 5 \times 10^{-10}$ cm²/s, respectively.[14,37,38] The inset demonstrates the same relationship displayed logarithmically.

2.3 Synthesis of Nanoscale Silicon and Germanium Electrode Materials

Nanoscale electrode materials of silicon and germanium can easily be prepared by a number of different chemical and physical vapor deposition methods. In the present work we prepared nanocrystalline silicon and germanium materials by gas-phase condensation and ballistic consolidation,[39] using a deposition chamber shown in Fig. 2.4a. A "forming gas" of 90% N_2 and 10% H_2 was streamed along the axis of the chamber with a differential pressure of 2 torr. An elemental charge was heated in a tungsten wire basket and evaporated into the gaseous atmosphere. The rate of evaporation was up to 10^{-3} g/cm^2/s (at approximately 1800°C). In the gas, the atoms are cooled rapidly, and nanocrystals are nucleated as the atoms collide. The nanocrystals move by Brownian motion in the gas, forming loose agglomerates when they meet. These particles, entrained in the gas, approach the sound velocity as the gas is drawn through a nozzle. The particles form a thin film of ballistically consolidated nanocrystals when they hit the substrate at high speed.

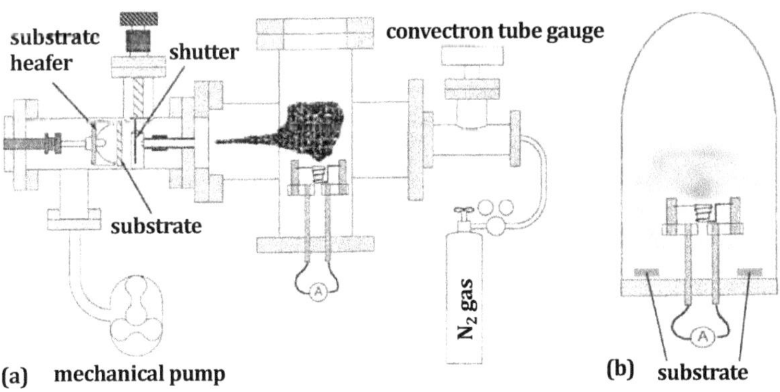

Figure 2.4 Schematics of the (a) gas-phase ballistic consolidation chamber and (b) bell jar evaporator.

Amorphous thin films of silicon and germanium were prepared by physical vapor deposition using a bell jar thermal evaporator (Fig. 2.4b). A charge of elemental material was evaporated under a vacuum of 6×10^{-6} torr in a tungsten wire heating basket. A nickel/copper substrate was placed directly below the tungsten basket, and the evaporated material was deposited onto the substrates in the form of a continuous thin film of approximately 100 nm in thickness (Fig. 2.4b).

In both cases the electrodes were intentionally made very thin, without binders or conductive additives, so that the intrinsic properties of nanostructured materials could be measured. Electrode substrates were prepared by first roughening the surface of the stainless steel plates (to be used as the spacers in 2016 coin cell) using 400 grit sandpaper. A thin nickel/copper coating (~100 nm) was then evaporated on the surface and finally, the silicon/germanium was deposited directly onto the metal-coated planar substrate.

2.4 Nanostructure and Electrochemical Performance

2.4.1 Nanocrystalline Silicon

The ballistically deposited silicon forms a web of interconnected nanocrystalline particles as shown by the bright-field transmission electron microscopy (TEM) image in Fig. 2.5a. The material has an open, high surface area structure with crystallites of typically 20–25 nm in diameter. Electrochemical cycling was performed in half-cells (using lithium metal as a counter electrode) with a standard electrolyte (1:1 EC/DMC with $LiPF_6$) and a 0.5 mm–thick fiberglass separator. The electrodes were cycled at a rate of approximately C/4 (equivalent to a 4 hour charge and a 4 hour discharge). The voltage profiles from electrochemical cycles 1, 25, and 50 are displayed in Fig. 2.6a. This electrode exhibited an initial discharge capacity of 2400 mAh/g during the first insertion of lithium, and a subsequent charge capacity of 1000 mAh/g, giving a columbic efficiency of 41% for the first cycle. This high irreversible capacity was limited to the first cycle, however. A more stable specific capacity of approximately 1000 mAh/g was observed on cycles 2–50. The capacity fade correlates inversely with the columbic efficiency, which was found to increase steadily up to 96% by cycle number 9. In this reversible region, the nanocrystalline electrode exhibited a mean capacity loss of approximately 20 mAh/g per cycle, leading to a final capacity of 525 mAh/g on cycle number 50.

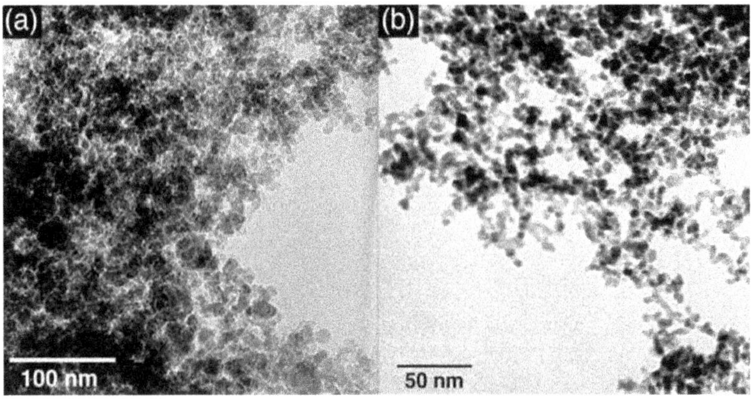

Figure 2.5 Bright-field TEM images of ballistically deposited (a) silicon and (b) germanium nanocrystals.

A close examination of the voltage profile can yield much information about the physical and chemical process occurring during lithiation. In addition to the lithium concentration, phase transitions are also easily identified by plateaus in the voltage profile. This is a consequence of the Gibbs phase rule, which states that the number of degrees of freedom (f) is a function of the number of components (c) and the number of phases (p) present:

$$f = c - p + 2. \tag{2.4}$$

Therefore, in a two-phase region, the chemical potential (or voltage) is fixed over changes in lithium concentration as one phase grows at the expense of the other.

The voltage profile of Fig. 2.6a exhibits a reasonably smooth slope on both the charge and discharge cycles. The differential capacity, $\mathrm{d}|x|/\mathrm{d}E$, where $|x|$ is the lithium concentration and E is the cell potential, is used to accentuate changes in the slope of the potential curve (Fig. 2.6b). Peaks in the differential capacity indicate regions of the potential where lithium ions are entering nearly equipotential sites. This plot suggests that the lithium insertion in nanocrystalline silicon occurs at 100 and 250 mV and lithium extraction occurs at 300 and 450 mV. The disparity is due to the over potential resulting from the constant current, or nonequilibrium state.

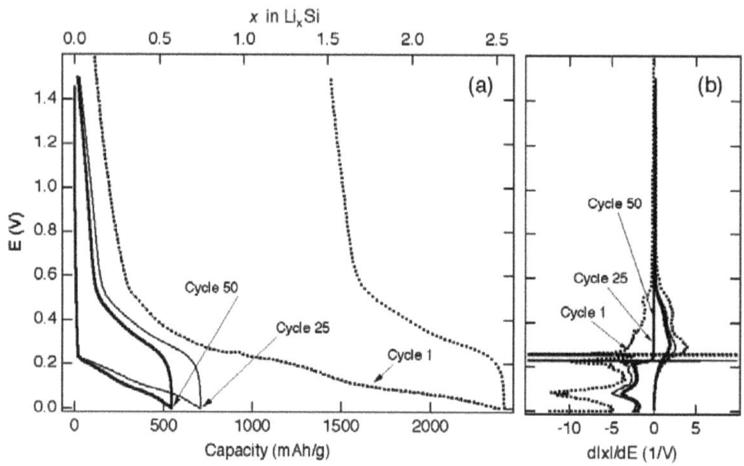

Figure 2.6 (a) Voltage profile and (b) differential capacity, $d|x|/dE$ for cycles 1, 25, and 50 from nanocrystalline silicon on a planar copper substrate.

The large native oxide on the surface of the silicon particles may be partially responsible for the reduced capacity observed in this material. Although SiO_2 does not typically react with Li, it may become more active in a disordered state and form either a lithiated phase, Li_4SiO_4, or Si and Li_2O during the first discharge.[40,41] This reaction is irreversible and typically only occurs during the early stages of the first discharge. In addition to the reduction of the disordered surface oxide, the low columbic efficiencies of the first few cycles can also be attributed to the formation of a solid–electrolyte interphase (SEI). This passivation layer is expected to form through a reaction of lithium with the electrolyte. Scanning electron microscopy images of the ballistically deposited silicon clusters after the first discharge (not shown here) revealed smooth electrode surfaces, in contrast with the fine-scale roughness of the as-deposited material. Although the change in the surface appearance could be partially attributed to the volume expansion during lithiation, there appears to be a contiguous passivation layer, which also emerges after the first lithium insertion. The formation of SEI leads to an irreversible capacity through two mechanisms, the loss of lithium to the formation of the SEI and an increase in the cell impedance. Since the sharp capacity fade is limited to the first few cycles, it is likely that the reactions contributing to the SEI layer occur mainly during the initial cycles.

2.4.2 Nanocrystalline Germanium

The ballistically deposited germanium also forms a low-density web of interconnected nanocrystalline particles (Fig. 2.5b) with a crystallite size of typically 5–10 nm in diameter. Electrodes of nanocrystalline germanium were prepared and cycled under the same conditions as previously described for silicon. A plot of the voltage profile from nanocrystalline germanium is shown in Fig. 2.7a. The most salient feature of this plot is the high reversible capacity indicating that the nanocrystalline electrode hosts up to 3.8 lithium atoms per germanium atom. A plot of the differential capacity, $d|x|/dE$, is displayed in Fig. 2.7b. This plot suggests that lithium insertion occurs at 190 and 350 mV and lithium extraction occurs at 400 and 550 mV.

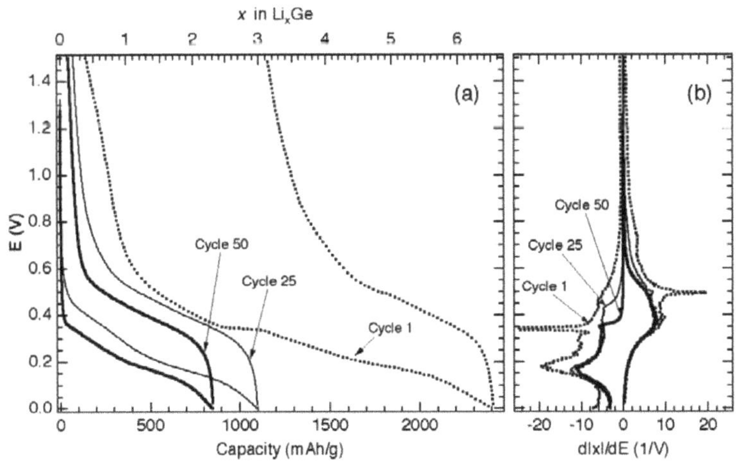

Figure 2.7 (a) Voltage profile and (b) differential capacity, $d|x|/dE$ for cycles 1, 25, and 50 from nanocrystalline germanium on a planar copper substrate.

The large irreversible capacity observed on the initial cycle may be attributed to a partial reduction of the native surface oxide and the formation of an SEI layer, as previously discussed for the silicon system. The peaks in the differential capacity [350 mV (insertion) and 500 mV (extraction)] in the first cycle disappear after the first few cycles and are indicative of the solid-electrolyte reaction to form the SEI layer. The first cycle capacity loss is approximately 70% of what was observed in the Li–Si system. The lower irreversible capacity on the first cycle is attributed to a reduced native oxide layer, which is only a few monolayers thick.

The reversible cycling of up to 3.8 lithium atoms per germanium atom in the nanocrystalline electrode is considerably larger than the capacities measured in the analogous silicon system, which exhibited a reversible capacity of 1.1 lithium atoms per silicon atom. The enhanced lithium uptake is attributed to the higher diffusivity of lithium in germanium at room temperature ($D_{Ge} = \sim 400\ D_{Si}$).

2.4.3 Effects of Lithiation on Nanoscale Silicon and Germanium

A bright-field TEM image and associated electron diffraction pattern from the nanocrystalline silicon before and after lithiation are displayed in Fig. 2.8. As expected, a significant volume expansion is evident with average particle size of 25–30 nm before lithiation and approximately 45–50 nm after lithiation. Other structural changes are clearly evident in these images. The electron diffraction pattern from the as-deposited silicon exhibits sharp rings indicative of a polycrystalline material. However, the broad diffuse rings seen in Li–Si diffraction pattern are indicative of an amorphous material. The crystalline silicon transforms into a metastable glassy phase at room temperature through electrochemically driven solid-state amorphization. Although the amorphous phase is not thermodynamically preferred, the sluggish mass transport at these temperatures seems to prevent any crystalline phases from forming. Evidence of the crystalline-to-amorphous transformation can also be observed in the voltage profiles of Fig. 2.6. The occurrence of a few broad peaks suggests that some local ordering does occur during cycling. However, the four sharp peaks expected from the four crystallographic phase transitions expected from the phase diagram are not observed during electrochemical lithiation at room temperature. It is important to note that the amorphization is driven by the room temperature thermodynamics and kinetics and is not attributed to the nanoscale microstructure.[42]

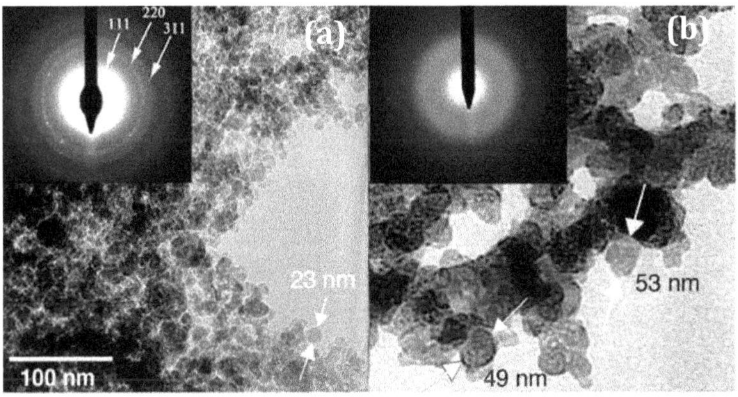

Figure 2.8 Bright-field TEM image and electron diffraction pattern (inset) of a silicon electrode (a) before and (b) after electrochemical lithiation.

Electrochemical lithiation of germanium at room temperature does not appear to undergo an electrochemically driven amorphous transition. A recent *ex situ* X-ray diffraction (XRD) study has demonstrated a number of crystallographic phase changes occurring during electrochemical lithiation.[43] The sequential formation of the following phases occurred during lithiation: Li_9Ge_4, Li_7Ge_2, and a mixture of $Li_{15}Ge_4$ and the fully lithiated phase $Li_{22}Ge_5$. But the more recent *in situ* XRD studies suggest that the end member of LiGe in the thin film electrodes system is $Li_{15}Ge_4$.[8]

2.4.4 Amorphous Silicon and Germanium

Structural phase transitions that occur during lithiation are typically undesirable since they often lead to slow kinetics and poor cycle life. In that respect, the electrochemically driven amorphization of silicon is advantageous since it allows lithiation to occur while bypassing multiple crystallographic transitions. However, the electrochemical properties of silicon and germanium could likely be further improved by initially preparing the electrodes in an amorphous state. Thin films of amorphous silicon and germanium (~100 nm in thickness) were prepared by evaporation as described above. Figure 2.9 shows the bright-field TEM images of the evaporated silicon and germanium films and the associated electron diffraction patterns. The absence of sharp peaks in the diffraction patterns suggests that the materials are entirely amorphous and the lack of structure in the bright-field images indicates that they are deposited as contiguous films, unbroken by grain boundaries, or cracks.

Figure 2.9 Bright-field TEM images of amorphous thin films of (a) silicon and (b) germanium prepared by evaporation. The corresponding electron diffraction patterns are shown in the insets.

Electrodes of silicon and germanium amorphous films were prepared by depositing the material directly onto a copper substrate. The electrochemical cells were prepared and cycled under conditions previously described for the nanocrystalline materials. The thin films exhibited excellent electrochemical performance as demonstrated in Fig. 2.10a and c, with voltage profiles obtained from cycles 1, 25, and 50 for silicon and germanium amorphous films, respectively. Plots of the differential capacity are shown in Fig. 2.10b and d, respectively.

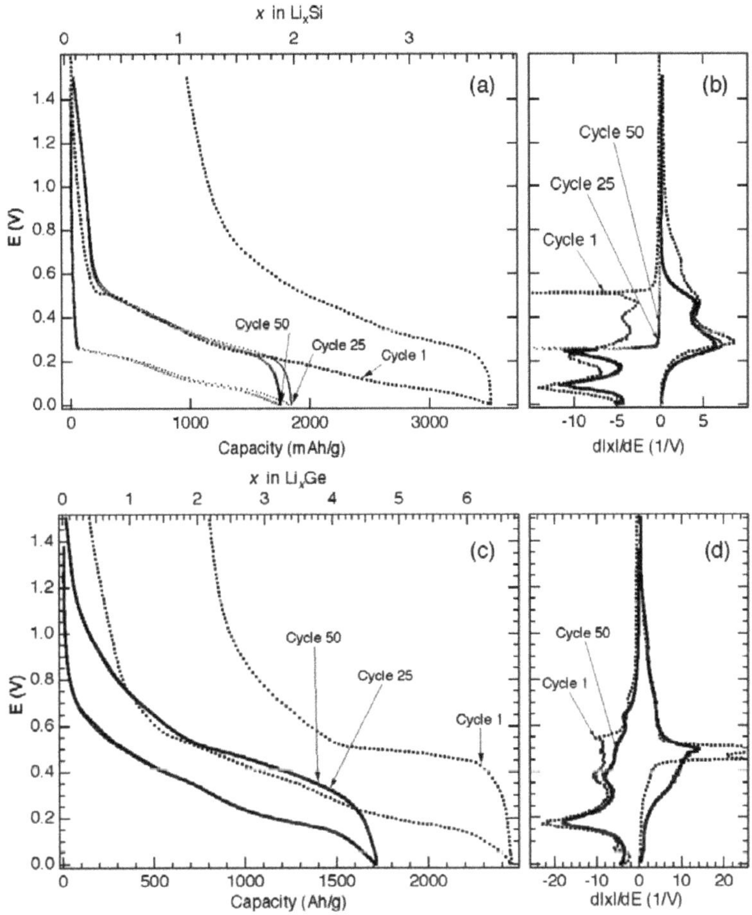

Figure 2.10 Voltage profile and differential capacity from amorphous thin films of silicon (a and b) and germanium (c and d) for cycles 1, 25, and 50.

For the silicon film, an initial discharge capacity of ~3500 mAh/g, and charge capacity of 2500 mAh/g were measured, which yields a columbic efficiency of 71% for the first cycle. Upon subsequent cycling, the electrode exhibited a rather stable specific capacity ~2000 mAh/g with a columbic efficiency of 98% on cycle 9. After 20 cycles, the amorphous thin film exhibited a mean capacity loss of only 8 mAh/g per cycle. The differential capacity plot (Fig. 2.10b) suggests that the lithium insertion occurs at 100 and 250 mV and lithium extraction occurs at 300 and 450 mV, similar to what was observed with the nanocrystalline silicon.

The evaporated amorphous germanium film exhibited exceptional electrochemical performance with small irreversible capacity on the first cycle. Upon subsequent cycling the film accommodates a theoretical maximum amount of lithium per germanium, which corresponds to roughly $Li_{22}Ge_5$. This is somewhat surprising since most studies have shown that electrochemical lithiation is only possible up to $Li_{15}Ge_4$. In this case, the amorphous thin film seems to remain amorphous during lithiation, avoiding the formation of the crystalline $Li_{15}Ge_4$ phase. A plot of the differential capacity (Fig. 2.10d) suggests that the lithium insertion occurs at approximately 190 mV and over the range of 350–600 mV and lithium extraction occurs over a range of 500–200 mV. It is interesting to note that the differential capacity plot from amorphous germanium shows a broad lithium insertion peak with a long tail that extends up to 600 mV. Lithium is inserted in a continuous process over a wide range of voltages. This is in contrast to the differential capacity plot from the nanocrystalline material (Fig. 2.7), which shows no lithium insertion above ~400 mV after the first few cycles.

The stability of the amorphous nanofilms during cycling is rather surprising.

Although these electrodes are thin, suggesting rapid lithium transport perpendicular to the films, the electrodes are attached to a rigid substrate. The large volume expansion that occurs during lithiation is expected to create considerable strain gradients as the lithium front propagates in and out of the film. The strains are more than sufficient to overcome the surface energy bonding the film to the substrate. However, there cannot be a complete decohesion of the film during cycling because the film remains electrically intact. It is possible that, during lithiation, there exist localized low energy sites that are preferentially filled. The preferential ordering is supported by the voltage profiles and differential capacity plots in Fig. 2.10. The local ordering of lithium would effectively reduce the mechanical strains of lithiation by allowing the volume changes to occur more homogeneously. Although the preferential ordering may reduce some of the alloying strains, it is also likely that the film is buckling and cracking in some regions. This would help dissipate much of the lattice stress while still allowing the film to maintain contact with the current collector.

2.4.5 Cycle Life

Cycle life plots from electrodes of bulk (>1 μm), nanocrystalline (<50 nm) and amorphous silicon and germanium are displayed in Fig. 2.11. The nanoscale electrodes (crystalline and amorphous) were cycled at a rate of approximately C/4 between 0 and 1.5 V, whereas the bulk electrodes were cycled at a much slower rate (~C/30) to maximize the specific capacity. Despite the gentle cycling conditions, bulk silicon and germanium exhibit extremely poor cycle characteristics, with nearly complete capacity loss by the seventh cycle. The nanocrystalline silicon and germanium electrodes exhibit a large first-cycle irreversible capacity, and subsequent constant capacity fade of approximately 10 mAh/g per cycle in both electrodes. A similar first cycle capacity loss is observed with the silicon and germanium amorphous films. However, both amorphous films exhibited a much longer cycle life with the germanium showing no appreciable capacity loss over 50 cycles, which may be due to its higher mechanical robustness.[44]

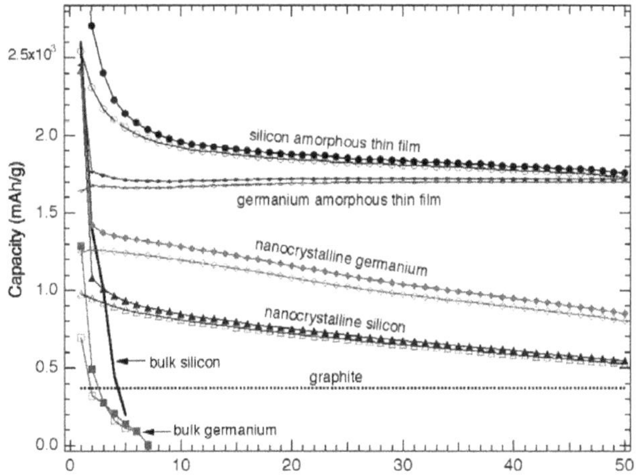

Figure 2.11 Gravimetric capacity of bulk, nanocrystalline, and amorphous films of silicon and germanium over 50 cycles. The dotted line represents the theoretical capacity for the commercially available lithium-intercalated graphite.

These results broadly demonstrate the utility of nanoscale electrodes, which yield greater capacities and cycle life than their bulk counterparts. We showed that improved electrochemical performance in the alloy electrodes could be achieved by preparing the electrodes in an amorphous rather than nanocrystalline state. Finally, we demonstrated that silicon and germanium are viable lithium electrodes and, when prepared with the proper nanostructure and morphology, can be cycled 50 times with little capacity loss.

2.5 Reaction of Lithium with Si: Nano *vs.* Bulk

Despite recent improvements, the alloying anodes, such as Li–Si and Li–Ge, continue to be plagued by poor cycling stability and slow kinetics. In order to better understand the mechanisms responsible for this degradation, more information is needed about how the local structure changes with lithiation and how it impacts the electrochemical behavior. Recently, the chemical and structural changes occurring in bulk and nanoscale silicon electrodes were investigated by *in situ* synchrotron XRD, *ex situ* TEM and electron energy-loss spectroscopy (EELS).[45] With the high-flux of the synchrotron beam, hundreds of XRD patterns were acquired over the first discharge (lithiation) to monitor the subtle structural changes occurring in the electrode over small voltage intervals (Fig. 2.12a). Elemental mapping (by energy-filtered imaging) with nanometer-scale resolution was performed at various stages of lithiation to interrogate where the lithium was going and how the particle morphology was changing at each stage. In contrast to bulk Si, which exhibits a linear decay of the diffraction intensity as a result of amorphization with lithiation,[46,47] the intensity of nano silicon falls more gradually at the early stage of lithiation (Fig. 2.12b), and drops rapidly toward the end. In addition, the reaction with nanoscale silicon starts at a higher potential than the bulk Si. This is attributed to the high surface area and abundant grain boundaries and other planar defects in the nanoscale Si. No $Li_{15}Si_4$ or other crystalline phases were detected by XRD in the nanoscale Si, over the full discharge (lithiation) process. This may be attributed to the nanoscale particle dimensions as reported for thin films and nanowires.[48,49] TEM and electron diffraction results clearly showed that in the fully lithiated state, the Si nanoparticles become mostly amorphous with lithiation, but some unreacted Si-rich 2–3 nm nanoclusters were also observed.

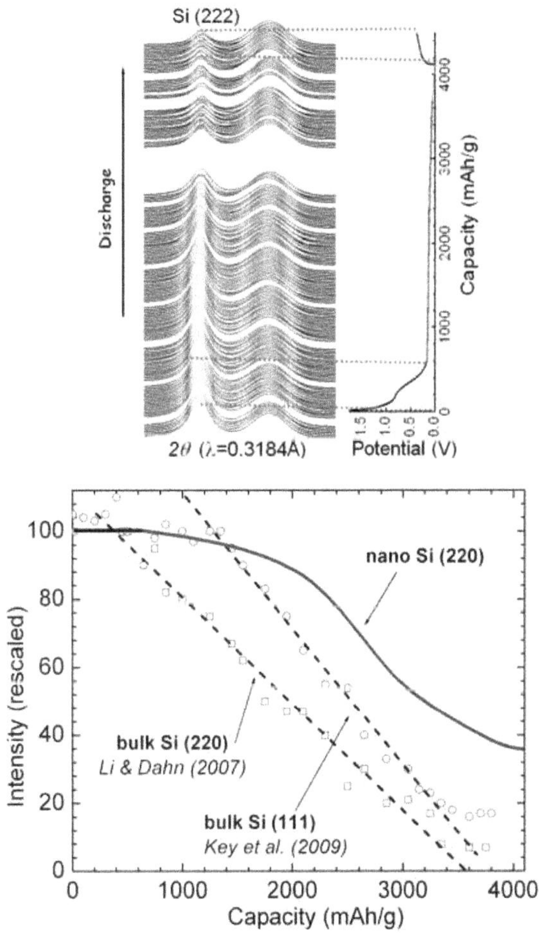

Figure 2.12 (a) Structural evolution of the silicon nanoparticles by *in situ* synchrotron XRD over the first discharge at a rate of C/100; (b) a comparison of the peak intensity of the nano-Si to the published results by Li and Dahn[46] and Key *et al.*[47] The broad peaks at ~10 degrees are due to the copper film (current collector).

2.6 Summary and Outlook

Over the past few years, intense research on the synthesis and characterization of nanostructured silicon and germanium electrodes has led to a much better understanding of the electrochemical reaction with lithium, but there is still a long way to go. A summary and outlook of a few key areas is given below:

(i) ***Fundamentals of the electrochemical reaction***: Much progress has been made on the phase transition behavior,[7,46–49] size dependence of mechanical fracture,[19,31] local alloying process during (de)lithiation and anisotropic expansion.[50–52] However, how nanostructure and morphology affect electrochemical performance, or conversely, how electrochemical cycling changes the nanostructure and morphology, is still not well understood. Further studies are necessary to determine how nanostructure and morphology can be manipulated to achieve optimal electrochemical performance. New tools need to be developed for nanoelectrochemistry where electrodes can be studied by a direct, *in situ* probe with nanoscale resolution during electrochemical cycling. These tools will reveal what types of structural changes occur and give us a better understanding of kinetics and mass transport, stress management and the role of defects in nanostructured materials. Real batteries utilizing nanoscale materials will likely require the development of new scaffolds or templates to maintain continuity with the current collector and act as a host matrix for the nanomaterials. Ultimately, new combinatorial methods will need to be developed, allowing us to scan through hundreds-to-thousands of electrode variations in order to optimize the structure, composition, density and morphology of new nanocomposite electrodes.[53]

(ii) ***Integrated Si or Ge electrodes for use in solid-state batteries***: The development of new electrode materials for rechargeable lithium batteries will likely continue to have a profound technological impact in the future. Advances in thin film solid-state cells will allow for the miniaturization and the integration of the lithium cell with system electronics (e.g., "battery on a chip").[53–55] Despite the large volume expansion in Si, the demonstration of high performance in all-solid-state batteries with silicon electrodes shows the promise of using Si in this type of application.[56] The resulting impact on microelectronics and wireless communications will lead to a number of improvements in existing technologies such as cellular phones, laptop computers and implantable medical devices.

(iii) ***Stabilized Si and Ge electrodes for a new generation of batteries***: One of the most common methods of reducing the problems associated with pulverization and decrepitation is to prepare the electrodes as composites.[57–61] Additional performance improvements have been achieved by engineering the materials into amorphous films,[19] columns,[62] and nanowires,[63] to better accommodate the large volume expansion. In addition, electronically conductive binders[64] and core-shell structures[65] have clearly demonstrated the increase of capacity retention during cycling. It is expected that, in the new generation of batteries, the energy density can be raised to 400 Wh/kg (cell level) by using a high-energy anode incorporating Si, coupled to a high-capacity, lithium-metal-oxide cathode.[66] It is likely that the high energy and power densities available in the next-generation lithium cells will lead to improved electrical energy storage systems for automotive applications, allowing us to shift away from petroleum-based energy toward more renewable sources.

References

1. R. A. Sharma, R. N. Seefurth, *J. Electrochem. Soc.*, **123**, 1763–1768 (1976).
2. C. J. Wen, R. A. Huggins, *J. Solid State Chem.*, **37**, 271–278 (1981).
3. C. van der Marel, G. J. B. Vinke, W. van der Lugt, *J. Solid State Comm.*, **54**, 917–919 (1985).
4. R. Nesper, H. G. von Schnering, *J. Solid State Chem.*, **70**, 48–57 (1987).
5. W. J. Weydanz, M. Wohlfahrt-Mehrens, R. A. Huggins, *J. Power Sources*, **81**, 237–242 (1999).
6. M. R. St. John, A. J. Furgala, A. F. Sammells, *J. Electrochem. Soc.*, **127**, C136–C137 (1980).
7. M.N. Obrovac, L. Christensen, *Electrochem. Solid State Lett.*, **7**, A93–A96 (2004).
8. L. Baggetto, P. H. Notten, *J. Electrochem. Soc.*, **156**, A169–A175 (2009).

9. J. O. Besenhard, H. P. Fritz, *J. Electroanal. Chem.*, **53**, 329–333 (1974).
10. J. O. Besenhard, H. P. Fritz, *J. Electroanal. Chem.*, **53**, 329–333 (1974).
11. R. Yazami, S. Genies, *Denki Kagaku*, **66**, 1293–1298 (1998).
12. J. O. Besenhard, J. Yang, M. Winter, *J. Power Sources*, **68**, 87–90 (1997).
13. M. Winter, J. O. Besenhard, M. E. Spahr, P. Novak, *Adv. Mater.* **10**, 725–763 (1998).
14. C. Fuller, J. Severiens, *Phys. Rev.*, **96**, 21–24 (1954).
15. J. Yang, M. Winter, J. O. Besenhard, *Solid State Ionics*, **90**, 281–287 (1996).
16. M. Winter, J. O. Besenhard, *Electrochim. Acta.*, **45**, 31–50 (1999).
17. S. Bourderau, T. Brousse, D. M. Schleich, *J. Power Sources*, **81**, 233–236 (1999).
18. H. Li, X. Huang, L. Chen, Z. Wu, Y. Liang, *Electrochem. Solid-State Lett.*, **2**, 547–549 (1999).
19. J. Graetz, C. Ahn, R. Yazami, B. Fultz, *Electrochem. Solid-State Lett.*, **6**, A194 (2003).
20. J. Graetz, J. Reilly, *Adv. Eng. Mater.*, **7**, 597 (2005).
21. G. W. Zhou, H. Li, H. P. Sun, D. P. Yu, Y. Q. Wang, X. J. Huang, L. Q. Chen, Z. Zhang, *Appl. Phys. Lett.*, **75**, 2447–2449 (1999).
22. J. Graetz, C. Ahn, R. Yazami, B. Fultz, *J. Electrochem. Soc.*, **151**, A698 (2004).
23. L. Y. Beaulieu, D. Larcher, R. A. Dunlap, J. Dahn, *J. Electrochem. Soc.*, **147**, 3206–3212 (2000).
24. A. Ulus, Y. Rosenberg, L. Burstein, E. Peled, *J. Electrochem. Soc.*, **149**, A635–A643 (2002).
25. D. Larcher, S. Beattie, M. Morcrette, K. Edstrom, J-C. Jumas, J-M. Tarascon, *J. Mater. Chem.*, **17**, 3759–3772, (2007).
26. W-J. Zhang, *J. Power Sources*, **196**, 13–24, (2011).
27. C. K. Chan, H. L. Peng, G. Liu, K. McIlwrath, X. F. Zhang, R. A. Huggins, Y. Cui, *Nat. Nanotechnol.*, **3**, 31–35, (2008).
28. G. X. Wang, L. Sun, D. H. Bradhurst, S. Zhong, S. X. Dou, H. K. Liu, *J. Power Sources*, **88**, 278–281 (2000).
29. C. P. Chen, H. Leipold, *Am. Ceram. Soc. Bull.*, **59**, 469–472 (1980).
30. J. Bagdahn, S. W. N., *Sens. Actuators, A*, **103**, 9–15 (2003).
31. X. H. Liu, L. Zhong, S. Huang, S. X. Mao, T. Zhu, J. Y. Huang, *ACS Nano*, **6**, 1522–1531 (2012).

32. S. W. Lee, M. T. McDowell, L. A. Berla, W. D. Nix, Y. Cui, *PNAS*, **11**, 4080–4085 (2012).
33. G. Che, B. B. Lakshmi, E. R. Fisher, C. R. Martin, *Nature.*, **393**, 346–349 (1998).
34. N. Li, C. R. Martin, B. Scrosati, *Electrochem. Solid-State Lett.*, **3**, 316–318 (2000).
35. N. Li, C. R. Martin, *J. Electrochem. Soc.*, **148**, A164–A170 (2001).
36. J. Patrissi, C. R. Martin, *J. Electrochem Soc.*, **148**, A1247–A1253 (2001).
37. K. Dokko, M. Mohamedi, Y. Fujita, T. Itoh, M. Nishizawa, M. Umeda, I. Uchida, *J. Electrochem. Soc.*, **148**, A422–A426 (2001).
38. J. Cho, H. Jung, Y. Park, G. Kim, H. Lim, *J. Electrochem. Soc.*, **147**, 15–20 (2000).
39. L. B. Hong, C. C. Ahn, B. Fultz, *J. Mater. Res.*, **10**, 2408–2410 (1995).
40. W.-S. Chang, C.-M. Park, J.-H. Kim, Y.-U. Kim, G. Jeong, H.-J. Sonh, *Energy Environ. Sci.*, **5**, 6895–6899 (2012).
41. B. Guo, J. Shu, Z. Wang, H. Yang, L. Shi, L. Chen, *Electronchem. Commun.*, **10**, 1876–1878 (2008).
42. P. Limthongkul, Y. Jang, N. Dudney, Y. Chiang, *Acta Materialia*, **51**, 1103–1113 (2003).
43. S. Yoon, C. Park, H. Sohn, *J. Electrochem. Solid State Lett.*, **11**, A42–A45 (2008).
44. X. H. Liu, S. Huang, S. T. Picraux, J. Li, T. Zhu, J. Y. Huang, *Nano Lett.*, **11**, 3991–3997 (2011).
45. F. Wang, L. Wu, X.-Q. Yang, Y. Zhu, J. Graetz (unpublished).
46. J. Li, J. R. Dahn, *J. Electrochem. Soc.*, **154**, A156–A161, (2007).
47. B. Key, R. Bhattacharyya, M. Morcrette, V. Seznec, J. M. Tarascon, C. P. Grey, *J. Am. Chem. Soc.*, **13**, 9239–9249 (2009).
48. T. D. Hatchard, J. R. Dahn, *J. Electrochem. Soc.*, **151**, A838–842 (2004).
49. J. Saint, M. Morcrette, D. Larcher, L. Laffont, S. Beattie, J.P. Peres, D. Talaga, M. Couzi, J. M. Tarascon, *Adv. Funct. Mater.*, **17**, 1765–1774 (2007).
50. X. H. Liu, H. Zheng, L. Zhong, S. Huang, K. Karki, L. Q. Zhang, Y. Liu, A. Kushima, W. T. Liang, J. W. Wang, *et al.*, *Nano Lett.*, **11**, 3312–3318 (2011).
51. S. W. Lee, M. T. McDowell, J. W. Choi, Y. Cui, *Nano Lett.*, **11**, 3034–3039 (2011).

52. J. L. Goldman, B. R. Long, A. A. Gewirth, R. G. Nuzzo, *Adv. Funct. Mater.*, **21**, 2412–2422 (2011).

53. M. Fleischauer, T. Hatchard, A. Bonakdarpour, J. R. Dahn., *Meas. Sci. Technol.*, **16**, 212–220 (2005).

54. N. Dudney, J. Bates, R. Zuhr, S. Young, J. Robertson, H. Jun, S. Hackney, *J. Electrochem. Soc.*, **146**, 2455–2464 (1999).

55. J. Whitacre, W. West, E. Brandon, B. Ratnakumar, *J. Electrochem. Soc.*, **148**, A1078–A1084 (2001).

56. V. P. Phan, B. Pecquenad, and F. Le Cras, *Adv. Funct. Mater.*, **22**, 2580–2584 (2012).

57. C. S. Wang, G. T. Wu, X. B. Zhang, Z. F. Qi, W. Z. Li, *J. Electrochem. Soc.*, **145**, 2751–2758 (1998).

58. H. Li, X. Huang, L. Chen, Z. Wu, Y. Liang, *Electrochem. Solid State Lett.*, **2**, 547–549 (1999).

59. N. Dimov, S. Kugino, M. Yoshio, *Electrochimica Acta*, **48**, 1579–1587 (2003).

60. U. Kasavajjula, C. S. Wang, A. J. Appleby, *J. Power Sources*, **163**, 1003–1039 (2007).

61. A. Magasinski, P. Dixon, B. Hertzberg, A. Kvit, J. Ayala, and G. Yushin, *Nat. Mater.*, **9**, 353–358 (2010).

62. M. D. Fleischauer, J. Li, M. J. Brett, *J. Elecgtrochem. Soc.*, **156**, A33–A36 (2009).

63. C. K. Chan, H. Peng, G. Liu, K. McIlwrath, X. F. Zhang, R. A. Huggins, Y. Cui, *Nat. Nanotech.* **3**, 31–35 (2008).

64. G. Liu, S. D. Xun, N. Vukmirovic, X. Y. Song, P. Olalde-Velasco, H. H. Zheng, V. S. Battaglia, L. W. Wang, W. L. Yang, *Adv. Mater.*, **23**, 4679–4683 (2011).

65. H. Wu, G. Chan, J. W. Choi, I. Ryu, Y. Yao, M. T. McDowell, S. W. Lee, A. Jackson, Y. Yang, L. Hu, Yi Cui, *Nat. Nanotech.*, **7**, 310–315 (2012).

66. M. M. Thackeray, C. Wolverton, E.D. Isaacs, *Energy Environ. Sci.*, in press (2012).

Chapter 3

Nano-Electrochemical Approach for Improvement of Lithium–Tin Alloy Anode

Tetsuya Osaka, Hiroki Nara, and Hitomi Mukaibo

Department of Applied Chemistry Waseda University, Ohkubo 3-4-1, Shinjuku-ku, Tokyo 169-8555, Japan

osakatets@waseda.jp

Tin (Sn) is promising material for the anode of lithium (Li) secondary batteries, because of its higher theoretical capacity than the conventional carbonaceous anodes. However Sn anode deteriorates with charge–discharge cycles due to its large volume change during the cycles. In this chapter, two methodologies with electrochemical process are introduced to overcome this problem. One is alloying with third metal which is not react with Li to suppress the volume change by surrounding Sn with third metal. The other is introduction of mesoporous structure to absorb the outward volume increase at charging: the mesopore could enable inward expansion of Sn during its lithiation. Additionally the introduction of mesoporous structure to electrodes increases the reaction efficiency, resulting in improvement of those rate properties. These findings can be applicable to the other materials like silicon.

Nanomaterials for Lithium-Ion Batteries: Fundamentals and Applications
Edited by Rachid Yazami

ISBN 978-981-4316-40-8 (Hardcover), 978-981-4364-23-2 (eBook)
www.panstanford.com

3.1 Introduction

With the advancement of portable electronic devices, further improvement of their power sources is required. Li-ion batteries possess the highest energy density among batteries in practical use, and the development of an improved anode material is considered as a breakthrough to meet such demand.

Presently, carbon is commercially used as the anode material for Li-ion batteries, and its capacity has already reached its theoretical value of 372 mAh/g. In 1994, Fuji Film Co. Ltd. filed a patent for Sn oxides as a novel anode material for Li-ion batteries with a higher theoretical capacity than carbon.[1] However, this material showed huge irreversible capacity, which was a fatal defect. Courtney and Dahn have developed a theory for the reaction mechanism of such materials.[2] Many attempts have been made to develop other Sn compounds in search of a suitable matrix that leads to high cyclability.[3–19] Osaka *et al.* have also investigated such candidates, and their work on SnS_2 has proved that this compound has a high potential as a novel anode material.[20] The effect of particle size had also been studied with this material. A matrix consisting of nanoparticles should have spaces between the particles, which may absorb the problematic volume change resulting from the reaction of Sn with Li during cycling. The large surface area of nanoparticles should also ease the stress accompanying such a volume change. The results have shown that by controlling the size of nanoparticles, improved cyclability can be achieved.[21]

The selection of an adequate matrix is the key to a successful Sn compound anode. Elements that are inactive against Li are assumed to suppress the volume change effectively without much irreversible capacity.[7] The use of Sn compounds with elements such as Fe,[10–14] Cu,[16–19] Mn,[11,15] and Co[11,15] has been investigated based on this assumption. Nickel (Ni) is a typical element which does not react with Li. Therefore, this element can be expected to serve as an appropriate matrix to improve the cyclability of the electrode without a high initial irreversible capacity. There have been studies on Sn–Ni compounds prepared by ball-milling[7,8] and electroplating,[9] but a capacity exceeding that of the current carbon material has not yet been reported. On the other hand, vacancy can be also expected to act as a matrix to absorb the volume change effect. In this chapter, Sn–Ni alloy thin films with various Ni/Sn ratios prepared by varying the ratio of Ni and Sn ions in the plating bath are discussed along with mesoporous Sn electrodeposited with lyotropic liquid crystal.

3.2 Electrodeposited Sn–Ni Alloy Thin Films as High-Capacity Anode Materials for Lithium-Ion Secondary Batteries

In this section, various Sn_xNi_ys electrodeposited were investigated on the stand point described above. Further investigation on Sn_xNi_y which indicated highest performance was discussed.

3.2.1 Preparation of Anodes with Various Compositional Ratios of Ni to Sn

The Sn–Ni alloy thin films were prepared by electrodeposition.[22] Thin films containing different Ni/Sn ratios were electrodeposited on Cu substrates from the baths listed in Table 3.1 with the current density of 5.0 mA/cm^2 with agitation by magnetic stirrer at 50°C. The samples were designated in terms of atomic percentage of Sn in the deposited film, which was determined by inductively coupled plasma (ICP) atomic emission spectroscopy. The samples obtained were as follows: $Sn_{54}Ni_{46}$, $Sn_{62}Ni_{38}$, $Sn_{84}Ni_{16}$, and $Sn_{92}Ni_8$, whose subscripts indicate their atomic ratio. The deposition time was 5 minutes for most samples except for those prepared for X-ray diffraction (XRD) analysis, which were deposited for 1 hour to obtain higher peak intensities and to suppress signals from Cu substrate.

Table 3.1 Bath compositions for electrodepositing Sn–Ni alloy thin films

	Amount of Sn in the samples/atom %			
	54	**62**	**84**	**92**
$NiCl_2 \cdot 6H_2O$/mol L^{-1}	0.125	0.075	0.063	0.050
$SnCl_2 \cdot 2H_2O$/mol L^{-1}	0.125	0.175	0.187	0.200
Potassium pyrophosphate ($K_4P_2O_7$)/mol L^{-1}	0.500	0.500	0.500	0.500
Glycine (H_2NCH_2COOH)/mol L^{-1}	0.250	0.125	0.100	0.075
Ammonium hydroxide (NH_4OH)/mol L^{-1}	5	5	5	5

Source: Reprinted with permission from *Electrochem. Solid-State Lett.*, **6**(10), A218 (2003). Copyright 2003, The Electrochemical Society.

3.2.2 Evaluation of Sn–Ni Alloy Thin Films with Various Compositional Ratios of Ni and Sn

XRD was used to evaluate the structural differences between the samples with various compositional ratios of Ni and Sn. Electrodeposited Sn–Ni film forms metastable phases that do not appear in the equilibrium phase diagram. Watanabe *et al.* carried out a detailed study of the structure of Sn–Ni films.[23] The peak assignments made for the samples shown in Fig. 3.1 are based on the result of their work. The metastable phase M1 is a Ni–Sn phase with a composition ratio of almost 50:50, and the metastable phase M2 is a phase where Ni is melted into Sn crystals. Samples with higher Sn contents were confirmed to have a structure similar to that of Sn. It is assumed that these electrodes with a high Sn content may show electrode properties similar to those of pure Sn electrodes, which means a high initial capacity, but a poor cyclability resulting from electrode degradation.

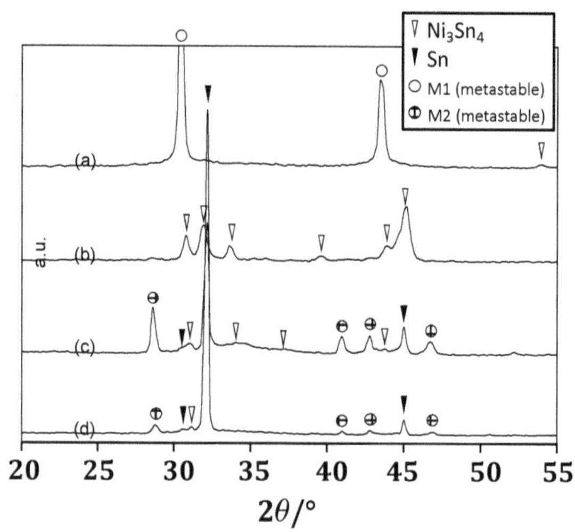

Figure 3.1 XRD patterns of Sn–Ni alloys obtained by electrodeposition for 1 hour (a) $Sn_{54}Ni_{46}$, (b) $Sn_{62}Ni_{38}$, (c) $Sn_{84}Ni_{16}$, and (d) $Sn_{54}Ni_{46}$. The XRD apparatus was equipped with Cu Kα radiation. These samples were electrodeposited for 1 hour and were not significantly different from those electrodeposited for 5 minutes except for their signal intensities of XRD peaks. Reprinted with permission from *Electrochem. Solid-State Lett.*, **6**(10), A218 (2003). Copyright 2003, The Electrochemical Society.

The first cyclic voltammogram (CV) cycles recorded with four different samples are shown in Fig. 3.2. It has been reported that in pure Sn electrode, three anodic peaks appear at 0.69, 0.61, and 0.45 V *vs.* Li/Li^+ in the anodic scan, and four peaks at 0.43, 0.57, 0.71, and 0.78 V *vs.* Li/Li^+ in the cathodic scan.[2] The high Sn content films with pure Sn phase showed CV characteristics close to those of pure Sn. The results are consistent with the XRD results.

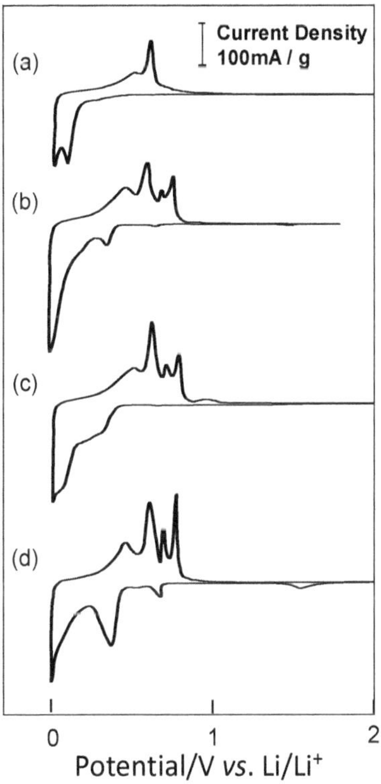

Figure 3.2 CVs of (a) $Sn_{54}Ni_{46}$, (b) $Sn_{62}Ni_{38}$, (c) $Sn_{84}Ni_{16}$, and (d) $Sn_{92}Ni_8$ with three-electrodes cell using Li foils as counter and reference electrode in the potential range of open-circuit potential (OCP) to 0.0 V *vs.* Li/Li^+ at the scan rate of 0.02 mV/s in 1 M $LiClO_4$/EC-PC (ethylene carbonate–propylene carbonates) (1:1 vol%). Reprinted with permission from *Electrochem. Solid-State Lett.*, **6**(10), A218 (2003). Copyright 2003, The Electrochemical Society.

Field emission scanning electron microscope (FE-SEM) images of $Sn_{62}Ni_{38}$ as deposited, after first charge until 0.0 V *vs.* Li/Li^+, and after first discharge until 3.0 V *vs.* Li/Li^+ are shown in Fig. 3.3. It is seen that the film as deposited consists of aggregates of small grains with an average size of 95 nm. After the first charge at a constant current, the grain size of the film increased to eight times as large (average size approx. 760 nm) as that of the as-deposited sample. With further cycling of the sample to the discharge potential of 3.0 V *vs.* Li/Li^+, the grains of the film decreased to an average of 150 nm, and they were aggregated to form a complicated three-dimensional porous structure. These changes in the electrode structure at different states of the initial cycle may be due to the insertion and desorption of Li.

Figure 3.3 FE-SEM images of $Sn_{62}Ni_{38}$ electrode (a) as deposited,(b) after 1st charge until 0.0 V *vs.* Li/Li^+, and (c) after 1st discharge until 3.0 V *vs.* Li/Li^+. Charge and discharge were performed with three-electrodes cell using Li foils as counter and reference electrode at the constant current density of 50 mA/g in 1M $LiClO_4$/EC-PC (1:1 vol%). Reprinted with permission from *Electrochem. Solid-State Lett.*, **6**(10), A218 (2003). Copyright 2003, The Electrochemical Society.

Figure 3.4 shows cycle performance of the electrodes with various Ni/Sn ratios. The cycle performance of $Sn_{62}Ni_{38}$ was outstanding among the four samples, showing the highest discharge capacity of *ca.* 650 mAh/g at about the 70th cycle. Under the assumption that all Sn reacts with Li to form $Li_{4.4}Sn$, the theoretical capacities of $Sn_{54}Ni_{46}$, $Sn_{62}Ni_{38}$, and $Sn_{84}Ni_{16}$ would be 700, 763, and 909 mAh/g, respectively. Nevertheless, when they were cycled, $Sn_{54}Ni_{46}$ and $Sn_{84}Ni_{16}$ showed a relatively low discharge capacity of *ca.* 300 and 272 mAh/g after 70 cycles, that is, the initial columbic efficiencies of the samples were as follows: $Sn_{54}Ni_{46}$ = 84.8%, $Sn_{84}Ni_{16}$ = 121.8%, and $Sn_{92}Ni_{8}$ = 121.7%. Under the same conditions, $Sn_{62}Ni_{38}$ showed the largest discharge capacity of 654 mAh/g, that is, the initial columbic efficiency of $Sn_{62}Ni_{38}$ was 105.0%. Another interesting feature of these samples is that they show an increase in capacity with cycling after first several cycles. $Sn_{62}Ni_{38}$ showed the biggest increase in 70 cycles (166 mAh/g). $Sn_{54}Ni_{46}$ also increased about 100 mAh/g of its capacity in 70 cycles, but the capacity of $Sn_{84}Ni_{16}$ started to decrease in about 35 cycles after 30 mAh/g increase of its capacity. The columbic efficiency exceeding 100% may be due to a side reaction (e.g., the elution of impurities within samples. The samples showed a stable columbic efficiency after two to three cycles, and the efficiencies after the 69th cycle were as follows: $Sn_{54}Ni_{46}$ = 92.2%, $Sn_{62}Ni_{38}$ = 99.3%, $Sn_{84}Ni_{16}$ = 97.4%, and $Sn_{92}Ni_{8}$ = 96.0%. The magnitude of the sharp drop in discharge capacity after the first cycle was greater for the higher Sn content samples. The fall of capacity of the $Sn_{84}Ni_{16}$ (after 40 cycles) and $Sn_{92}Ni_{8}$ (after 35 cycles) samples may indicate the deterioration of the pure Sn phase in those electrodes. This result is consistent with an assumption that the samples with a high Sn content are easier to deteriorate. A unique characteristic observed in Fig. 3.4 is the increase in capacity with the number of cycles for all samples. The $Sn_{62}Ni_{38}$ showed the highest capacity increase of all, which was approximately 140 mAh/g. This may be due to the changes in surface condition accompanying a volume change during cycling (as seen in Fig. 3.3), which should lower the resistance to Li diffusion within the sample. Overall, the cyclability of each sample was adequate after the first several cycles, indicating the effectiveness of the Ni matrix. It is concluded that the $Sn_{62}Ni_{38}$ is the most appropriate composition with a structure leading to a small initial degradation, good cyclability, and the highest capacity.

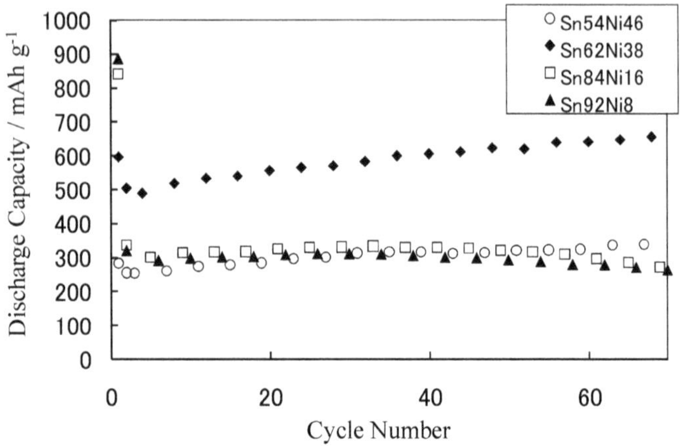

Figure 3.4 Cycle performance of electrodes with various Ni/Sn ratios at the constant current density of 50 mA/g with three-electrodes cell using Li foils as counter and reference electrode in the potential range of 0.0 to 3.0 V *vs.* Li/Li^+ in 1M $LiClO_4$/EC-PC (1:1 vol%). Reprinted with permission from *Electrochem. Solid-State Lett.*, **6**(10), A218 (2003). Copyright 2003, The Electrochemical Society.

To evaluate these phenomena furthermore, structural and morphological changes with cycling (repeated Li^+ insertion and extraction) were investigated by XRD and SEM. Since the results of $Sn_{92}Ni_8$ were similar to that of $Sn_{84}Ni_{16}$, this chapter will only discuss the results of the latter case.

Figure 3.5 illustrates the structure and surface morphology of $Sn_{54}Ni_{46}$, $Sn_{62}Ni_{38}$, and $Sn_{84}Ni_{16}$ before cycling. The peak assignments of the XRD are based on the results of Watanabe *et al.*[23] Metastable phases are phases that do not appear in the equilibrium phase diagram and that formed due to the electrodeposition process. The metastable phase (M1) is a NiSn phase with a composition ratio of almost 50:50, and the metastable phase (M2) is a phase where Ni is melted into pure Sn crystal lattice. M1 phase is the main phase confirmed from the XRD of $Sn_{54}Ni_{46}$ and not identified in $Sn_{62}Ni_{38}$ or $Sn_{84}Ni_{16}$. For $Sn_{62}Ni_{38}$, Ni_3Sn_4 phase is the only identified phase. The assumption that all Ni atoms within $Sn_{62}Ni_{38}$ form Ni_3Sn_4 alloy with Sn leaves at least 19 at % Sn in excess. Nevertheless, presence of distinct pure Sn phase could not be distinguished from Fig. 3.5. On comparing the peaks attributed to Ni_3Sn_4 in the results with those indicated in the JCPDS data, the peaks appeared slightly shifted, indicating that the Sn atoms may exist melted within the Ni_3Sn_4 lattice or as a nanocrystalline or an amorphous phase in the film. The XRD peaks of $Sn_{84}Ni_{16}$ are assigned to Ni_3Sn_4, M2, and pure Sn phases. Peaks at 34.5°, 35.0°, 42.2°, and 46.7° could not be identified. From SEM observation, the majority particles size of $Sn_{54}Ni_{46}$, $Sn_{62}Ni_{38}$ and $Sn_{84}Ni_{16}$ was 20–24, 25–29, and 40–44 nm, respectively, and films with higher Sn content lead to rougher, more granular surface.

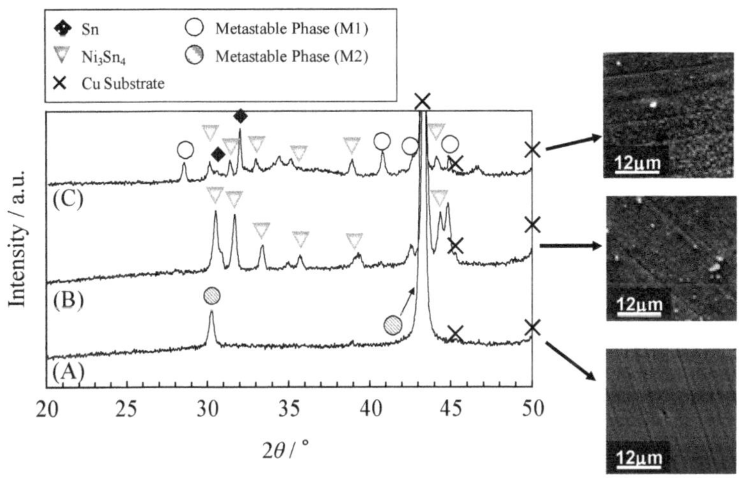

Figure 3.5 XRD patterns and SEM images of (A) $Sn_{54}Ni_{46}$, (B) $Sn_{62}Ni_{38}$, and (C) $Sn_{84}Ni_{16}$ before charge-discharge cycles. Reprinted with permission from *J. Power Sources*, **146**, 457 (2005). Copyright 2005, Elsevier.

Figure 3.6 shows the results of the XRD patterns and SEM images after the first charging (Li insertion). Upon charging, the peak attributed to M1 phase of $Sn_{54}Ni_{46}$ at 30.3° decreases its intensity. This suggests that part of the phase reacted with Li^+, resulting in a state of nano-crystalline or amorphous phase that could not be detected by the XRD. It should also be noted that no peaks assigned to Li–Sn alloy phase were confirmed at this charged state. This indicates that no XRD-detectable Li–Sn binary alloy phases were formed by the reaction of Sn with Li^+. On the contrary, $Sn_{62}Ni_{38}$ and $Sn_{84}Ni_{16}$ showed new peaks indicating the presence of Li–Sn alloys. As it is well known, there are several types of Li–Sn alloys with different composition.[24-26] However, the results here show them as two broad peaks, and since many peaks of various Li–Sn composition is overlapped in this range, it was not possible to attribute them to a certain Li–Sn phase. This indicates that lithiation reaction of Sn had taken place during charging in $Sn_{62}Ni_{38}$ and $Sn_{84}Ni_{16}$ thin films. Simultaneously, most of the peaks due to Ni_3Sn_4, M2, and Sn phase became unidentifiable. The peaks attributed to pure Ni were not identified. This will be discussed further with the results shown in Fig. 3.7. The SEM images of each sample after Li^+ insertion showed a bumpier surface compared to those images before cycling (Fig. 3.5).

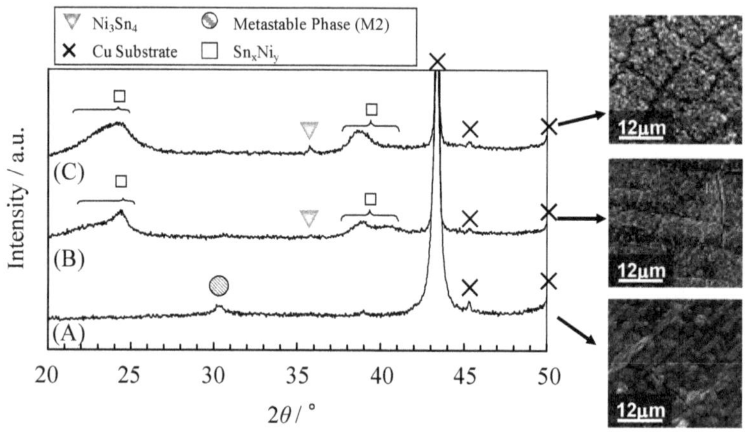

Figure 3.6 XRD patterns and SEM images of (A) $Sn_{54}Ni_{46}$, (B) $Sn_{62}Ni_{38}$, and (C) $Sn_{84}Ni_{16}$ after initial charging. These Sn–Ni electrodes were charged until 0.01 V *vs.* Li/Li^+ at the constant current density of 250 mA/g with three-electrodes cell using Li foils as counter and reference electrode in 1 M $LiClO_4$/EC-PC (1:1 vol%). Reprinted with permission from *J. Power Sources*, **146**, 457 (2005). Copyright 2005, Elsevier.

After the first discharge process where Li^+ was extracted from the sample (Fig. 3.7), the XRD patterns suggest a partial recovery of the electrode structure. For $Sn_{54}Ni_{46}$, the peak intensity of the metastable phase at 30.3° grows higher again, indicating that the reaction between the M1 phase and Li^+ is reversible. The peak is also slightly broader than it was before cycling (Fig. 3.5), suggesting the decrease in the crystal size. In the case of $Sn_{62}Ni_{38}$, although all the peaks attributed to Ni_3Sn_4 in Fig. 3.5 were confirmed again, their peak intensity had become relatively weaker. From the XRD pattern of $Sn_{84}Ni_{16}$, only the peak at 45.0° was identified for M2 phase. Recovery of all the peaks of Ni_3Sn_4 and Sn were confirmed, but

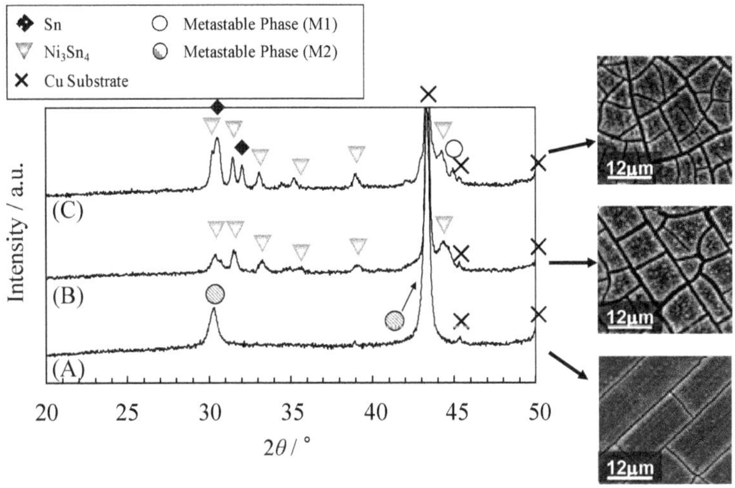

Figure 3.7 XRD patterns and SEM images of (A) $Sn_{54}Ni_{46}$, (B) $Sn_{62}Ni_{38}$, and (C) $Sn_{84}Ni_{16}$ after first discharging. These Sn–Ni electrodes were cycled in the potential range of 0.01–1.20 V *vs.* Li/Li^+ at the constant current density of 250 mA/g with three-electrodes cell using Li foils as counter and reference electrode in 1 M $LiClO_4$/EC-PC (1:1 vol%). Reprinted with permission from *J. Power Sources*, **146**, 457 (2005). Copyright 2005, Elsevier.

their peak intensity was different from Fig. 3.5 without regularity. Specifically, compared with the peak intensities of Fig. 3.5, the peaks at 30.3° (Ni_3Sn_4), 30.7° (Sn), 31.5° (Ni_3Sn_4), and 44.2° (Ni_3Sn_4) showed increased intensity, while the peak at 33.1° (Ni_3Sn_4) and 45.0° (Sn) showed almost the same intensity, and peaks at 32.0° (Sn), 35.8° (Ni_3Sn_4), and 39.0° (Ni_3Sn_4) showed decreased intensity. The peaks attributed to Li–Sn alloy phases were no longer detectable for both $Sn_{62}Ni_{38}$ and $Sn_{84}Ni_{16}$. Assuming from the fact that the Ni–Sn alloy phase has recovered in both $Sn_{62}Ni_{38}$ and $Sn_{84}Ni_{16}$ with Li^+ extraction, Ni may have existed within the film in the charged state (Fig. 3.6) as nanocrystalline or amorphous phase; hence their phases could not be identified in XRD results. From the SEM images, formation of surface cracks with Li^+ extraction was confirmed. Relatively, the cracks after discharging are smaller for $Sn_{54}Ni_{46}$. This indicates that the reaction of M1 phase and Li^+ results in smaller volume change compared with the reaction between Ni_3Sn_4 and Li^+. This may be considered to be consistent with the smaller amount of Li insertion (lower discharge capacity) indicated in Fig. 3.7. Also, these cracks confirmed on the electrode surface can be expected to lower the diffusion resistance of Li^+ into the inner electrode. This generation of cracks may be one of the reasons for the rise in the capacity with cycling seen in the cycle performance of each electrodeposited Sn–Ni alloy anode (Fig. 3.7).

Figure 3.8 shows the XRD results and SEM images of each sample after the 10th discharge (Li extraction). For $Sn_{54}Ni_{46}$ and $Sn_{62}Ni_{38}$, no more formation of new phases or disappearance of existing phases can be confirmed compared with the result for the sample after first discharge (Fig. 3.7), while both samples showed lower peak intensity. Especially for $Sn_{62}Ni_{38}$, it may be noted that the XRD result does not indicate aggregation of pure Sn phase with cycling as reported for Sn oxide materials.[27] For $Sn_{84}Ni_{16}$, all the peaks attributed to Sn showed increase in its intensity that may be indicating the aggregation of Sn phase with cycling. Furthermore, this sample showed small peak that was attributed to Li–Sn alloy phase at the lower degree (around 25°). This suggests that with repeated cycling, some Li gets trapped within the Li–Sn alloy and is unable to dealloy during the discharge process. The SEM images show cracked surface morphology similar to Fig. 3.7.

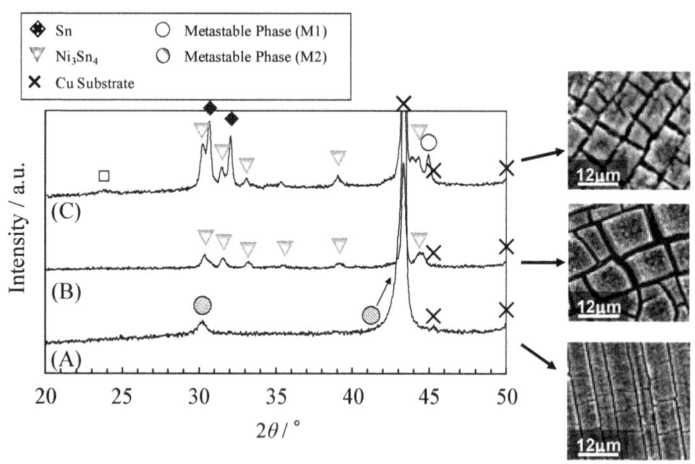

Figure 3.8 XRD patterns and SEM images of (A) $Sn_{54}Ni_{46}$, (B) $Sn_{62}Ni_{38}$, and (C) $Sn_{84}Ni_{16}$ after 10th discharging. These Sn-Ni electrodes were cycled in the potential range of 0.01–1.20 V *vs.* Li/Li^+ at the constant current density of 250 mA/g with three-electrodes cell using Li foils as counter and reference electrode in 1 M $LiClO_4$/EC-PC (1:1 vol%). Reprinted with permission from *J. Power Sources*, **146**, 457 (2005). Copyright 2005, Elsevier.

The change of the electrode crystallinity with cycling was also evaluated using μ-electron diffraction (Fig. 3.9). Clear Laue patterns can be seen from the as-deposited image and from the first discharge image, indicating its crystallinity. But after the first charge, broad halo can be identified, suggesting that the Li insertion has a significant effect on lowering the crystallinity of the electrode structure in the nanometer scale.

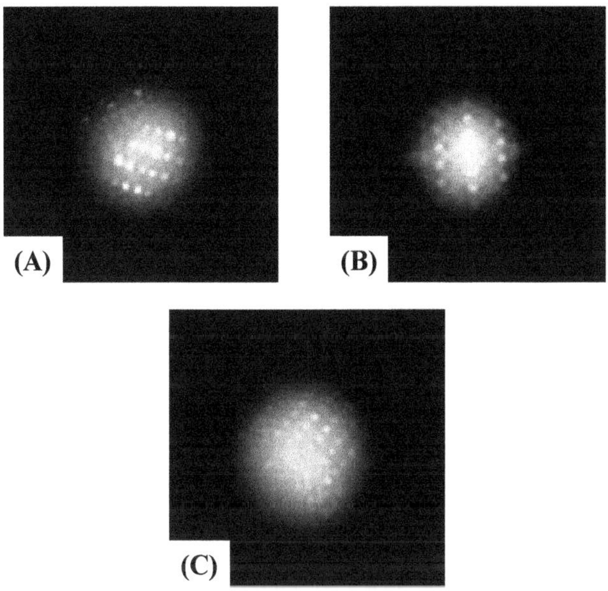

Figure 3.9 μ-electron diffraction pattern of $Sn_{62}Ni_{38}$ (A) as deposited, (B) after the first charge, and (C) after the first discharge. The $Sn_{62}Ni_{38}$ electrode was cycled in the potential range of 0.01–1.20 V *vs.* Li/Li^+ at the constant current density of 250 mA/g with three-electrodes cell using Li foils as counter and reference electrode in 1 M $LiClO_4$/EC-PC (1:1 vol%). The acceleration voltage was 300 kV and the diameter of the beam for ED was 2 nm. Reprinted with permission from *J. Power Sources*, **146**, 457 (2005). Copyright 2005, Elsevier.

Overall, the obtained results suggest that a reaction mechanism of Sn–Ni alloys is different to the ones considered for the Sn oxide anodes.[2] With the Li^+ insertion during the charge process, Sn atoms segregate from the Sn–Ni alloy structure and alloys with Li to form the Li–Sn alloy phase. This reaction should be reversible. With discharge, the Li^+ is extracted from the Li–Sn alloy phase, and the dealloyed Sn atom gets absorbed into the Ni matrix again forming the Sn–Ni alloy phase.

By considering this reaction mechanism with the gained results on the cycle performance and structural changes of the electrodes, it may be suggested that the structure of $Sn_{54}Ni_{46}$ was unable to free the Sn from the metastable Sn–Ni alloy crystal to allow its full alloying with Li. This could also be supported by the fact that the initial discharge capacity was 597 mAh/g for $Sn_{62}Ni_{38}$ and 841 mAh/g for $Sn_{84}Ni_{16}$ but 334 mAh/g for $Sn_{54}Ni_{46}$. In the case of $Sn_{84}Ni_{16}$, pure Sn phases were observed which clearly did not alloy with Ni, hence leading to the large capacity drop after the second cycle. The rechargeable capacity also begins to drop from the 35th cycle indicating its relatively rapid degradation compared to $Sn_{54}Ni_{46}$ and $Sn_{62}Ni_{38}$ (Fig. 3.4).

It was indicated that the structure that allows Li and Sn, Sn and Ni to reversibly alloy/dealloy is the key to gain high-capacity long-life anode materials. Ni_3Sn_4 phase may have been able to realize such reversible reactions, hence resulting in the high capacity of $Sn_{62}Ni_{38}$.

To obtain further information on the changes induced by cycling, cross-sectional transmission electron microscope (TEM) pictures were taken on $Sn_{62}Ni_{38}$ (Fig. 3.10), which showed the highest performance as Li-ion battery anodes during galvanostatic cycling test.[28] The film thickness of $Sn_{62}Ni_{38}$ before cycling was *ca.* 0.6 μm, and the crystal particle size was 0.05–0.13 μm. The film thickness and crystal particle size varied largely upon cycling, and formation of cracks and pores were seen with cycling. The results are summarized in Table 3.2. With the first charging discharging process, the dense film became porous, and the thickness expanded. The surface layer seen on the film upon charging disappeared with discharging. This may indicate the growth of Li oxides during the procedure to prepare TEM samples by exposure to the air. Or it may indicate some decomposition of the organic electrolyte during charging. The decrease in the crystal size of the film was also confirmed with cycling. This may be due to the reaction where Li comes into the Ni_3Sn_4 phase to generate Ni-rich and Li–Sn alloy phases, and to the converse reaction where Li goes out from the Li–Sn alloy and Ni-rich phases and generate Ni_3Sn_4 phases. This reversible reaction could be contributing to the good anode property of $Sn_{62}Ni_{38}$.

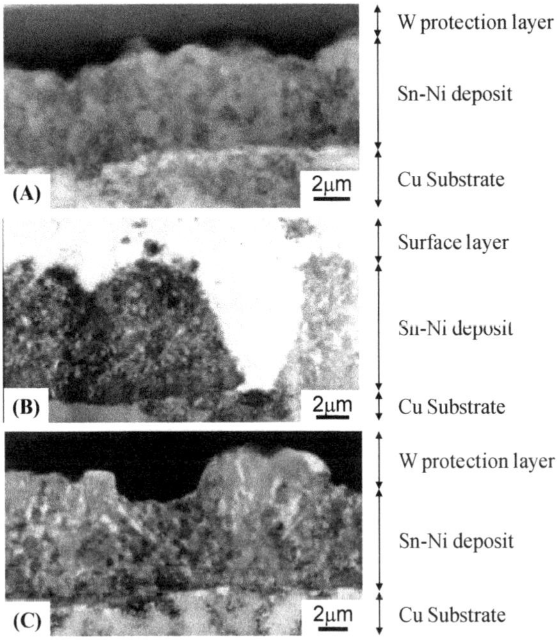

Figure 3.10 TEM images of the $Sn_{62}Ni_{38}$ (A) as deposited, (B) after the first charge, and (C) after first discharge. The $Sn_{62}Ni_{38}$ electrode was cycled in the potential range of 0.01–1.20 V *vs.* Li/Li^+ at the constant current density of 250 mA/g with three-electrodes cell using Li foils as counter and reference electrode in 1 M $LiClO_4$/EC-PC (1:1 vol%). The acceleration voltage was 300 kV and the diameter of the beam for ED was 2 nm. Reprinted with permission from *J. Power Sources*, **146**, 457 (2005). Copyright 2005, Elsevier.

Table 3.2 Summary of the TEM results in Fig. 3.10

State of sample	Film thickness/ μm	Crystal particle size/μm	Film state	Surface film
Before cycling	0.6	0.05–0.13	Dense	Not confirmed
After 1st charge	1.25	0.01–0.13	Porous	Confirmed
After 1st discharge	0.75	0.01–0.18	Porous	Not confirmed

Source: Reprinted with permission from *J. Power Sources*, **146**, 457 (2005). Copyright 2005, Elsevier.

3.2.3 Evaluation of $Sn_{62}Ni_{38}$ Showed the Highest Performance as Li-Ion Battery Anodes during Galvanostatic Cycling Test

Moreover, $Sn_{62}Ni_{38}$, which showed the highest performance as Li-ion battery anodes during galvanostatic cycling test, and pure Sn electrode as a comparison, was especially evaluated electrochemically.

At first, a pure Sn electrode was electrochemically characterized. The charge (lithiation) and discharge (delithiation) profiles of the electrodeposited Sn anode in the first two cycles are shown in Fig. 3.11. During the first charge, the potential of the Sn anode dropped rapidly to about 0.4 V *vs.* Li/Li^+, stabilized, and then gradually changed toward cathodic values again. The plateau at 0.4 V *vs.* Li/Li^+ is considered to result from the reaction of Li–Sn with metallic Li.[25] The gradual decrease of the voltage results from the coexistence with several LiSn alloy phases in the active material. During discharging, multiple plateaus were clearly observed. By the second charge, the capacity dramatically dropped to values less than 40 mAh/g and was worse in the subsequent cycles. The profiles are similar to those reported by Tamura *et al.* for an electrodeposited Sn on Cu foil using different bath compositions.[18] The significant drop in the capacity was considered by Tamura *et al.* to result from a lack of interface strength between the entire part of the active material and the Cu current collector.[18] It is well known that the electrochemical reduction of a Sn electrode in a Li^+-containing electrolyte leads to the subsequent formation of a number of intermetallic phases Li_xSn.[2] On the other hand, the Sn pulverizes rapidly when it experiences full lithiation due to the large volume increase leading to the electrode failure in the subsequent cycles. Indeed, only a visual observation of the electrode showed that the electrodeposited Sn has detached from the Cu substrate.

A point briefly discussed here is the appearance of the abnormal high-voltage plateau during the charge of pure Sn electrode. This phenomenon has not been often investigated and was reported only by Tamura *et al.*[18] and by Beaulieu *et al.*[29] In this experiments, when the Sn electrode was charged from 2.5 to 0.0 V *vs.* Li/Li^+, a voltage plateau close to 1.5 V *vs.* Li/Li^+ could be seen as shown in the insert of Fig. 3.11. The occurrence of this plateau is believed to be due to the formation of a thin film resulting from a catalytic decomposition of electrolyte by the pure Sn electrode.[18,29] It was thus confirmed here that the occurrence of the high-voltage plateau is not an experimental artifact in agreement with Tamura *et al.*[18] and Beaulieu *et al.*[29] Consequently, the abnormal high-voltage behavior of Sn certainly deserves further investigations onto the nature of the catalytic Sn surface.

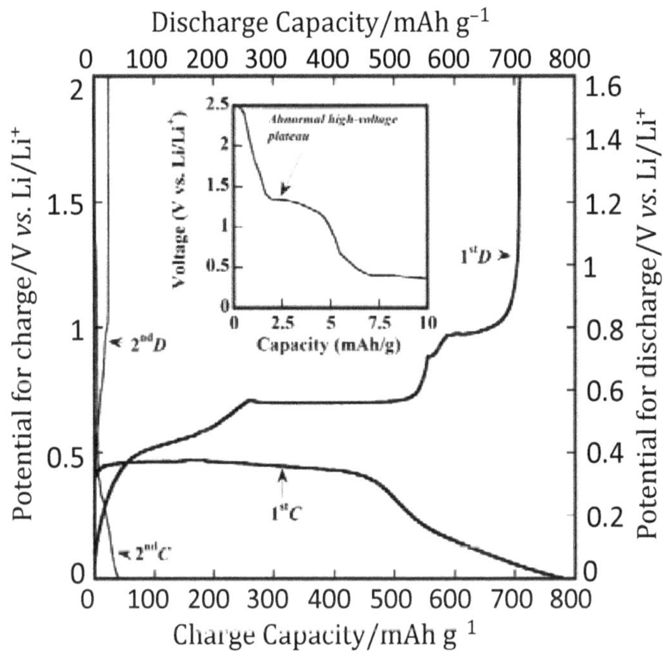

Figure 3.11 Potential *vs.* capacity profiles of the electrodeposited Sn on Cu substrate. The Sn electrode was cycled in the potential range of 0.01–2.0 V *vs.* Li/Li⁺ at the constant current density of 50 mA/g with three-electrodes cell using Li foils as counter and reference electrode in 1 M $LiClO_4$/EC-PC (1:1 vol%). The insert shows the second potential curve at charging illustrating the presence of the abnormal high-potential plateau. The letters *C* and *D* in the figure stand for charge and for discharge. Reprinted with permission from *J. Electrochem. Soc.*, **152**(3), A560 (2005). Copyright 2005, The Electrochemical Society.

Electrochemical characterization of $Sn_{62}Ni_{38}$ electrode was characterized electrochemically. Figure 3.12 shows the charge–discharge curves of electrodeposited $Sn_{62}Ni_{38}$ alloy in the first, second, tenth, and fiftieth cycles against Li metal. The first charge and discharge capacities are both the same, *ca.* 660 mAh/g, which shows that the initial cycling efficiency is 100 %. A very interesting feature of Fig. 3.12 is that the discharge capacity decreased as cycling extended to 10 cycles and started to increase for higher cycles. The 50th cycle is illustrated in the figure for clarity, indeed the recovery of the capacity was observed by the 10th cycle as shown later in this chapter.

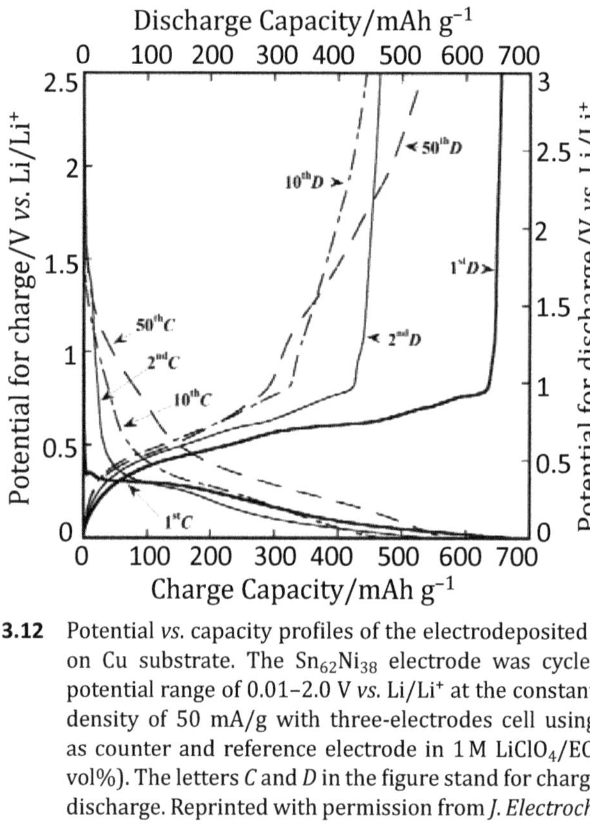

Figure 3.12 Potential *vs.* capacity profiles of the electrodeposited $Sn_{62}Ni_{38}$ on Cu substrate. The $Sn_{62}Ni_{38}$ electrode was cycled in the potential range of 0.01–2.0 V *vs.* Li/Li^{+} at the constant current density of 50 mA/g with three-electrodes cell using Li foils as counter and reference electrode in 1 M $LiClO_4$/EC-PC (1:1 vol%). The letters *C* and *D* in the figure stand for charge and for discharge. Reprinted with permission from *J. Electrochem. Soc.*, **152**(3), A560 (2005). Copyright 2005, The Electrochemical Society.

As to the anomalous high-voltage, one expects that alloying other metal with Sn would likely avoid the occurrence of the abnormal high-voltage irreversible capacity.[30] Figure 3.13, illustrates the second and third charge curves related to the $Sn_{62}Ni_{38}$ alloy electrode. It clearly shows the high-voltage plateau at the second charge process, whereas the plateau disappeared at the third charge process. The high-voltage plateau was observed with $Sn_{84}Ni_{16}$ and $Sn_{92}Ni_8$ but not with the $Sn_{54}Ni_{46}$ alloy. It is quite clear from Fig. 3.13 that Sn alloyed with another metal (for instance, Ni) does exhibit a high-voltage plateau for a Sn content higher than 62 atom%. As for the Sn electrode, these phenomena require further investigations.

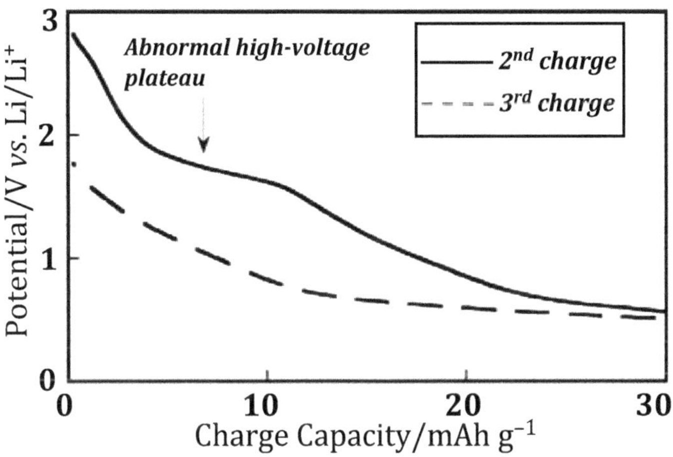

Figure 3.13 Potential *vs.* capacity profiles of the electrodeposited $Sn_{62}Ni_{38}$ on Cu substrate at second and third charging. The $Sn_{62}Ni_{38}$ electrode was cycled in the potential range of 0.01–2.0 V *vs.* Li/Li^+ at the constant current density of 50 mA/g with three-electrodes cell using Li foils as counter and reference electrode in 1 M $LiClO_4$/EC-PC (1:1 vol%). Reprinted with permission from *J. Electrochem. Soc.*, **152**(3), A560 (2005). Copyright 2005, The Electrochemical Society.

The CVs for the $Sn_{62}Ni_{38}$ film for the first three cycles at 0.02 mV/s are shown in Fig. 3.14. The samples were cycled between the OCP and 0.0 V *vs.* Li/Li^+. During the first reduction process, the voltammograms are characterized by two reduction peaks at 0.65 and 0.34 V. The peak at 0.65 V was found to depend on the Sn content, that is, its maximum current increased when the Sn content in the alloy increased.[28] The peak at 0.65 V has been observed during the second and subsequent cycles with SnO materials;[31] however, it has not been observed during the first reduction of SnO_2,[32] nor with Sn–Fe-based alloys,[10–12] and may therefore be associated with the Li–Sn alloying process.

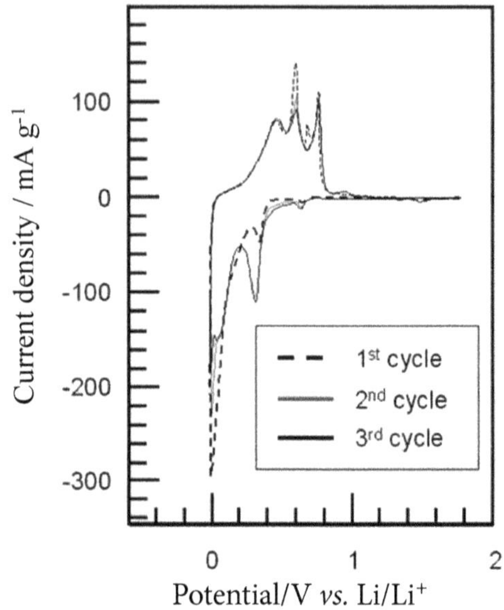

Figure 3.14 The first three CVs of the electrodeposited $Sn_{62}Ni_{38}$ on Cu substrate. The $Sn_{62}Ni_{38}$ electrode was cycled in the potential range of 0.01–1.8 V *vs.* Li/Li^+ at the scan rate of 0.02 mV/s with three-electrodes cell using Li foils as counter and reference electrode in 1 M $LiClO_4$/EC-PC (1:1 vol%). The dash line represents the first cycle, the solid lines represent the second and third cycles. Reprinted with permission from *J. Electrochem. Soc.*, **152**(3), A560 (2005). Copyright 2005, The Electrochemical Society.

Figure 3.15a shows a family of four Nyquist plots of $Sn_{62}Ni_{38}$ thin film electrode subjected to up to 10 cycles of Li insertion. Figure 3.15b displays an enlarged view of their high-frequency domain. The impedance spectra of Fig. 3.15 are basically the typical Nyquist plot of an ion-insertion thin film electrode. There is one semicircle in the high-frequency region, and a steep sloping line at the lower frequencies due to the onset of the finite length effect. The high-frequency semicircle was found to be dependent upon cycling, that is, it decreased in size from the first to the tenth cycle. It is interesting to note that no diffusion impedance is clearly seen at the first three charging processes. However, Warburg diffusion impedance appeared at the tenth cycle in the middle frequency region (see insert in Fig. 3.15b). This is naturally ascribed to solid-state diffusion of Li-ion into the bulk anode material. The appearance of the Warburg impedance should be due to an increase in the electrode thickness. Indeed, the theoretical analysis of the complex impedance for insertion electrodes predicts that the diffusion control expands with increasing film thickness.[33] For very thin films, the kinetic control region may overlap with the charge saturation region and the diffusion region may appear in a very narrow range of frequencies. As the film thickness gets bigger, charge saturation dominates only at the very low frequencies, and diffusional control is observed over a large frequency range. The steep sloping line reflects a capacitive behavior, which is due to accumulation of the intercalant (Li) into the bulk anode.

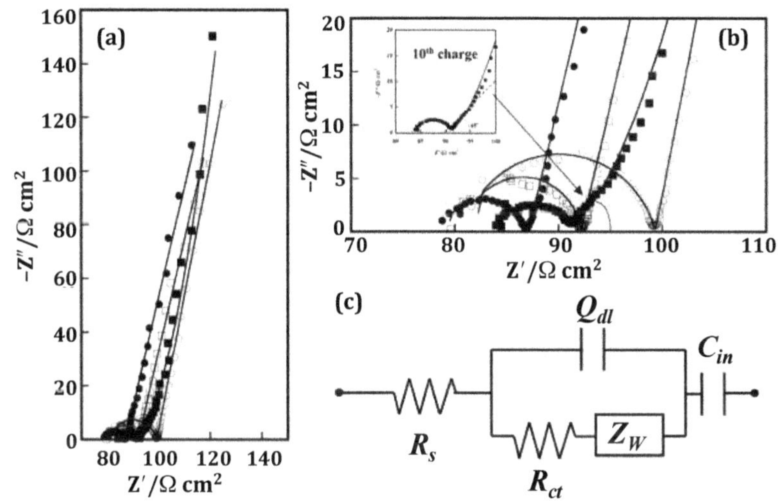

Figure 3.15 Nyquist plots of the electrodeposited $Sn_{62}Ni_{38}$ on Cu substrate measured after (○) 1st, (□) 2nd, (●) 3rd, and (■) 10th charge. (a) Full scale, (b) enlarged view of the high-frequency domain of the impedance spectra, and (c) modified Randles–Ershler model used to simulate the impedance spectra. Solid lines in Fig. 3.15a,b represent the simulation results. The impedance were measured at OCP (after charge) with the amplitude of 10 mV in the frequency range from 20 kHz to 10 mHz. The $Sn_{62}Ni_{38}$ electrode was cycled in the potential range of 0.01–2.0 V *vs.* Li/Li^+ at the current density of 50 mA/g with three-electrodes cell using Li foils as counter and reference electrode in 1 M $LiClO_4$/EC-PC (1:1 vol%). The dash line represents the first cycle, the solid lines represent the second and third cycles. Reprinted with permission from *J. Electrochem. Soc.*, **152**(3), A560 (2005). Copyright 2005, The Electrochemical Society.

According to the solid electrolyte interface (SEI) model, one would expect to observe two semicircles in the impedance spectrum, one semicircle reflecting the charge-transfer process at the $Sn_{62}Ni_{38}$/electrolyte interface, and one related to the formation of SEI. However, if their time constants are similar, the semicircles could overlap one another and become irresolvable. Moreover, it has been demonstrated with mesocarbon microbeads single-particle anode (no binder and no conductive agent) that the SEI has to be aged or forcibly grown for long hours at its formation potential to be able to see two separate semicircles in the impedance spectra.[34,35]

Features of the impedance spectra of Fig. 3.15a may be modeled by a simple modified Randles–Ershler equivalent circuit shown in Fig. 3.15c. In this model, R_s is the solution resistance, and R_{ct} is the charge-transfer resistance at the electrode/electrolyte interface. A constant phase element (CPE) was used instead of a double-layer capacitance to take into account the surface roughness of the particle. C_{in} is the insertion capacitance, and Z_W is the Warburg impedance that corresponds to the solid-state diffusion of the Li-ion into the bulk anode. The Warburg element was used only for impedance data obtained at the tenth charge. The electrical components of the surface film which is likely formed on the electrode were disregarded, because no time constant related to this process could be seen in the electrochemical impedance spectroscopy (EIS) spectra. It was also checked that their inclusion in the model of Fig. 3.15c does not improve the fit.

The solid lines in Fig. 3.15a,b represent the simulation results, and these show an excellent fitting obtained with the simple model of Fig. 3.15c over a six-decade range of frequency. The chi-square quality of fitting parameter was within 10^{-6}–10^{-5} range, and the estimated standard deviations of the individual electrical components were less than 10%. Parameters of interest obtained from the nonlinear least square fitting procedure are listed in Table 3.3. The solution resistance was more or less constant during the first three cycles of charge, whereas it slightly increased at the tenth cycle. The charge-transfer resistance, R_{ct} decreased as the cycling increased indicating an improvement in the performance of the electrode. Moreover, it points out that the contribution of the SEI to the Li insertion and extraction kinetics is not significant at these first ten Li insertion stages, otherwise the charge-transfer process would become difficult, and one would observe an increase in its resistance value, since the growth of a dense SEI layer would causes an electronic insulation and the Li ions would have to first migrate through the SEI layer to reach the anode. Nonetheless, if a pinhole is present in the SEI layer, electrons will flow relatively unimpeded from the SEI layer to the anode.

Table 3.3 Parameters of interest determined from fitting the experimental impedance spectra of Fig. 3.15a to the equivalent circuit of Fig. 3.15c

	$R_s/\Omega\ cm^2$	$R_{ct}/\Omega\ cm^2$	$C_{dl}/\mu F\ cm^{-2}$	$D_{Li}/cm^2\ s^{-1}$
1st charge	80.7	18.6	22.0	6.70×10^{-10}
2nd charge	80.1	11.2	25.0	6.10×10^{-10}
3rd charge	79.0	8.3	32.0	6.50×10^{-10}
10th charge	83.5	7.7	140	8.04×10^{-10}

Note: Impedance spectra were modeled using complex nonlinear least-squares (CNLS) fitting program EQUIVCRT developed by B. A. Boukamp. The Chi-square (x^2) function gives a good indication of the quality of the fit; a value of 10^{-4}–10^{-5} or less for x^2 indicates a reasonably good fit.

Source: Reprinted with permission from *J. Electrochem. Soc.*, **152**(3), A560 (2005). Copyright 2005, The Electrochemical Society.

The CPE exponent, *n*, ranged between 0.85 and unity and thus may be assimilated to the double-layer (dl) capacitance, C_{dl}. The double-layer capacitance was found in the range of 22–32 mF/cm^2 for the first three charges but increased to about 140 mF/cm^2 at the tenth charge, which clearly indicates changes in the electrode surface condition. The insertion capacitance C_{in} remained practically unchanged from the first to the third charge as follows from the fitting results. This means that the same level of Li insertion is obtained. In contrast, C_{in} increased at the tenth charge. The element R_s should not only reflect the solution resistance but also electronic extension from the current collector to the electrode bulk. The electronic resistance of the electrode active material may depend on cycling owing to changes in its composition; the resistance of the other passive component (current collectors) is expected to be invariant with the state of charge.

If the homogeneous diffusion ($\omega \gg 2D_{Li}/l^2$) is observed (last part of the diagram), the current is approximately 90° out of phase with the voltage. Under this condition, the real part of the impedance is independent of the frequency. The extrapolated real intercept of the straight-line region gives rise to the limiting low-frequency resistance, R_L, defined by Ho *et al.*[33]

$$Z' = R_L = \frac{V_m}{nFA}\frac{dE}{dx}\frac{l}{3D_{Li}} \tag{3.1}$$

V_m is the molar volume of the host structure, dE/dx the slope of the coulometric titration curve *vs.* mobile ion concentration x at each x value, l the maximum length of the diffusion pathway (l can be assimilated to the electrode film thickness), and n the number of equivalent.

The imaginary part of the impedance as

$$Z'' = \frac{1}{\omega C_{in}} = \frac{V_m}{nFA}\frac{dE}{dx}\frac{1}{\omega l} \tag{3.2}$$

$$\frac{1}{C_{in}} = \frac{dZ''}{d\omega^{-1}} \tag{3.3}$$

Thus, in the range where the impedance phase angle approaches $\pi/2$, the value of the intercalation capacitance, C_{in} can be determined graphically from the slope of Eq. 3.3.

Therefore, D_{Li} may be deduced from Eq. 3.1 through Eq. 3.3.

$$D_{Li} = \frac{l^2}{3R_L C_{in}} \tag{3.4}$$

Although C_{in} can be graphically determined with good precision, the determination of R_L by extrapolation is not precise and an inaccuracy is introduced in the calculation of the diffusion coefficient using Eq. 3.3. Consequently, R_L and C_{in} were determined by using the complex nonlinear least-squares (CNLS) fitting program. The diffusion coefficient value calculated thus at various cycle of charge are listed in Table 3.3. There is no significant change in the D_{Li} value from the first to the third cycle of Li insertion and sustained at an average value of 6.43×10^{-10} cm^2/s. However, at the tenth cycle, D_{Li} increased to 8.04×10^{-10} cm^2/s.

Additionally, structural and morphological changes of $Sn_{62}Ni_{38}$ electrode upon cycling was also investigated. *Ex situ* XRD analysis and morphological changes by SEM were performed at the $Sn_{62}Ni_{38}$ subjected to several cycles of charge and discharge.

Figure 3.16 displays the progression of XRD patterns for the $Sn_{62}Ni_{38}$ at the first charge and after further cycling up to ten cycles. X-ray peaks for the as-deposited electrode are also reported for reference. Shown in Fig. 3.17 is the evolution of the surface morphology of the electrode at some stages of cycling. The as-deposited $Sn_{62}Ni_{38}$ had an average particle size (for 100 particles) of 30 nm (Fig. 3.17a). Figure 3.16 shows that the Ni_3Sn_4 was the main phase confirmed from the as-deposited state. The Bragg peaks from the Ni_3Sn_4 alloy disappeared at the first charge, indicating that that a reaction between the NiSn alloy and Li is occurring. This proves that during the first charge Ni_3Sn_4 is active. The reaction product of the first charge shows two broad X-ray peaks at

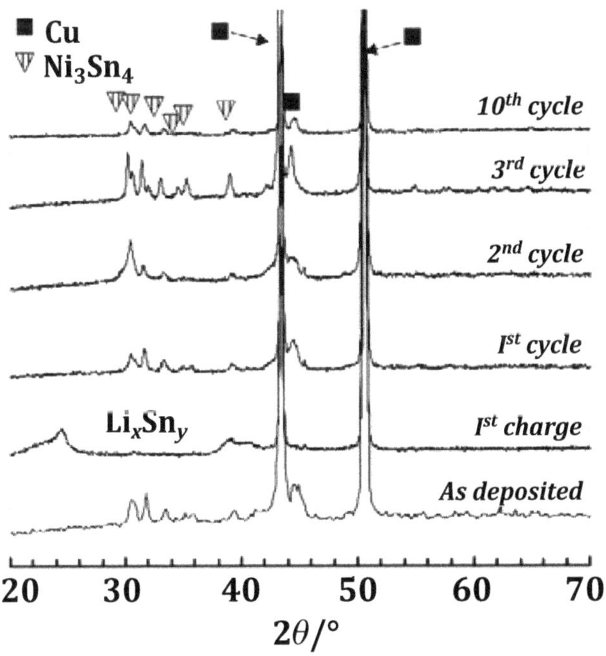

Figure 3.16 *Ex situ* XRD pattern taken at different stages during cycling of the electrodeposited $Sn_{62}Ni_{38}$ on Cu substrate. The $Sn_{62}Ni_{38}$ electrode was cycled in the potential range of 0.01–1.8 V *vs.* Li/Li$^+$ at the scan rate of 0.02 mV/s with three-electrodes cell using Li foils as counter and reference electrode in 1 M $LiClO_4$/EC-PC (1:1 vol%). Reprinted with permission from *J. Electrochem. Soc.*, **152**(3), A560 (2005). Copyright 2005, The Electrochemical Society.

$2\theta \approx 24$ and 38°, respectively. Courtney and Dahn identified these two peaks as those of Li–Sn alloy phases.[2] At the following cycle, peaks of Ni_3Sn_4 reappeared, whereas those of Li–Sn alloy phases disappeared. The formation of Li-rich alloy phases leads to fast mechanical degradation of the electrode due to a large volume change during cycling. This is supported by the SEM image of the electrode, which clearly shows expansion in the grains resulting in a bumpy surface with Li^+ insertion and formation of surface cracks with Li^+ elimination (Fig. 3.17c). It should be noted that the Ni_3Sn_4 phase is maintained even after extended cycling indicating the reversibility of the reaction between Ni_3Sn_4 and Li^+.

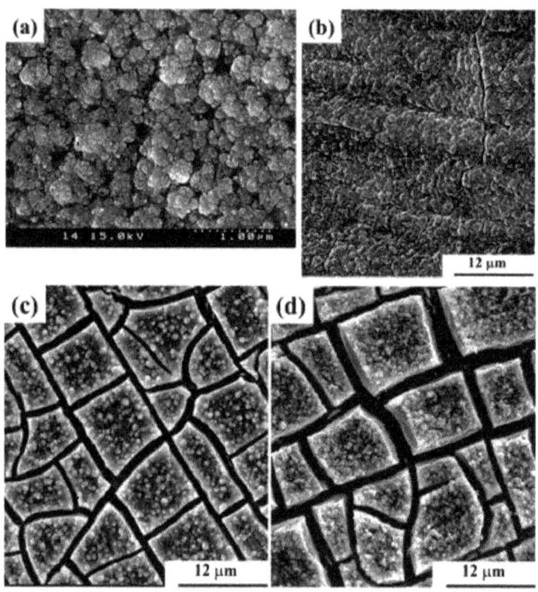

Figure 3.17 (a) High-resolution FE-SEM image of the as-deposited $Sn_{62}Ni_{38}$. (b), (c), and (d) are *ex situ* SEM images taken at 1st charge, 1st discharge, and 10th discharge, respectively. The $Sn_{62}Ni_{38}$ electrode was cycled in the potential range of 0.01–1.8 V *vs.* Li/Li^+ at the scan rate of 0.02 mV/s with three-electrodes cell using Li foils as counter and reference electrode in 1 M $LiClO_4$/EC-PC (1:1 vol%). Reprinted with permission from *J. Electrochem. Soc.*, **152**(3), A560 (2005). Copyright 2005, The Electrochemical Society.

An additional study was made to evaluate the structure and morphology changes of $Sn_{62}Ni_{38}$ cycled for a longer period.

Figure 3.18 displays the capacity *vs.* cycle number profile upon 150 cycles of charge and discharge. The profile shows an outstanding performance of the $Sn_{62}Ni_{38}$ alloy, that is, after a drop in the capacity to around 460 mAh/g, this one has markedly recovered and sustained at around an average value of 600 mAh/g over the 150 cycles tested. It is noted that the capacities of other Sn–Ni compositions tested ($Sn_{54}Ni_{46}$, $Sn_{84}Ni_{16}$, and $Sn_{92}Ni_{8}$) although stable, were very much inferior and were less than 400 mAh/g upon cycling.[28] With regard to the reproducibility of the remarkable performance of the $Sn_{62}Ni_{38}$, the average charge for various experimental trials was about 625 mAh/g. The recovery in the capacity observed with this material is a very interesting phenomenon and has been observed by Mao and Dahn with the FeSn system, which showed an increase in capacity with cycle number.[11] The authors postulated that this is caused by the breakdown of the Fe skin that impedes access to the interior of the grains of this phase by the repeated volume changes in the electrode.

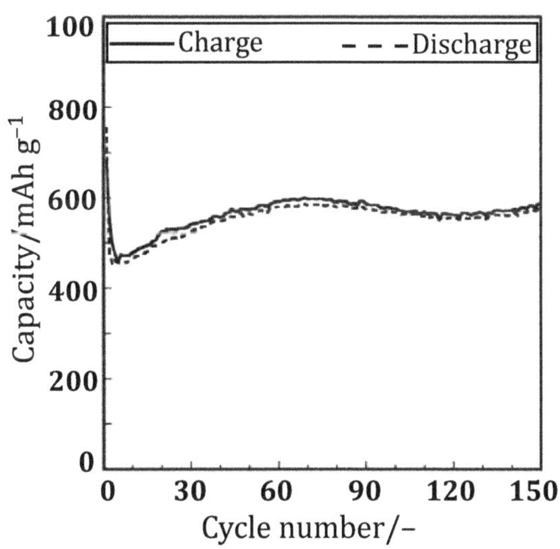

Figure 3.18 Cycle-life performance of the electrodeposited $Sn_{62}Ni_{38}$ on Cu substrate. The $Sn_{62}Ni_{38}$ electrode was cycled in the potential range of 0.01–2.0 V *vs.* Li/Li^+ at the constant current density of 50 mA/g with three-electrodes cell using Li foils as counter and reference electrode in 1 M $LiClO_4$/EC-PC (1:1 vol%). Reprinted with permission from *J. Electrochem. Soc.*, **152**(3), A560 (2005). Copyright 2005, The Electrochemical Society.

Figure 3.19 also reports few SEM images taken at the 50th and 100th cycles. The cracks observed in the tenth cycle (Fig. 3.17d) have coalesced in the 50th and the 100th cycle, and higher roughness can be noticed from the surface morphology. These images are evidence that alterations in the surface condition accompanying volume changes during cycling were responsible of the $Sn_{62}Ni_{38}$ behavior.

Figure 3.19 Cycle-life performance of the electrodeposited $Sn_{62}Ni_{38}$ on Cu substrate. The $Sn_{62}Ni_{38}$ electrode was cycled in the potential range of 0.01–2.0 V *vs.* Li/Li^+ at the constant current density of 50 mA/g with three-electrodes cell using lithium foils as counter and reference electrode in 1 M $LiClO_4$/EC-PC (1:1 vol%). Reprinted with permission from *J. Electrochem. Soc.*, **152**(3), A560 (2005). Copyright 2005, The Electrochemical Society.

It is obvious that the electrode does not return to the original volume upon lithium extraction. The major increase in volume happens in the first alloying. This volume enlargement of the host matrix causes an increase in the electrode/electrolyte interface area. The repeated formation of cracks and pores has expanded the volume of the electrode so much that it has covered up the cracks formed during the earlier cycling. The resulting expanded electrode offers a higher porosity than the original material, and provided that there is still electronic contact within the host after this activation, Li diffusion is facilitated with lower volume expansions in the following cycles. This might be one of the reasons for the rise in the capacity seen in the cycle performance of each sample.

The other probable positive changes upon cycling are improvements in the impregnation of the porous electrode with the solution and reorganization of the surface films (better cohesion and adhesion), which improves their transport properties (Li-ion migration). SEM images for the aged (long-term cycled) electrode showed no sign of major failure mechanisms due to exfoliation, cracking, or any other bulk destruction of the SnNi particles. However, the images of the aged electrode clearly show that it is covered by granular surface films. The crack formation exposes new surfaces to the electrolyte, which will be followed by massive precipitation of surfaces species. This explains the shift in the solution resistance observed by impedance. The observed surface films rearrange and refine, and their coherence and adherence to the anode increases which causes a decrease in the resistances for Li-ion migration through the surface layer for Li-ion transfer at the anode. The new layer will serve as a protective overlayer and enhances the long-term cycling efficiency and stability of the $Sn_{62}Ni_{38}$ anode.

3.3 *In situ* Stress Transition Observation of Electrodeposited Anode Materials

To obtain further interpretation of Sn-based electrodes during its lithiation–rest–delithiation–rest process, an *in situ* measurement of the stress transition based on the optical cantilever method was carried out. The stress transition during the reaction of Li (the first element) and Sn (the second element) is discussed in detail, along with the effect of alloying Sn with the third element (Ni). The other object was to develop a useful stress-measurement methodology for *in situ* stress measurement during charge/discharge cycles.

3.3.1 Stress Calculation

The electrode mechanical behavior during stress development was modeled under the following assumptions and definitions 1–11. The abbreviations and the physical constants that were used for the calculations are listed in Table 3.4.

1. The cantilever can be modeled as a wide plate with dimensions of length (*l*), width (*W*), and thickness (*t*), where $t << W < l$ as shown in Fig. 3.20.
2. The stress in the direction of *y*-axis (along the beam) is the only component that is relevant.
3. The physical constants of the active material are similar to those of the bulk material.
4. The active material and the substrate are totally elastic and incompressible.
5. One side of the substrate is uniformly electrodeposited with the active material.
6. The adhesion of the active material thin film electrode to the substrate is complete. (There is no delamination.)
7. The only reactive material is Sn, which interacts with Li.
8. The thin film reactions are as follows: (i) Sn reacts completely with Li to form $Li_{22}Sn_5$ [volume increase factor (VIF): $V_{Li_{22}Sn_5}/V_{Sn} = 3.592$] and (ii) $Sn_{62}Ni_{38}$ reacts completely with Li to form Ni and Li_7Sn_2[36] (VIF: $V_{Li_7Sn_2}/V_{Sn} = 2.992$).
9. The volume expansion is isotropic, unless it is confined by the substrate.
10. The active material is void-free after lithiation.
11. The stress generated at the contact interface between the active material and the substrate is the subject of calculation.

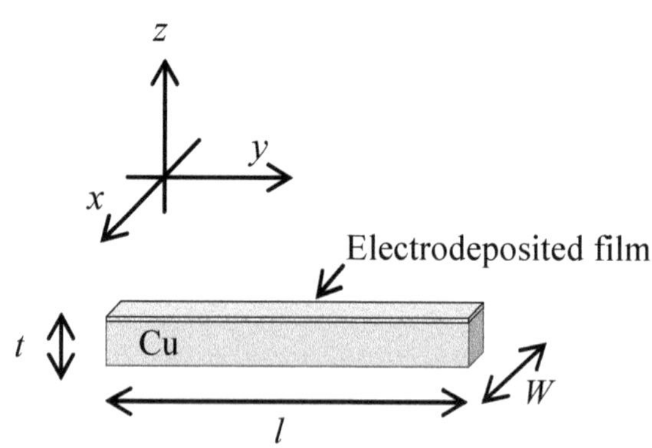

Figure 3.20 Definitions of *x*-, *y*-, and *z*-axes and abbreviations used in the model. Reprinted with permission from *Electrochem. Solid-State Lett.*, **10**(3), A570 (2007). Copyright 2007, The Electrochemical Society.

Table 3.4 Abbreviations and physical constants used for calculation

Abbreviations	**Meanings**	**Physical value**
σ_{Sn}, $\sigma_{Sn_{62}Ni_{38}}$	Stress in the direction of *y*-axis generated at the contact interface	–
Y_{Sn}	Young's modulus of Sn	50×10^9 (Pa)
Y_{NiSn}	Young's modulus of Ni_3Sn_4	133×10^9 (Pa)
Y_{Cu}	Young's modulus of Cu	130×10^9 (Pa)
ν_{Cu}	Poisson ratio of Cu	0.343
ε_y	The strain in *y*-axis direction (in-line)	–

Source: Reprinted with permission from *Electrochem. Solid-State Lett.*, **10**(3), A570 (2007).

The theoretical value of the stress generated at the contact interface between the electrodeposited active material and substrate was calculated to obtain the order of magnitude of the stress generated by lithiation. As it has been reported from previous studies,[25,37,38] the actual electrode is not a uniform material and is expected to contain and/or form voids and cracks with lithiation. It has also been shown by *in situ* atomic force microscopy (AFM) observations that the electrode experiences plastic deformation, and the thickness of the electrode changes irreversibly to some extent with the lithiation and delithiation process.[29] Because of such indeterminate nature of the assumptions, the absolute value of the calculated stress is not the subject of discussion.

In the case of pure Sn, the VIF after the lithiation is 3.59. When the active material is free to expand without stress, the linear increase factor in the direction of *y*-axis is

$$\varepsilon_{Sn_y} = \left[(l - \Delta l/l)\right]_{Sn} - 1 = 3.59^{1/3} - 1 = 1.53 - 1 = 0.53$$

Therefore, when the active material is completely bound to the substrate and $l = 9 \times 10^{-3}$ m, the stress that is developed at the contact interface to counterbalance the volume expansion is

$$\sigma_{Sn} \sim Y_{Sn} \times \varepsilon_{Sn_y} = 27(\text{GPa})$$

where σ_{Sn} is the stress in the contact interface between the Cu substrate and the electrodeposited Sn, Y_{Sn} is the Young's modulus of Sn, and $\varepsilon_{\mathrm{Sn}_y}$ is the strain of the electrodeposited Sn in the *y*-axis direction (along the beam). Note that this is only an approximation because for such large strain, the thin film is most likely to be beyond the critical yield point, where the material begins to plastically deform. This strain value was used to compare between the strains developed in the Sn to the $Sn_{62}Ni_{38}$ alloy.

In the case of the $Sn_{62}Ni_{38}$ alloy, the volumetric ratio of Sn atoms in the active material is approximately equal to 0.77, which was calculated from the atomic radii of Sn (1.58 Å) and Ni (1.25 Å) and the composition of the active material Sn/Ni = 62:38. Therefore, the VIF of the total active material is 2.53. For pure Sn, the linear increase factor is $\varepsilon_{\mathrm{NiSn}_y} = [(l + \Delta l)/l - 1]_{\mathrm{NiSn}} = 2.53^{1/3} - 1 = 0.36$. Therefore, when $l = 9 \times 10^{-3}$ m, the stress at the contact interface is $\sigma_{\mathrm{NiSn}} \sim Y_{\mathrm{NiSn}} \times \varepsilon_{\mathrm{NiSn}_y} = 48$ GPa, where σ_{NiSn} is the stress in the contact interface between the Cu substrate and the electrodeposited $Sn_{62}Ni_{38}$, Y_{NiSn} is the Young's modulus of Ni_3Sn_4 alloy (which is the main structural phase of $Sn_{62}Ni_{38}$(28)), and $\varepsilon_{\mathrm{NiSn}_y}$ is the strain of the electrodeposited $Sn_{62}Ni_{38}$ in the *y*-axis direction (along the substrate like beam).

The results of the above calculations indicate that the $Sn_{62}Ni_{38}$ electrode, which shows a longer cycle life than the Sn electrode, is more highly stressed. Note that these calculations are tentative and assume linear strain–stress relation, which is probably incorrect for such large strains. However, this rough calculation demonstrates the very basic fact that Young's modulus of the thin film plays an important role in the electrochemical behavior of the electrode and the subsequent electrode reliability.

3.3.2 *In situ* Stress Measurement

Figure 3.21 illustrates the structure of the cell used for the *in situ* stress measurement. The active material was electrodeposited on only one side of the electrode. Hence the electrode acted as a cantilever which is deflected by the stress generated at the interface between the active material and the substrate. With the volume expansion of the active material, tensile stress was generated on the Cu substrate, whereas the volume contraction would result in compressive stress. A laser beam was irradiated on the Cu (non-electrodeposited) side of the electrode to detect the deflection using a deflection measuring system.

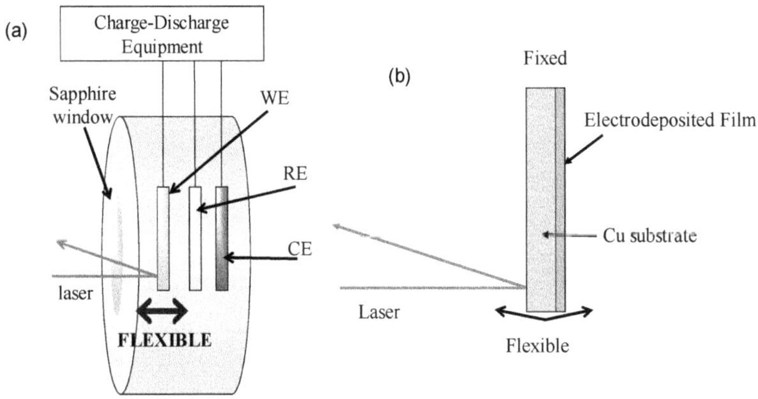

Figure 3.21 Schematic illustration of the experimental setup used for *in situ* stress transition measurement. (a) The electrochemical cell and (b) the enlarged view of the cantilever-like working electrode. Reprinted with permission from *Electrochem. Solid-State Lett.*, **10**(3), A570 (2007). Copyright 2007, The Electrochemical Society.

3.3.3 *In situ* Stress Transition Observation of Electrodeposited Sn-Based Anodes

The *in situ* stress transition of the electrodeposited Sn and $Sn_{62}Ni_{38}$ anodes were evaluated by means described above. Figure 3.22a,b shows the results of *in situ* stress measurements of the Sn and the $Sn_{62}Ni_{38}$ electrodes, respectively. The negative values of the deflection profile represent compressive stress (expansion of the active material). Tests of repeated charge–discharge cycles showed that the deflection response of the electrode continued, and it was confirmed that the substrate elasticity was retained during the first cycle. It is difficult to obtain quantitative reproducibility due to the individual textures that differ with each electrode (such as their cracks and pores). However, the qualitative reproducibility which indicates that the electrode shows a tendency that is dependent on the electrode composition has been confirmed.

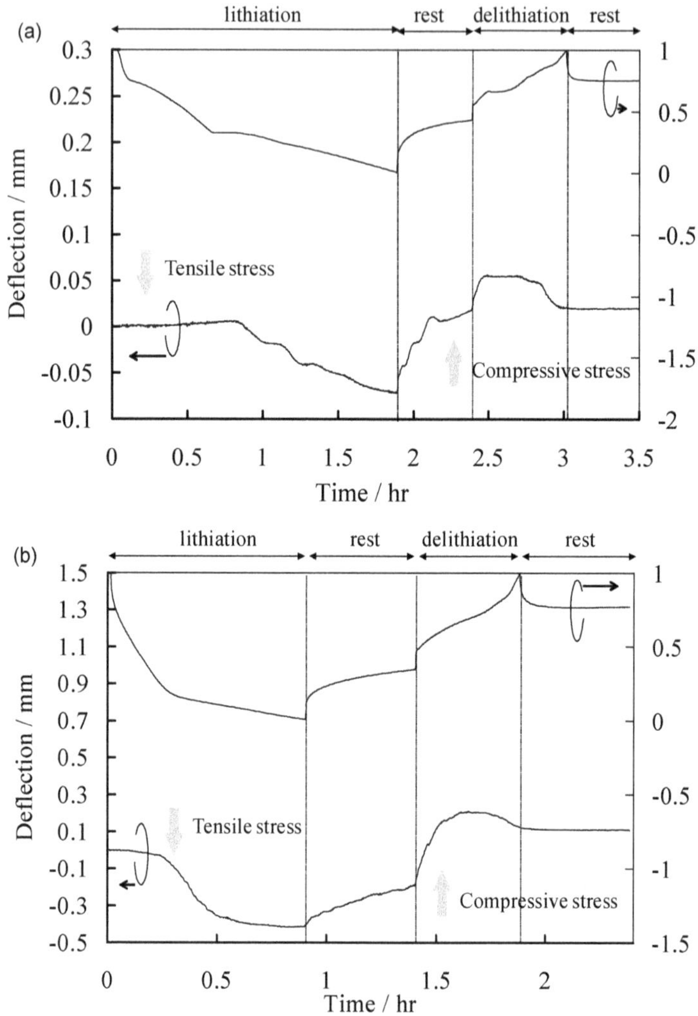

Figure 3.22 The deflection and potential profiles *vs.* time during the 1st cycle of (a) Sn electrode and (b) $Sn_{62}Ni_{38}$ electrode. The electrodes were charge and discharged from the potential of OCP to 0.01 and then from 0.01 to 1.0 V *vs.* Li/Li^+ at the constant current density of 250 mA/g with three-electrodes cell using Li foils as counter and reference electrode in 1 M $LiClO_4$/EC-PC (1:1 vol%). Reprinted with permission from *Electrochem. Solid-State Lett.*, **10**(3), A570 (2007). Copyright 2007, The Electrochemical Society.

From the potential profile of the Sn electrode, two descending plateaus which start around 0.78 and 0.32 V *vs.* Li/Li^+, respectively, are observed during the lithiation process (Fig. 3.22a). According to the reports by Winter *et al.*[25] the first plateau is attributed to the gradual lithiation of Sn to Li_2Sn_5 and LiSn and the second plateau to the further lithiation of the active material to phases such as Li_7Sn_3, Li_5Sn_2, and $Li_{13}Sn_5$. An increase in the tensile stress is first detected from the deflection profile, when the potential reaches 0.32 V *vs.* Li/Li^+. The VIF of each Li_xSn_y alloy [$VIF(Li_xSn_y) = V_{Li_xSn_y}/V_{Sn}$] has been calculated by Dahn *et al.* in Table 3.5.[2] Considering the difference in the VIF(LiSn):1.53 and VIF(Li_7Sn_5):2.28, this result indicates that there is a critical point between these volume changes where the generated stress becomes large enough to affect the Cu substrate. However, the succeeding flexion point of the deflection profile at 0.24 and 0.19 V *vs.* Li/Li^+ do not agree with the potentials where distinct phase transformations of Li–Sn alloys were reported to occur.[25] This result indicates that the 100 atom % Sn electrode undergoes "stress accumulation" (during the plateau of the deflection profile) and "stress release" (during the large gradient of the deflection profile) which is independent of the crystal-structure transformations of the electrode. This phenomenon could be due to the high-charge current density that leads to a rapid volume change of the active material and prevents the Cu substrate from accurately responding to the actual stress transition.

Table 3.5 VIF of each Li_xSn_y

	Li_2Sn_5	LiSn	Li_7Sn_3	Li_5Sn_2	$Li_{13}Sn_5$	Li_7Sn_2	$Li_{22}Sn_5$
VIF	1.23	1.53	2.28	2.76	2.81	2.99	3.59

In the case of the $Sn_{62}Ni_{38}$ electrode, the potential profile shows a slope that starts from 0.80 V *vs.* Li/Li^+ and a plateau from 0.24 V *vs.* Li/Li^+. From the potential profile of the Sn electrode, the latter plateau can be assumed to be the formation of Li-rich Sn–Li alloy phases. 0.24 V *vs.* Li/Li^+ is lower than the corresponding lithiation potential of the Sn electrode, which could indicate the higher overpotential associated with the electrochemical reaction to form Li-rich Sn–Li alloy phases in the $Sn_{62}Ni_{38}$ electrodes. The deflection profile shows only one flexion point of tensile stress increase which starts at 0.24 V *vs.* Li/Li^+. Unlike Sn, flexion points of the deflection profile which could not be attributed to the structural transformations were not observed. This could be due to the effect of the third element (Ni), which increases the Young's modulus of the electrode and increases the stress generated from the volume change, hence leading to a better response from the Cu substrate. The higher magnitude of deflection also indicates the larger stress in this electrode compared with the Sn electrode, which is consistent with the results indicated from the calculation.

When the electrode reaches 0.01 V *vs.* Li/Li^+ and the current is turned off, the potential of both electrodes initially rises rapidly, and after a transient state, it slowly draws near to a steady state. There has been no clear explanation on the phenomena within the electrode at this transient stage during the open circuit (*i.e.*, the "rest" condition). In the deflection profile, both electrodes show a rapid drop in the tensile stress starting at the moment when applied current was cut off. This indicates that there is some excess stress in the electrode during the lithiation process, where a large change took place in the amount of Li at the surface of the electrode. This deviation of Li could have been rapidly relaxed by the fast diffusion of the Li into the active material after the electric field was turned off. Similar to the initial lithiation process, the stepwise stress transition from the Sn electrode deflection profile and the smooth transition from the $Sn_{62}Ni_{38}$ electrode deflection profile could be reflecting the difference in the magnitude of the generated stress. $Sn_{62}Ni_{38}$, which has a larger Young's modulus, could have a larger effect on the Cu substrate compared with the Sn.

During the subsequent discharge process, the potential rises with accordance to the delithiation of Li–Sn alloys. *Ex situ* surface morphology observed in previous reports show that the electrodes crack during this delithiation process.[37,38] In the deflection profile, a further drop in stress was observed for both electrodes. This is assumed to be due to cracking of the electrode, which is expected to relax the stress more quickly than the delithiation process does. Here, the phase transformation does not seem to cause any distinct effect on the stress transition. As the delithiation process was allowed to continue, the rapid relaxation of compressive stress stopped, and the stress became tensile.

During the rest period after the delithiation, the potential dropped to a constant value with a transient step much shorter than the rest after lithiation. This indicates little change in the electrode state, which is consistent with the deflection profile that remains constant during this step. The deflection profile also indicates that there was constant residual tensile stress in the electrode after the delithiation. This indicates that the initial, as-deposited electrode had an inherent stress, which probably was originated from the electrodeposition process. Apparently, this initial inherent stress was released as a result of the cracking of the electrode during delithiation.

Finally, magnitudes of the stress values of both samples, (a) the Sn and (b) the $Sn_{62}Ni_{38}$, were calculated from the deflections measured at 0.01 V *vs.* Li/Li^+ using Stoney's equation. As a result, the stress of Sn (σ_{Sn}) and $Sn_{62}Ni_{38}$ (σ_{NiSn}) were σ_{Sn} = 52 [MPa] and σ_{NiSn} = 490 [MPa].

As it has been indicated during the discussion of the lithiation process, the relation between the two stress values ($\sigma_{Sn} < \sigma_{NiSn}$) is consistent with the results of the calculation. However, the absolute values of the stress are smaller by an order of 10^2–10^3 than those derived by the calculation. This discrepancy may be partly due to incomplete lithiation of the active material and partly due to inadequacy of the assumptions made for the calculation, such as the formation of voids and cracks, as described in the previous discussion. These phenomena would significantly release the excessive stress that is induced during the reaction of the electrode. The difference between the calculation and the experimental results also indicate that the stress-release effect of the Sn electrode is larger than the $Sn_{62}Ni_{38}$ electrode. This could be due to the different critical yield point of the materials, where Sn could accommodate the stress more than the $Sn_{62}Ni_{38}$ by plastic deformation.

3.4 Mesoporous Sn Anode Improved Electrochemical Properties

Many mesoporous materials with specific structural features (e.g., uniform mesopore size and high surface area) have extensively been investigated. In particular, mesoporous metals with high electroconductivity are very promising for various electrochemical applications. Since the first report by Attard *et al.*,[39] several mesoporous metals including Sn[40–42] have been prepared by the reduction of the corresponding metal ions in the presence of lyotropic liquid crystals (LLC) made of nonionic surfactants.[43] The mesoporous metals have been mainly prepared by electrodeposition for the reduction of metal ions to form thin films on conductive substrates.

Here a mesoporous Sn from LLC was prepared and investigated on its cycle and rate properties as an anode for Li-ion secondary batteries. The mesoporous structure should suppress the influence of the volume change during the charge and discharge cycling of Li-ions, leading to a longer life. Also, the mesoporous structure with high surface area can provide a low diffusion resistance of Li-ions into the electrode, which should contribute to an increase of the reaction rate.

3.4.1 Preparation of Mesoporous Sn Anode

Mesoporous Sn was prepared on a copper foil by electrodeposition. LLC including Sn ions was prepared by mixing 0.300 g nonionic surfactant, octaethyleneglycol monohexadecyl ether ($C_{16}EO_8$), and an aqueous Sn solution (0.300 g) which was prepared by dissolving both 1.69 g $SnCl_2 \cdot 2H_2O$ and 9.15 g of H_2SO_4 aq. (18.3 mol/L) into water and fixing the volume to 50 mL. The LLC takes a 2-D hexagonal symmetry, as is confirmed by low-angle XRD analysis. The electrodeposition was carried out at room temperature and H_2O saturated atmosphere with a constant voltage of −100 mV till the current of 1.5 C/cm^2 was passed. Sn plate was used as the counter electrode.

3.4.2 Electrochemical Property of Mesoporous Sn Anode

Figure 3.23a shows the TEM image of the as-prepared sample and its selected electron diffraction (ED) pattern in a 100 nm region. A disordered mesoporous structure was observable over the entire area (Fig. 3.23a). The bright and dark areas indicate pores and deposited Sn, respectively. The ED patterns are assignable to (111) diffraction of the α-Sn structure and (200), (101), (220), and (211) diffractions of the β-Sn structure. The ring pattern indicates that the deposited Sn is composed of very minute polycrystals.

Figure 3.23 Bright-field TEM images and selected ED patterns in a 100 nm region. (a) As-prepared mesoporous Sn. (b) Mesoporous Sn anode after the 100th cycle. The sample was scratched from the substrate. The powdery samples were dispersed in ethanol and mounted on a TEM microgrid. The sample was cycled in the potential range of 0.01 to 1.0 V *vs.* Li/Li^+ at the constant current density of 994 mA/g (1 C) with three-electrodes cell using Li foils as counter and reference electrode in 1 M $LiClO_4$/EC-PC (1:1 vol%). Reprinted with permission from *Chem. Lett.*, **37**(2), 142 (2008). Copyright 2008, The Chemical Society of Japan.

Figure 3.24 shows the discharge capacity of the mesoporous Sn anode and that of a Sn anode prepared using electroplating bath without nonionic surfactant,[38] called the "dense Sn anode," for comparison. The discharge capacities of both the mesoporous and dense Sn anodes increased at the initial stage (0–5th cycle). This is because the electrochemically active surface area increased due to crack formation by the volume change with cycles.[45] At the second stage (*ca.* 5th–20th cycle), the discharge capacity of the dense Sn anode was dramatically decreased due to electrical disconnection. The disintegration of the dense anode from the Cu foil was visually observed. On the other hand, in the case of the mesoporous Sn anode, such a phenomenon was not confirmed. The mesoporous structure suppresses the influence of the volume change during the cycling to some extent, which can prevent the disintegration of the Sn anode from the substrate. At the following stage (over 20th cycle), the discharge capacities of both Sn anodes were maintained with small changes. After the 100th cycle, the discharge capacity of the mesoporous anode was 425 mAh/g, while that of the dense anode was only 46 mAh/g. For comparison, when nanosized carbon particles were cycled at a constant current of 100 mA/g, the discharge capacity after the 100th cycle was 290 mAh/g.[44] Consequently, the introduction of the mesoporous structure is effective to improve the cycle durability.

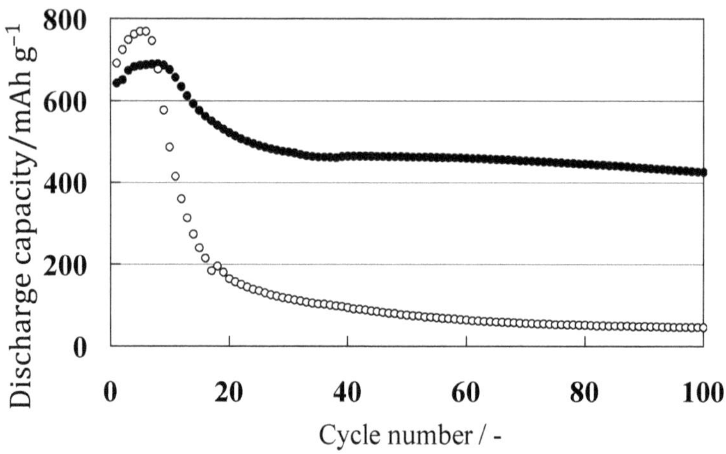

Figure 3.24 The discharge capacities of the mesoporous Sn (○) and the dense Sn anodes (●). The charge–discharge current density was 994 mA/g (1 C rate) in the potential range of 0.01 to 1.0 V *vs.* Li/Li$^+$ with three-electrodes cell using Li foils as counter and reference electrode in 1 M $LiClO_4$/EC-PC (1:1 vol%). Reprinted with permission from *Chem. Lett.*, **37**(2), 142 (2008). Copyright 2008, The Chemical Society of Japan.

Figure 3.23b shows the TEM image of the mesoporous Sn anode after the 100th cycle and its selected-ED pattern. From Fig. 3.23b, the formation of Sn grains of 10 nm size was observable, indicated by the arrow (Fig. 3.23b). The ring pattern including the intense spots (Fig. 3.23b) shows the presence of both very minute polycrystalline phase and crystal Sn grains. It has been previously shown that an amorphous-like phase should be effective to improve the cycle durability of Sn anode.[32] Therefore, the very minute polycrystalline phase should contribute to the improvement of the cycle durability of the mesoporous Sn anode. When the dense Sn anode was used, larger Sn crystals grow easily, after the charge/discharges of Li ions are repeated for several times.[45] On the contrary, the mesoporous Sn anode still retains its very minute polycrystalline phase even after the 100th cycle. Therefore, the presence of the mesoporosity prevented the larger Sn crystal growth during cycling, leading to a longer life.

Figure 3.25 shows the discharge capacity at the 1st cycle with various C rates. With increasing C rate, the discharge capacity decreased for both anodes. At lower C rate, the difference between mesoporous and dense Sn anodes was hardly observed. However at higher C rate, a remarkable difference was observed; only the mesoporous Sn anode operated at as high as 17 C rate. This result is ascribable to the alleviation of voltage loss due to the moderation of actual current density and should be ascribed to the shortening of the solid-state diffusion distance of Li. As the result, rate property improved on introducing the mesoporous structure.

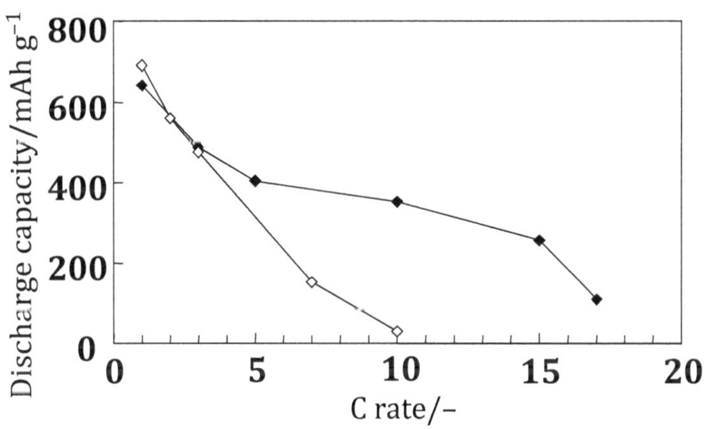

Figure 3.25 The discharge capacity at the first cycle of the mesoporous Sn anode (◆) and the dense Sn anode electrodeposited with a conventional bath (◇). The charge–discharge was carried out at various current density of 0.994 to 1.69 A/g (1–17 C rate) in the potential range of 0.01 to 1.0 V *vs.* Li/Li^+ with three-electrodes cell using Li foils as counter and reference electrode in 1 M $LiClO_4$/EC-PC (1:1 vol%). Reprinted with permission from *Chem. Lett.*, **37**(2), 142 (2008). Copyright 2008, The Chemical Society of Japan.

3.5 Conclusions and Outlook

In this chapter, Sn-based anodes for Li-ion secondary batteries were discussed mainly to improve cycle properties in terms of suppression of volume-change effect, alloying with inactive metal with Li or introduction of mesoporous structure. As these results show, the cycle properties were significantly improved. Moreover, in case of mesoporous Sn anode, high-rate property was also achieved.

The demand for energy sources with better performance is increasing with the rapid development of electronic devices. Although the fuel cells are attracting strong attention as the next-generation power supply sources, they suffer from poor power output. Improvement of conventional Li-ion batteries is an urgent and essential issue to be realized, especially for the applications in portable electronic devices which require high power. Under such circumstances, the substitution of the anode material from the conventional carbonaceous materials to the Sn-based materials is very promising in terms of enhancement of the electrode capacity.

Indeed, the Sn-based materials suffer short cycle life. However, the studies indicated in this chapter proposes that by understanding the electrochemical phenomenon induced within the electrode during cycling with diversified viewpoints, Sn-based materials can be designed as the novel anode electrode for high-energy Li-ion battery.

Acknowledgment

This work was financially supported in part by a grant-in-aid by 21st Century COE Program (Practical Nano-Chemistry) and COE Research (Molecular Nano-Engineering), Encouraging Development Strategic Research Centers Program "Establishment of Consolidated Research Centers Institute for Advanced Science and Medical Care," and the Global COE Program "Practical Chemical Wisdom" from The Ministry of Education, Culture, Sports, Science and Technology, Japan.

References

1. Fuji Photo Film Co. Ltd., *European Patent No.* 0,651,450,A1 (1995).
2. I. A. Courtney, J. R. Dahn, *J. Electrochem. Soc.*, **144**, 2045 (1997).
3. I. A. Courtney, J. R. Dahn, *J. Electrochem. Soc.*, **144**, 2943 (1997).
4. J. Morales, L. Sanchez, *J. Electrochem. Soc.*, **146**, 1640 (1999).
5. J. Santos-Pena, T. Brousse, D. M. Schleich, *Solid State Ionics*, **135**, 87 (2000).

6. X. W. Zhang, C. S. Wang, A. J. Appleby, F. E. Little, *J. Power Sources*, **109**, 136 (2002).
7. J. H. Ahn, Y. J. Kim, G. X. Wang, M. Lindsay, H. K. Liu, S. X. Dou, *Mater. Trans.*, **43**, 63 (2002).
8. G. M. Ehrlich, C. Durand, X. Chen, T. A. Hugener, F. Spiess, S. L. Suib, *J. Electrochem. Soc.*, **147**, 886 (2000).
9. O. Crosnier, T. Brousse, X. Devaux, P. Fragnaud, D. M. Schleich, *J. Power Sources*, **94**, 169 (2001).
10. O. Mao, J. R. Dahn, *J. Electrochem. Soc.*, **146**, 414 (1999).
11. O. Mao, J. R. Dahn, *J. Electrochem. Soc.*, **146**, 423 (1999).
12. O. Mao, R. A. Dunlap, J. R. Dahn, *J. Electrochem. Soc.*, **146**, 405 (1999).
13. O. Mao, R. L. Turner, I. A. Courtney, B. D. Fredericksen, M. I. Buckett, L. J. Krause, J. R. Dahn, *Electrochem. Solid-State Lett.*, **2,** 3 (1999).
14. D. Larcher, L. Y. Beaulieu, O. Mao, A. E. George, J. R. Dahn, *J. Electrochem. Soc.*, **147**, 1703 (2000).
15. L. Beaulieu, D. Larcher, R. A. Dunlap, J. R. Dahn, *J. Alloys Compd.*, **297**, 122 (2000).
16. K. D. Kepler, J. T. Vaughey, M. M. Thackeray, *Electrochem. Solid-State Lett.*, **2**, 307 (1999).
17. D. Larcher, L. Y. Beaulieu, D. D. MacNeil, J. R. Dahn, *J. Electrochem. Soc.*, **147**, 1658 (2000).
18. N. Tamura, R. Ohshita, M. Fujimoto, S. Fujitani, M. Kamino, I. Yonezu, *J. Power Sources*, **107**, 48 (2002).
19. G. X. Wang, L. Sun, D. H. Bradhurst, S. X. Dou, H. K. Liu, *J. Alloys Compd.*, **299**, L12 (2000).
20. T. Momma, N. Shiraishi, A. Yoshizawa, T. Osaka, A. Gedanken, J. J. Zhu, L. Sominski, *J. Power Sources*, **97–98**, 198 (2001).
21. H. Mukaibo, A. Yoshizawa, T. Momma, T. Osaka, *J. Power Sources*, **119–121**, 60 (2003).
22. A. Ito, H. Enomoto, *J. Met. Finish. Soc. Jpn.*, **36**, 466 (1985).
23. T. Watanabe, T. Hirose, K. Arai, M. Chikazawa, *J. Japan Inst. Metal*, **63**, 496 (1999).
24. C. J. Wen, R. A. Huggins, *J. Electrochem. Soc.*, **128**, 1181 (1981).
25. M. Winter, J. O. Besenhard, *Electrochim. Acta*, **45**, 31 (1999).
26. J. Chouvin, J. Olivier-Fourcade, J. C. Jumas, B. Simon, O. Godiveau, *Chem. Phys. Lett.*, **308**, 413 (1999).

27. I. A. Courtney, W. R. McKinnon, J. R. Dahn, *J. Electrochem. Soc.*, **146**, 59 (1999).
28. H. Mukaibo, T. Sumi, T. Yokoshima, T. Momma, T. Osaka, *Electrochem. Solid-State Lett.*, **6**, A218 (2003).
29. L. Y. Beaulieu, S. D. Beattie, T. D. Hatchard, J. R. Dahn, *J. Electrochem. Soc.*, **150**, A419 (2003).
30. S. D. Beattie, T. Hatchard, A. Bonakdarpour, K. C. Hewitt, J. R. Dahn, *J. Electrochem. Soc.*, **150**, A701 (2003).
31. J. Z. Li, H. Li, Z. X. Wang, X. J. Huang, L. Q. Chen, *J. Power Sources*, **81–82**, 346 (1999).
32. M. Mohamedi, S. J. Lee, D. Takahashi, M. Nishizawa, T. Itoh, I. Uchida, *Electrochim. Acta*, **46** 1161 (2001).
33. C. Ho, I. D. Raistrick, R. A. Huggins, *J. Electrochem. Soc.*, **127**, 343 (1980).
34. K. Dokko, Y. Fujita, M. Mohamedi, M. Umeda, I. Uchida, J. R. Selman, *Electrochim. Acta*, **47**, 933 (2001).
35. M. Umeda, K. Dokko, Y. Fujita, M. Mohamedi, I. Uchida, J. R. Selman, *Electrochim. Acta*, **47**, 885 (2001).
36. P. E. Lippens, J. Olivier-Fourcade, J. C. Jumas, *IMLB, Biarritz, France* (2006).
37. L. Y. Beaulieu, K. W. Eberman, R. L. Turner, L. J. Krause, J. R. Dahn, *Electrochem. Solid-State Lett.*, **4**, A137 (2001).
38. H. Mukaibo, T. Momma, M. Mohamedi, T. Osaka, *J. Electrochem. Soc.*, **152**, A560 (2005).
39. G. S. Attard, C. G. Goltner, J. M. Corker, S. Henke, R. H. Templer, *Angew. Chem., Int. Ed. Engl.*, **36**, 1315 (1997).
40. G. S. Attard, J. M. Elliott, P. N. Bartlett, A. Whitehead, J. R. Owen, *Macromol. Symp.*, **156**, 179 (2000).
41. A. H. Whitehead, J. M. Elliott, J. R. Owen, *J. Power Sources*, **81–82**, 33 (1999).
42. A. H. Whitehead, J. M. Elliott, J. R. Owen, G. S. Attard, *Chem. Commun.*, **4**, 331 (1999).
43. Y. Yamauchi, T. Yokoshima, H. Mukaibo, M. Tezuka, T. Shigeno, T. Momma, T. Osaka, K. Kuroda, *Chem. Lett.*, **33**, 542 (2004).
44. Z. H. Yi, Y. G. Liang, X. F. Lei, C. W. Wang, J. T. Sun, *Mater. Lett.*, **61**, 4199 (2007).
45. H. Mukaibo, T. Momma, T. Osaka, *J. Power Sources*, **146**, 457 (2005).

Chapter 4

Alloy Electrode and Its Breakthrough Technology

Kiyotaka Yasuda
SILX Project, Corporate Technology Center, Mitsui Mining & Smelting Co. Ltd., 1333-2 Haraichi, Ageo-shi, Saitama 362-0021, Japan
yasuda_k@mitsui-kinzoku.co.jp

Lithium-ion batteries are the state-of-the-art rechargeable power sources and have been widely used not only for portable equipment such as cellular phones, personal computers, and digital cameras, but also power tools and industrial applications like robot and electric vehicles. With the improvement of material and battery technologies, the battery capacity has doubled since its commercialization in the early 1990s while its basic concept and chemistry has remained unchanged. Recently we have seen this improvement slow down, as it is approaching the theoretical limit of the system. A new battery system is necessary to meet the rigorous demands of energy and power increments for the progress of multi-functional mobile equipment and to secure sufficient mileage of an electric vehicle per charge. Under these circumstances, alloy anode materials are expected to be promising candidates to remarkably increase the

Nanomaterials for Lithium-Ion Batteries: Fundamentals and Applications
Edited by Rachid Yazami

ISBN 978-981-4316-40-8 (Hardcover), 978-981-4364-23-2 (eBook)
www.panstanford.com

capacity of lithium-ion batteries. In this chapter, the potential and benefits of these alloy anode materials and recent breakthroughs that have been achieved are introduced.

4.1 Introduction

Lithium metal forms a highly attracting anode that makes a high-energy density battery system due to its low potential and intrinsically high gravimetric electrochemical capacity.[1] However, lithium metal electrode has poor electrochemical cycle efficiency, resulting in lower cycle performance, and it also been facing serious battery safety problem, as lithium metal tends to form dendrite during the charging process and, therefore, its practical use seems to be difficult in current technologies. Some elements and alloys have been known to react with lithium.[2,3] Much effort and different strategies have been put into lithium alloys in order to overcome these problems of lithium metal anode and deal with the volume changes that occur during cycling.[4–21] Despite much effort, none of them has succeeded in obtaining the electrode that has sufficient cycle number without sacrificing high capacities of alloy electrodes.

In the meantime, it was demonstrated that lithium can be reversibly inserted into graphite at room temperature in an organic electrolyte in 1983.[22] Lithium-ion battery was first commercialized with this carbon-based anode in 1991. Graphite has a capacity of 372 mAh/g, corresponding to the intercalation of one lithium atom per six carbon atoms.[23–27] Though carbon has rather lower capacity than lithium metal and lithium alloy anodes, volume change was small and represented longer cycle performance. After this commercialization, many researchers and companies have put their effort on new carbonaceous anodes.

The advantage of lithium-ion battery is their high energy density. They can store more energy than conventional nickel/cadmium and nickel/metal hydride batteries. Lithium-ion batteries have clearly established their position as a rechargeable battery technology of choice for cellular phones and notebook computers. And now they are also taking over the world of power tools. Most representative commercialized lithium-ion batteries use a layered compound, such as lithium cobalt oxide, as the positive electrode and graphite as the negative electrode. The battery energy density has doubled so far with the improvement of material and battery technologies without changing its basic concept and chemistry and has been approaching the theoretical limit of the system.

Further energy and power improvement is, however, necessary to meet the rigorous demand of the state-of-the-art functional equipment for future mobile society. Besides, global attempts and development have been energetic to use lithium-ion batteries for hybrid, plug-in hybrid, and electric vehicles toward future extinction of oil resources. For hybrid and plug-in hybrid vehicle market, higher output/input characteristic is necessary for the battery system to assist the engine and accept the recovering energy in a short time in a wide range of temperatures. Especially for hybrid electric vehicle application, graphite electrode has a potential safety hazard for lithium deposition due to its kinetic limitation of lithium acceptability at lower temperatures and the lower potential that is too close to that of elemental lithium. For electric vehicle application, immediate representative chemistry is the lithium manganese oxide spinel cathode/natural carbon anode to ensure the battery power and lower the cost of materials. However, superior new system is intrinsically reputed in securing the sufficient mileage of an electric vehicle per charge and power at wide range of temperatures. Thus, the applications do not stop at current existing technologies and remarkable improvement of energy and power densities for lithium-ion batteries is an important theme on which the market can develop.

From the energy increment point of view, it is important and desirable to improve both cathode and anode materials. There are a few cathode materials that have higher capacity than lithium cobalt oxide, but their expected capacity progress is in the range of tens of percents at most. On the other hand, there are some candidates for anode materials to have very high specific capacities. Consequently research interest is being focused on alloy anode system and considerable research and development works have been conducted on lithium alloy systems over the last decades. After 2000, nanotechnology has been in the limelight in global technologies, and nano-lithium or nano-alloy anodes have also been energetic.

In this chapter, the issues of lithium and the potential of alloy anode materials are briefly touched upon along with recent breakthroughs that have been achieved. Future development trend and its challenge have also been stated.

4.2 Lithium Metal Electrode

The half cell reaction of lithium can be expressed through the following equation:[28]

$$Li^{+} + e^{-} \Leftrightarrow Li\ (-3.045\ V\ vs.\ NHE;\ T = 25?)$$

Lithium metal has a very low-standard electrochemical potential and a theoretical capacity of 3860 mAh/g, and therefore, it is the most attractive anode candidate among all elements. Practically, lithium metal electrodes are widely used for lithium primary cells. Nevertheless, lithium metal anode has not been utilized for rechargeable batteries so far. The major drawback is the poor cycle efficiency due to unexpected lithium metal formation, that is, dendrite formation during cycling.[29,30] Considering the discharging process of this dendrite lithium metal, dissolution of the lithium near the current collector is advantageous rather than dissolution of dendrite that results in the formation of isolated and dead lithium during the repetition of charge/discharge cycles. This dendrite formation results in lower electrochemical efficiency and directly affects the cycle performance of the battery system. Moreover, dendrite may penetrate the separator and may lead to an internal short-circuit of the cell. Dendrite formation is therefore the most critical problem of lithium metal electrode for rechargeable batteries.

Much work has been done to recognize the formation mechanism of lithium dendrite. It has been observed that surface layer on the surface of lithium metal strongly affects the dendrite formation. Several investigations have been performed to clarify the composition and characteristics of these surface films by using *in situ* infrared absorption spectroscopy and X-ray photoelectron spectroscopy.[31-36] In those investigations, it is generally reported that it tends to generate dendrite in case the structure and chemical composition of the surface film are not homogeneous. Furthermore, the composition of surface films is also dependent upon the selection and composition of electrolyte. Selection of solvent additives also affects the cycle performance.

Several studies are still been being performed to control the lithium metal precipitation and prolong the cycle life of the lithium metal electrode. Further improvement seems to be still necessary to improve the cycle efficiency for the lithium metal electrode.

4.3 Lithium Alloy Electrodes

4.3.1 A Brief History of Lithium Alloy Electrodes

Research on lithium metal–iron sulfide molten salt battery has been performed for electric vehicle in the United States of America in 1960s. This battery is composed of lithium metal anode and iron sulfate cathode in a LiCl/KCl molten salt and operates at temperatures around 400 degrees Celsius. A number of binary alloy system was investigated in high temperature systems, including Li-Al,[37–42] Li-Si,[43–47] Li-Mg,[48] Li-Sn,[49–52] Li-Bi,[53,54] Li-Sb,[53–55] Li-In,[56] and Li-Ga.[57] In 1970s the research on alloying reactions of these elements in organic electrolyte was also performed[58–64] and it succeeded other investigations including several ternary systems.[2,65–68] First commercialized lithium alloy electrode was an alloy of Bi–Pb–Cd–Sn, the so called Wood's metal, and it employed in the coin-type cells by Matsushita Battery[69,70] in the 1980s. The shortcoming of this alloy electrode was the large volume change of the alloy itself during the charge–discharge process, and reliable cycle life was only obtained with the shallow depth of charge (typically <10%). Thus, the specific capacity density of the commercialized alloy electrode was very low and not comparable to the capacity expected to lithium alloy electrodes. After the commercialization of lithium-ion battery system by Sony using the present graphite-base negative electrode in the beginning of 1990s,[71,72] the lithium-ion battery became the mainstream of high-capacity rechargeable batteries and most researchers have put their interest and effort on new carbonaceous anodes. The carbonaceous anode, such as graphite has now established the leading position of lithium-ion battery technology. Carbonaceous anode has a long cycle life because of the lithium intercalation reaction, but their drawback is its far lower theoretical capacity density compared to lithium alloy anodes. It is also noticeable that there intrinsically remains safety concern for carbonaceous anode at the end of charging, particularly at high rate of charging, since the potential of carbonaceous anodes tends to reach the lithium potential. Consequently, continuous effort is being made on lithium alloy systems to meet the demand for energy and power increments.

In the middle of the 1990s, some lithium transition metal nitrides were reported that showed good reversibility with high specific capacity densities.[73-76] This system doubled the capacity of graphite system and indicated fairly better cycle life. However, it also led to higher discharge potential and faced severe difficulty from a practical point of view, for example, it shad to be treated in dry atmosphere. In 1996, Fujifilm Celltech announced their development of new lithium-ion battery technology with an amorphous tin-base composite metal oxide system (TCOs) as anode.[77-79] This active material had higher reversible specific capacity of over 600 mAh/g and 2200 mAh/cm^3, which corresponded to nearly twice the level of those for graphite. This electrode indicated a good cycle life of over 300 cycles and the swelling of the electrode was well-suppressed. It was interpreted that nanosized $Li_{4.4}Sn$ particles produced from tin oxide at the initial charging process were well-dispersed in inert glassy matrix oxides. This system has, however, not been commercialized because of its huge irreversibility at initial charging process. But this research thereafter led to vigorous research field on the alloy systems.[3,10,12,14,17,19,80-108]

Cycle performance of electrodes is dependent not only upon the active material but also strongly upon the morphology of the electrode as a whole including copper foil. One idea is to directly unify the lithium alloy active materials onto copper current collector. Electrodeposited tin-base electrodes[91,109-112] and vacuum-deposited electrodes, such as tin,[113,114] silicon,[115-117] and germanium[118] have fairly been investigated. It appeared that cycle performance was getting worse with thickness increment, so the thickness of active layers should be designed to 5 μm at the most. It was also demonstrated that copper foil with rough surface and/or annealed copper foil were fairly effective to prolong the cycle performance of the tin- or silicon-base film electrodes. However, these attempts were not so realistic since the production of the copper foil with both-side rough surface needs much complicated process and extra cost, and annealing of electrode unfortunately softens copper foil itself and results in lower strength of copper foil current collector. Therefore, despite these considerable works and efforts, the development of lithium alloy electrode is still on the way to commercialization. Another attempt was to form composite lithium alloy with carbon-base anode.[119-121] Composite electrode has better cycle performance but the capacity is rather lower than those expected for lithium alloy electrode. Sony started to commercialize a new lithium-ion battery, using carbon-alloy composite, Sn–Co–C, amorphous alloy anode in 2005. Yet these composite alloy anode systems do not meet the demand and requirement for a sufficient energy increment.

Tracking the number of new technologies and ideas for the last two decades to overcome these two conflicting facts of capacity and cycle life on lithium alloy electrodes, it is inferred that lithium alloy electrodes only work with both macroscopic design of the whole electrode structure and microscopic nanotechnology at each interface. Much attention will be given further and effort will be made on lithium alloy systems for the next generation lithium battery.

4.3.2 Classification of Lithium Alloy Systems

Lithium reacts with numerous metals M (M = Mg, Ca, Al, Si, Ge, Sn, Pb, As, Sb, Bi, Pt, Ag, Au, Zn, Cd, Hg, etc.) at room temperature if the metal is polarized to adequate potentials in organic electrolyte containing lithium ions. The cell reaction of so-called lithium alloys can be expressed according to the following equation:

$$M + xLi^{+} + xe^{-} \Leftrightarrow Li_{x}M$$

Table 4.1 shows the theoretical specific capacity densities (including lithium) and average potentials together with volume changes for representative alloy systems. As the energy is estimated by multiplication of voltage and capacity of the cell system, an anode with lower potential has an advantage in energy point of view. All lithium insertion potentials for alloys are higher compared with graphite. Anode materials with higher specific capacity with lower potential are preferable, and silicon and tin are attractive candidates. However, large volume expansion by lithium accommodation should also be taken into account (LiAl: 97%,[6] $Li_{4.4}Sn$:358%,[9] $Li_{4.4}Si$:323%[6]). Mechanical stress related to the volume change induces a rapid decay in mechanical stability. The electrode suffers from cracking and crumbling as well as from consequent loss of electronic interparticle contact. Without appropriate design of the whole electrode, as referred in the following sections, the lithium alloy electrode typically deteriorates after only a few charge/discharge cycles as shown in Fig. 4.1.

Table 4.1 Comparison of the specific capacities and average potentials of some lithium alloy electrodes

Material	Average voltage (V) *vs.* Li	Specific capacity(Ah/kg)
Li	0	3,860
$Li_{4.4}Si$	0.2	2,010
$Li_{4.4}Sn$	0.5	790
LiAl	0.3	790
Li_3Sb	0.9	564
LiC_6	0.1	340

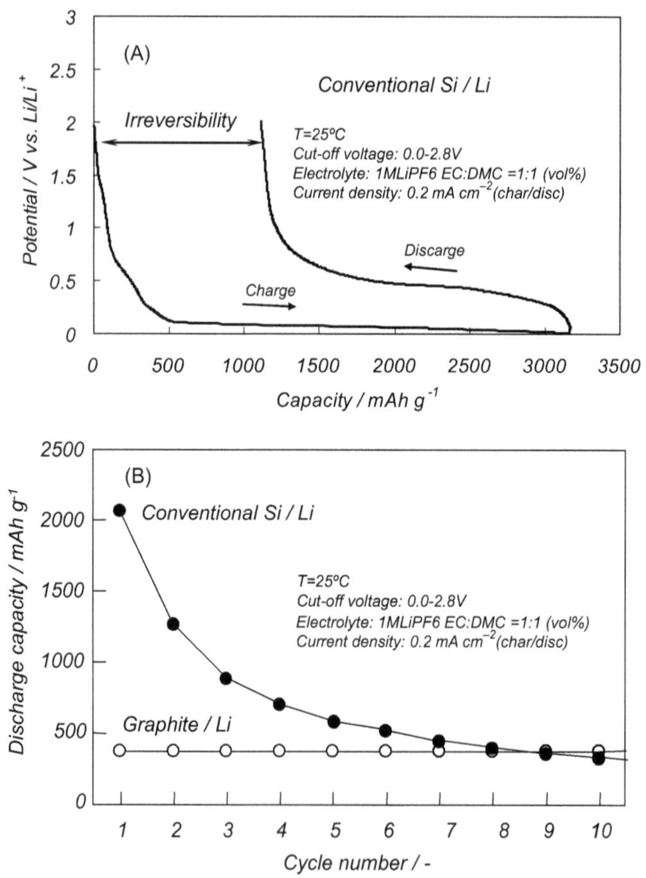

Figure 4.1 Cycle performance of typical conventional lithium alloy electrode: (A) Potential profile of Si at initial charge/discharge process, (B) Capacity retention during cycling for Si and C electrodes.

Intermetallic compounds have also been intensively investigated. Among those, the antimonides Zn_4Sb_3,[122] InSb,[19] SnSb,[87,123] $TiSb_2$,[124] VSb_2,[124] $CrSb_2$,[125] $CoSb_3$,[126] Cu_2Sb,[127] and $CoFe_3Sb_{12}$,[92] the stannides $MnSn_2$,[123] $Mn_{1.77}Sn$,[85] Mn_3Sn,[85] Mn_3SnC,[84,85] FeSn,[9] $FeSn_2$,[124,128] $CoSn_2$,[124] Ni_3Sn_2,[16] Cu_6Sn_5,[15,86,129] CaSn,[130] and $CeSn_3$[50] and silicides CrSi,[131] NiSi,[132,133] FeSi,[133] and Mg_2Si[134] have been given the most attention so far. However, the cycle performance was poor and less than 100 charge/discharge cycles for most of all systems. Pulverization due to the large volume change by lithium accommodation lead to the lack of contact either among particles or between active material and current collector, resulting in lower cycle performance of the alloy electrodes.

There seems to be a limit to improve the cycle performance by only adjusting the alloy composition and their structures. The idea or breakthrough to control the swelling of these alloy systems and improve their cycle performance is given in the later part of the chapter.

4.3.3 Reaction Mechanisms of Lithium Alloy Systems

Lithium can react with metals and intermetallic compounds, for example, binary systems in various ways.[94,135,136]

(i) Internal displacement of the atoms within a parent host structure, referred to as absorption and reconstitution reaction:

$$M+Li \Leftrightarrow Li_xM$$

$$MM'+Li \Leftrightarrow Li_xMM'$$

Unlike the hydrogen storage materials, topotactic reaction is usually limited to only small amount of lithium since ionic radius of lithium is ca. 90 pm, much larger than that of hydrogen of 40 pm, and there are hardly any sufficient interstitial spaces in all host metals for lithium accommodation. A few intermetallic compounds have been proposed to reversibly absorb lithium in their interstitial space without changing their structures, for example, Mg_2Ge[14,20,137] and $La_3Ni_2Sn_7$,[138] but the specific capacities of these intermetallic compounds are rather lower than the other materials. Generally, alloying usually occurs for a larger lithium insertion in most of all intermetallic compounds and considerable atomic displacement and structure rearrangement are necessary for this reaction and usually lead to a poor cycle performance.

(ii) Complete phase separation of one of the metals by initial lithium insertion yields a new binary phase, referred to as displacement reaction:

$$MM' + Li \rightarrow Li_xM + M'$$

In this reaction, inactive element M′ is extruded from intermetallic compound during the first charging process, leading to a finely dispersed complex matrix structure. Lithium reacts with finely dispersed nanoparticle M thereafter. In this case, though the cycle performance can be improved to some extent, the volume change of the lithium-active element M is not easily relieved. Examples are FeSi[133] and NiSi.[133]

(iii) Substitution to form a ternary phase by lithium insertion may involve solid--solution behavior, referred to as mixed reaction:

$$MM' + Li \Leftrightarrow Li_xMM'_{1-y} + yM'$$

or if the displacement is irreversible:

$$MM' + Li \rightarrow Li_xMM'_{1-y} + yM' \Leftrightarrow Li + MM'_{1-y}$$

At initial charging process, a part of inactive element M′ is put out from the intermetallic compound and the remaining component reversibly reacts with lithium thereafter. This two-step process is decently advantageous to as it stabilizes the electrode structure and prolongs the cycle performance. This reaction leads to *in situ* composite matrix of finely inter-dispersed lithium-active and inactive phases, and the inactive phase can buffer the volume change of the active phase by lithiation to some extent and also provide a better conductive environment to the electrode system. Therefore, the total volume change of the electrode could be alleviated effectively. Good examples are Sn_2Fe[128] and Cu_6Sn_5.[13,15]

Each reaction above is not strictly distinguished from other reactions and the borders of those reactions are usually floating and depend on the electrochemical conditions such as charging rates, depth of charge, cut-off voltages, and also the morphology of starting materials, electrolytes, additives, etc.

4.3.4 Approach to Improve the Cycle Performance

Since there is no direct evidence, two conclusions can be accepted to relieve the large volume change of lithium alloy anode. First, nanosized metal or alloy can keep rather stable microstructure during charge/discharge cycling since it has high endurance to volume change due to its superplasticity and high ductility.[7,139,140] Second, two-step process for intermetallic compounds leads to the active/inactive composite matrix in which nanoscale active material is uniformly dispersed into matrix of inactive component, obviously to improve the stability of the electrode as discussed in the previous section. Owing to these beneficial factors, nanosized alloys,[87,93] nanocomposite materials[121,141,142] or nanostructured compounds[84] have been investigated.

However, there seem to be a limit to improve the stability and reliability of the lithium alloys by simply modifying the composition and morphology of these materials. One issue is the segregation during charge/discharge cycling. For instance, nanoparticles less than 100 nm resulted in serious aggregation.[143] Segregation further led to the isolation of active materials and lack of electroconductivity between interparticle and copper current collector.

Attempt from a different view has been reported, namely unifying the active materials with current collector such as electroplating of tin[109] or tin-base alloys[110–112] or vapor deposition of silicon[116,117] or silicon alloys on nickel[144] or copper foils. Table 4.2 summarizes the several attempts proposed so far to improve lithium alloy electrodes.

In regards to production methods, the most frequently used method to synthesize alloy electrodes seems to be mechanical alloying by ball-milling. Ball-milling can be suitable to synthesize nanostructured materials but this method should be carefully examined for mass production. There seems to be no better solutions to apply ball-milling technology in a mass plant level. An alternative is the synthesis by precipitation in chemical bathes and the precipitation might be used for mass production. However, as nano-material has extremely high surface energy and lightly reacts with oxygen, the suppression of surface oxidation after mechanical milling and precipitation should be strongly controlled. Thus, much care and new technology are necessary for mechanical alloying and precipitation technologies to bring them into practice in a mass production level. Vapor deposition or sputtering is also a possible process but large investment is definitely necessary and therefore unlikely to get into practice. Continued electroplating of tin-base alloy onto foils can be mostly established and therefore expected to be a rather favorable process for mass production.

Table 4.2 Attempts to improve the lithium alloy electrode

(i)	Two-step process using intermetallic compound
	• SnO
	1st step: $SnO_x + 2xLi^+ + 2xe^- \rightarrow xLi_2O + Sn$
	2nd step: Sn (nano-partical) + $yLi^+ + ye^- \Leftrightarrow Li_ySn$ (nano-partical)
	• Cu_6Sn_5
	1st step: $Cu_6Sn_5 \Leftrightarrow Li_2CuSn + Cu \Leftrightarrow Li_2CuSn + Cu \Leftrightarrow Li_{4.4}Sn + Cu$
	2nd step: Li_2CuSn (nano-partical) + Li $\Leftrightarrow Li_{4.4}Sn$ (nano-partical) + Cu
(ii)	Nano-material/composite (20~200 nm)
	• Mechanical alloying (many intermetallic/composite materials)
	• Chemical precipitation (Sn–Cu–B)
	• Sputtering
(iii)	Thin film and uniformed with current collector
	• Electrochemical plating on copper foil (Sn, Sn–Zn, Sn–Cu)
	• Evaporation/Sputtering on copper foil (Si, Mg–Si)

4.3.5 Problem of Thin-Film Electrode Technologies

Silicon and tin are the most promising candidates among all elements as described in the previous sections.

Tin-base alloys have been widely investigated so far and the cycle performance is being improved, but the drawback that still remains is their higher initial irreversibility, which is in the range of 15% to 30% for mechanical alloying or nanopowers. The main reasons for this larger irreversibility are (i) lithium is fixed by the reaction with surface oxygen of tin-base nanopowers and (ii) active phase is apart from current collector and isolated at initial charging process. For the electroplated tin on copper foil, the irreversibility is well improved to less than 10%.[109] It was reported that electroplated tin on the copper foil with rough surface and further annealed, which leads to the Cu–Sn alloy during annealing and microcolumnar structure at first charging, are effective to improve the cycle performance of electroplated tin-base electrode. However, considering the mass production, the time of annealing should be carefully considered to secure the Cu–Sn alloying. In addition, annealing leads to the softening of the copper foil and loses its strength.

Silicon has the highest capacity among all possible candidates as lithium alloy anode, but its shortcoming is its inferior conductivity to others like tin or aluminum. Silicon powder electrode rapidly collapses and loses its capacity because of its lower conductivity. Alloying or mixing with conductive materials is an effective way of giving better conductivity but usually loses the capacity advantage of silicon. Thereupon, direct thin-film formation on copper foil has been proposed and investigated.[116,117] Structure of silicon thin film is well-characterized to reflect the roughness of foil as a current collector. Columnar structure of amorphous silicon is obtained by using copper foil with rough surface and indicates a rather stable cycle performance.[116] Over 200 charge/discharge cycles was reported for the electrode with 50 to 500 nm thick silicon film on nickel film with rough surface treatment.[117] However, much thicker silicon film is theoretically required to suit the capacity of cathode, for example, 3 to 5 μm thick silicon is a necessary equivalent to the practical capacity of 2 to 3 mAh/cm^2, but the cycle performance of the thicker silicon film electrodes is still poor.

The cycle performance of tin and silicon thin-film electrodes can be improved in a similar manner, that is by treating the rough surface of copper foils as a current collector is a crucial and indispensable component. Modification of surface roughness of copper foils is widely used, for example, coarse surface copper foil is effective to strengthen the adhesion to resin, and smooth surface is used for fine-pitch etching in electronic industries. However, those roughness changes are all executed only on one side of the copper foil. For battery application, roughness treatment is necessary for both sides of copper current collector, but no practical process seems to exists.

Considering these problems, it seems difficult to bring these thin film technologies into practice.

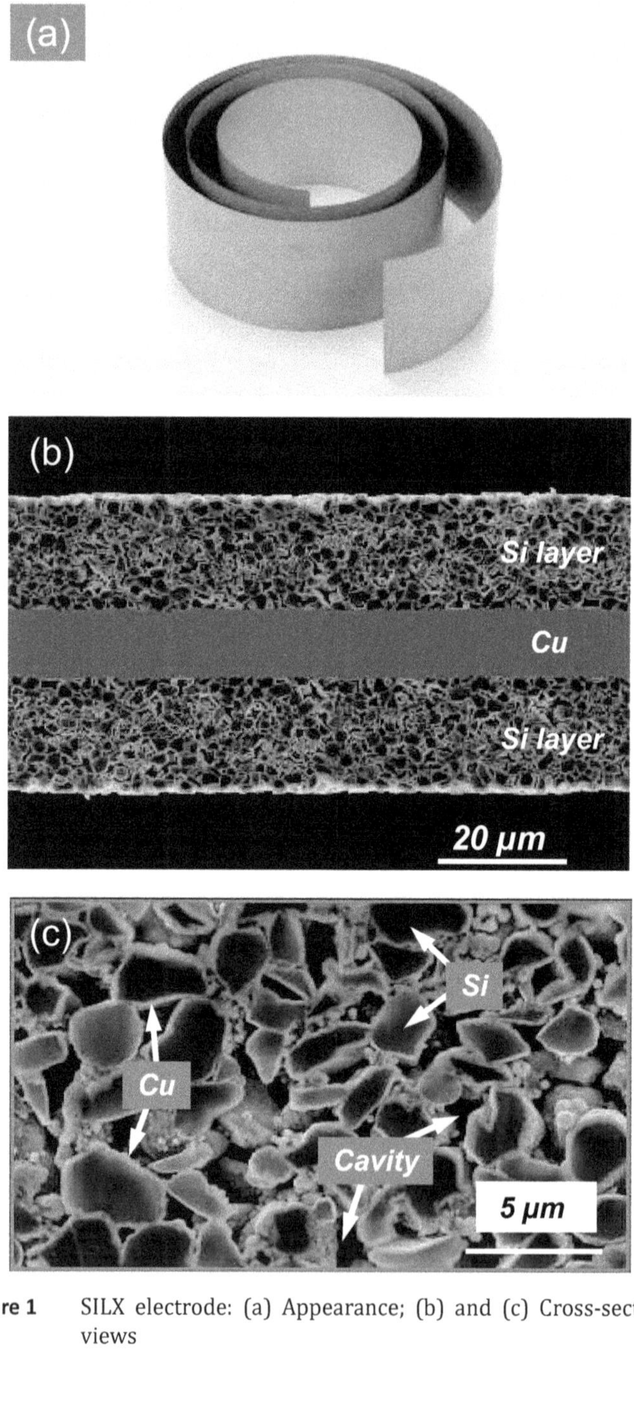

Picture 1 SILX electrode: (a) Appearance; (b) and (c) Cross-sectional views

4.3.6 Breakthrough Technology with Nanostructure Design Integrated with Current Collector

Authors have been developing a silicon-based anode technology, SILX, composed of a unique Si–Cu powder/substrate–united structure. Picture 1 shows the scanning electron microscope image of the cross-sectional view of SILX. In this example, pure silicon of over 99.0% purity with 2.5 μm in average diameter is used. Each silicon particle is covered with 100 to 200 nm thin layer of copper, built up with primary 5 to 50 nm of crystallite, to provide conductivity, and all powders are connected each other to sustain the network structure. About 30% of cavity or internal space is evident to give room to electrolytes and also to accommodate the natural expansion of silicon by lithium accommodation to some extent. The thickness of the active material layer is about 20 μm, which is about one-fourth of the current existing graphite systems in the same capacity range. Structure designs, such as active material, particle diameter, or thickness, can be changed in the range of several tens of micrometers without great efforts.

Figure 4.2 shows the potential profile for SILX electrode. Some polarization can be seen at the initial charging process to activate each silicon particle, and the electrode can be utilized thereafter without any big polarization. The initial irreversibility is rather lower, in this example 7%, compared to other silicon or tin-base electrodes where typically it is seen over 20%. High reversibility gives a high-energy system and several choices of cathode selection since the counter cathode does not loose the unnecessary lithium and damage of cathode becomes lower. This high reversibility is interpreted mainly for two reasons: (1) lower oxygen contents, oxygen is getting rid of from silicon surface during the SILX production process and (2) lower bare silicon surface, as a decent part of silicon surface is covered with copper, and this is one of the significant advantages of SILX technology. The capacity retentions of SILX and conventional silicon powder electrodes are shown together with conventional graphite electrode in Fig. 4.3. SILX has apparently sufficient superior capacity to graphite and stable cycle performance to conventional silicon electrode.

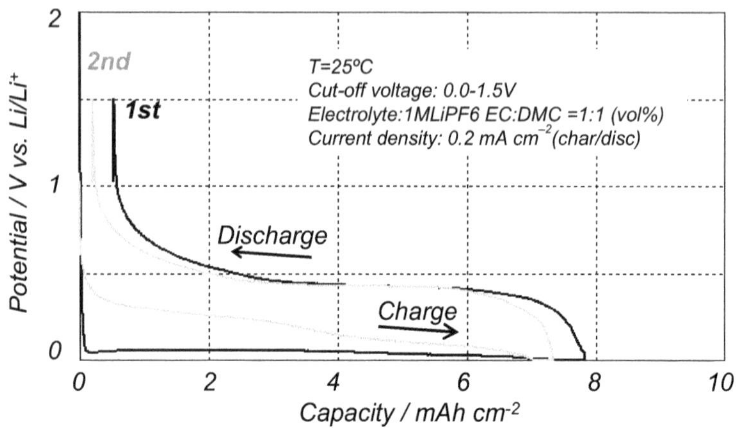

Figure 4.2 Potential profiles at initial and second charge/discharge process.

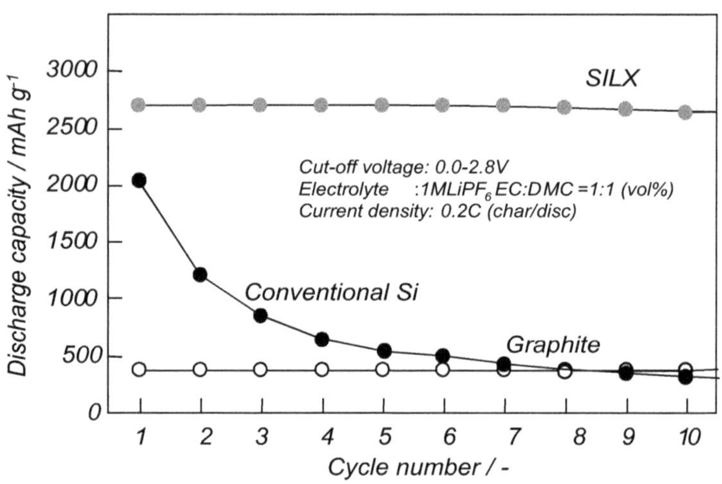

Figure 4.3 Capacity retentions of SILX together with conventional Si and graphite.

Before discussing the cell design and total energy of the battery system using SILX anode, the relationship between silicon utilization and its retention after charge/discharge cycling are examined (Fig. 4.4). In this example, a SILX electrode described in Picture 1 is matched with a layered structured material of $LiMn_{1/3}Co_{1/3}Ni_{1/3}O_2$ in 1 M $LiPF_6$ EC/DEC electrolyte with some fluorinated EC as additive. The volume capacity density after charge/discharge cycling tends to decrease with the silicon utilization of ca. 55% or more, while the capacity retention indicates a similar trend to the volumetric capacity density, namely the deterioration gets larger with the increasing silicon utilization over 55%. Following experimental data for the SILX battery system is for the silicon utilization of 50%. Of course, the limitation of silicon utilization varies depending not only on SILX, such as, silicon active materials and structure, but also strongly on other components, such as, electrolyte, cathodes, etc.

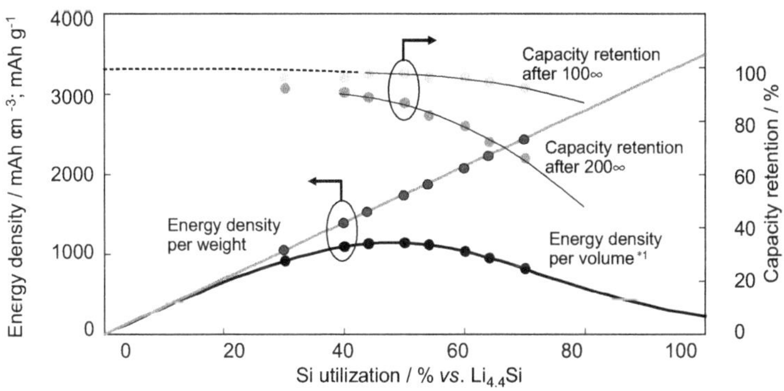

Figure 4.4 Relationship between silicon utilization and its capacity retention. *1 Calculated based on the volume after 100 cycles.

Authors have made prismatic cells from SILX and conventional graphite systems. In this case, LMCO ($Li_xCo_{1-x}Mn_xO_2$), which is suitable for SILX and has a capacity equivalent to $LiCoO_2$, is matched as a cathode. The energy density of SILX system has 30% more than the graphite system (Fig. 4.5).

One of the limiting factors for graphite-base anode is its rate performance especially at lower temperatures and it is very essential for applications like power tools and HEV. There is a potential hazard to deposit lithium metal on the surface of graphite electrode at high-rate charging and low temperatures. Commercialized cells for power applications are designed to lower the capacity densities by thinning the active layer and adding the conductive elements to secure the rate performance. Figure 4.6 shows the discharge capabilities of conventional graphite and SILX systems at different temperatures. Capacities for both systems are standardized to 100% to ease the recognition of difference. All cells are charged at 0.5 C and discharged at 1 C. The capacity retentions are ca. 20% for graphite system, while ca. 90% at 253K and still 60% at 243K for SILX stem. Though this data is to indicate the discharging characteristics, apparent advantage of SILX system can also be seen for charging characteristic. The superior rate performance is interpreted to be due to the low reaction impedance at the interfaces between the surface of silicon and electrolyte with the result of electrochemical impedance measurements.

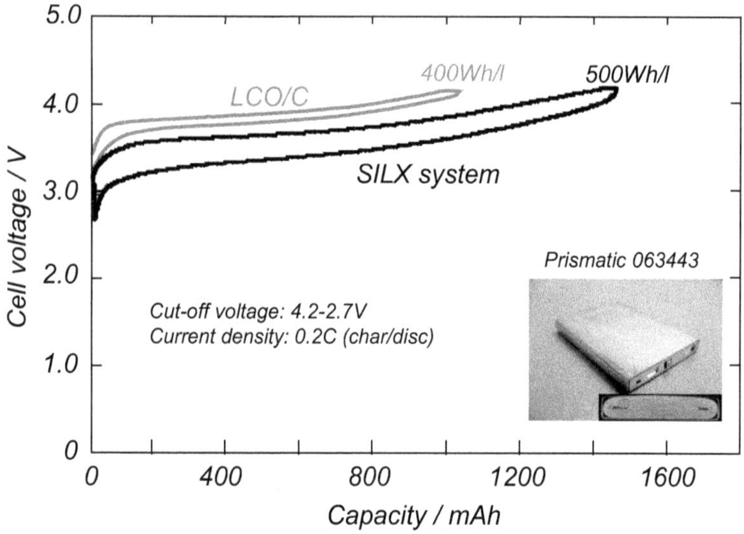

Figure 4.5 Potential profiles and energy densities of SILX and graphite systems.

The big concern is to control the swelling of the silicon electrode during charge/discharge cycling and to prolong the cycle life. The cycle performance of the SILX system measured with the coin-type cell is indicated in Fig. 4.7. The potential profile after 500 cycling has rather big polarization but the capacity of the cell is still retained over 80%. From this stable cycle performance, it can be interpreted that the nanodesigned surface copper plays several significant functions to stabilize the electrode: (1) giving conductivity to each silicon particle, (2) preventing silicon particles from isolation and keeping the better conductivity through whole electrode even after cycling, and (3) protecting unexpected chemical reaction of silicon with electrolyte. The cycle performances within restricted capacity

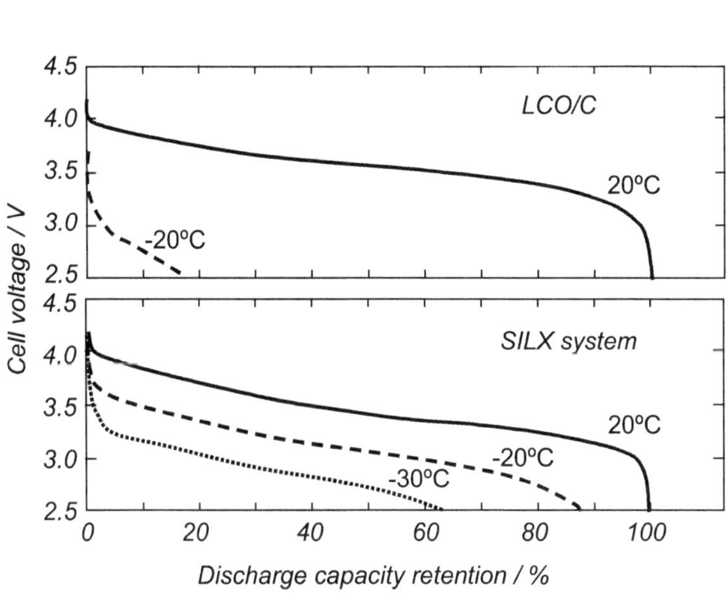

Figure 4.6 Discharge capabilities of conventional graphite and SILX systems at different temperatures.

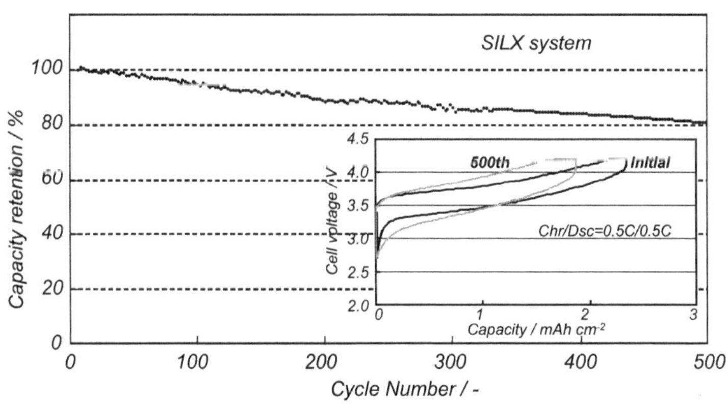

Figure 4.7 Cycle performance of the SILX system measured with coin-type cell.

ranges are also shown together with potential profiles in Fig. 4.8, assuming the use of SILX system for HEV or PHEV application. Considering the capacity of SILX system, the range of 10% is enough to estimate the cycle life for HEV application. Potential curves have gradually been changing with increasing cycle numbers, the cells are still working better after thousands of cycling. Cycle performance will be improved by researching the relationship of silicon electrode and other components, especially the electrolyte.

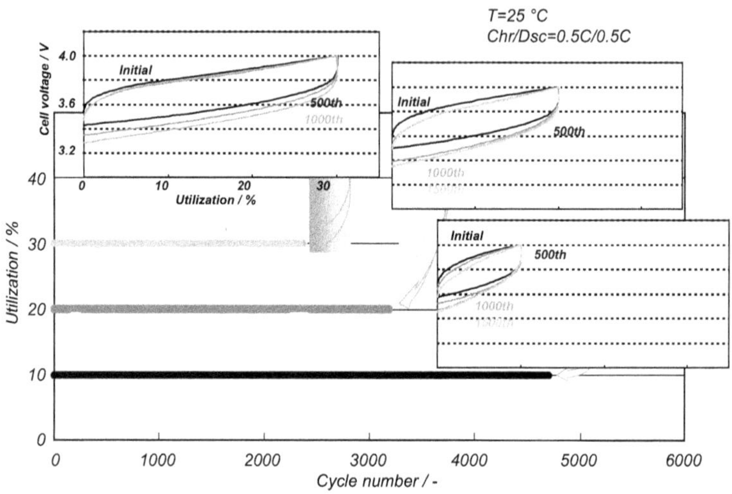

Figure 4.8 Cycle performance of SILX system within restricted capacity ranges.

In summarizing the research data so far, the SILX is the promising breakthrough technology to meet the performance requirements of increasing energy density and power. Low-temperature performance is especially one of the key issues for lithium-ion batteries used in HEV and SILX.

4.4 Conclusion and Outlook

Lithium alloy anodes have far higher advantages in capacity compared with current graphite-base anode. Strong points of alloy electrodes are also their superior rapid-charging and low-temperature characteristics, and therefore, they are expected to spread not only over small mobile applications but also out to HEV use. Although there are still issues that have to be overcome for commercializing the lithium alloy system, they can be solved with new attempt and invention.

The unique unified structure with current collector, SILX, has better stability as an alloy electrode and promising possibility to take over current existing technology. Future improvements are necessary and will be explored definitely with applying nanotechnologies, such as nanopowders or nanofilms. Moreover, collaboration to explore the synthetic research spreading over suitable copper foil, electrolyte, and cathode materials are strongly expected to commercialize the next-generation lithium-ion battery system taking over and leaping current lithium-ion batteries.

Acknowledgments

This work was done as the CTO project in Mitsui Mining & Smelting Co. Ltd., Japan. I would like to thank my colleagues who worked with me on SILX technology, especially Y. Sakaguchi for his earlier work and H. Ide and Y. Todorof for the valuable discussions and comments and preparation of the materials.

References

1. J. P. Gabano (ed.), *Lithium Batteries* (Academic Press 1983).
2. J. Wang, I. D. Raistrick, R. A. Huggins, *J. Electrochem. Soc.* **133** 457 (1986).
3. R. A. Huggins, *J. Power Sources* **81–82** 13 (1999).
4. A. N. Dey, *J. Electrochem. Soc.* **118** 1547 (1971).
5. A. Anai, R.A. Huggins, *J. Power Sources* **38** 351(1992).
6. D. Fauteux, R. Koksbang, *J. Appl. Electrochem.* **23** 1 (1993).
7. J. Yang, M. Winter, J. O. Besenhard, *Solid State Ionics* **90** 281 (1996).
8. J. O. Besenhard, J. Yang, M. Winter, *J. Power Sources* **68** 87 (1997).
9. O. Mao, J. R. Dahn, *J. Electrochem. Soc.* **146** 414 (1999).
10. J. O. Besenhard, M. Wachtler, M. Winter, R. Andreas, I. Rom, W. Sitte, *J. Power Sources* **81/82** 268 (1999) 268
11. J. Yang, Y. Takeda, N. Imanishi, O. Yamamoto, *J. Electrochem. Soc.* **146** 4009 (1999).
12. W. J. Weydanz, M. Wohlfahrt-Mehrens, R. A. Huggns, *J. Power Sources* **81/82** 4401 (1999) 4401
13. K. D. Kepler, J. T. Vaughey, M. M. Thackeray, *Electrochem. Solid-State Lett.* **2** 307 (1999).

14. H. Sakaguchi, H. Honda, T. Esaka, *J. Power Sources* **81/82** 229 (1999).
15. D. Larcher, L. Y. Beaulieu, D. D. MacNeil, J. R. Dahn, *J. Electrochem. Soc.* **147** 1658 (2000).
16. G. M. Ehrlich, C. Durand, X. Chen, T. A. Hugener, F. Spiess, S. L. Suib, *J. Electrochem. Soc.* **147** 886 (2000).
17. H. Kim, Y.-J. Kim, D.G. Kim, H.-J. Sohn, T. Kang, *Solid State Ionics.* **144** 41 (2001).
18. K. C. Hewitt, L. Y. Beaulieu, J. R. Dahn, *J. Electrochem. Soc.* **148** A402 (2001).
19. J. T. Vaughey, J. O. O'Hara, M. M. Thackeray, *Electrochem. Solid-State Lett.* 3 13 (2000).
20. H. Honda, H. Sakaguchi, T. Fukunaga, T. Esaka, *Electrochemistry* **70** 99 (2002).
21. H. Sakaguchi, H. Honda, Y. Akasaka, T. Esaka, *J. Power Sources* **119–121** 50 (2003) 50
22. R. Yazami, P. Touzain, *J. Power Sources* **9** 365 (1983).
23. R. Fong, U. V. Sacken, *J. Electrochem. Soc.* **137(7)** 2009 (1990).
24. N. Takami, A. Satoh, M. Hara, T. Ohsaki., *J. Electrochem. Soc.* **142 (2)** 371 (1995).
25. K. Tatsumi, N. Iwashita, H. Sakaebe, H. Shioyama, S Higuchi, A. Mabuchi, H. Fujimoto, *J. Electrochem. Soc.* **142(3)** 716 (1995).
26. W. Xing, J. R. Dahn, *J. Electrochem. Soc.* **144(4)** 1195 (1997).
27. K. Sato, M. Noguchi, A. Demachi, N. Oki, M. Endo, *Science* **264** 556 (1994).
28. A. J. Bard, L. R. Faulkner, *Electrochemical Methods* (John Wiley & Sons, Inc. 1980).
29. I. Yoshimatsu, T. Hirai, J. Yamaki, *J. Electrochem. Soc.* **135** 2422 (1988).
30. M. Arakawa, S. Tobishima, Y. Nemoto, M. Ichimura, J. Yamaki, *J. Power Sources* **4344** 27 (1993).
31. E. Peled, *J. Electrochem. Soc.* **126** 2047 (1979).
32. J. G. Thevenin, R. H. Muller, *J. Electrochem. Soc.* **134** 273 (1987).
33. M. Garreau, *J. Power Source* **20** 9 (1987).
34. D. Aurbach, M. L. Daroux, P. W. Faguy, E. Yeager, *J. Electrochem. Soc.* **134** 1611 (1987).
35. D. Aurbach, O. Chusid, *J. Electrochem. Soc.* **140** L155 (1993).
36. D. Aurbach, O. Chusid, *J. Electrochem. Soc.* **140** L1 (1993).

37. C. E. Johnson, M.S. Foster, *J. Electrochem. Soc.* **116** 1612 (1969).
38. N. P. Yao, L. A. Heredy, R. C. Saunders, *J. Electrochem. Soc.* **118** 1039 (1971).
39. E. C. Gey, D. R. Vissers, F. J. Martiono, K. E. Andersen, *J. Electrochem. Soc.* **124** 1160 (1977) 1160
40. C. A. Melendres, *J. Electrochem. Soc.* **124** 651 (1977).
41. J. R. Selman, D. K. DeNuccio, C. J. Sy, R. K. Steunenberg, *J. Electrochem. Soc.* **124** 1161 (1977) 1161
42. C. A. Melendres, C. C. Sy, *J. Electrochem. Soc.* **125** 727 (1978).
43. S. Lai, *J. Electrochem. Soc.* **123** 1196 (1976).
44. R. A. Sharma, R. N. Seefurth, *J. Electrochem. Soc.* **123** 1763 (1976).
45. R. N. Seefurth, R. A. Sharma, *J. Electrochem. Soc.* **124** 1207 (1977).
46. C. J. Wen, B. A. Boukamp, R. A. Huggins, W. Weppner, *J. Electrochem. Soc.* **126** 2258 (1979).
47. R. N. Seefurth, R. A. Sharma, *J. Electrochem. Soc.* **127** 1101 (1980).
48. C. S. Tedmon Jr, W. C. Hagel, *J. Electrochem. Soc.* **115** 151 (1968).
49. W. Mueller, H. Schaefer, *Z. Naturforsch.* **28b** 246 (1973).
50. M. S. Foster, C. E. Crouthamel, S. E. Wood, *J. Phys. Chem.* **70** 3042 (1966).
51. C. J. Wen, R. A. Huggins, *J. Solid State Chem.* **35** 376 (1980).
52. C. J. Wen, R. A. Huggins, *J. Electrochem. Soc.* **128** 1181 (1981).
53. W. Weppner, R. A. Huggins, *J. Electrochem. Soc.* **124** 1569 (1977).
54. W. Weppner, R. A. Huggins, *J. Solid State Chem.* **22** 297 (1977).
55. W. Weppner, R. A. Huggins, *J. Electrochem. Soc.* **125** 7 (1978).
56. C. J. Wen, R. A. Huggins, *Mater. Res. Bull.* **15** 1225 (1980).
57. C. J. Wen, R. A. Huggins, *J. Electrochem. Soc.* **128** 1636 (1981).
58. J. O. Besenhard, H. P. Fritz, *Electrochim. Acta* **20** 513 (1975).
59. M. M. Nicholson, *J. Electrochem. Soc.* **121** 735 (1974).
60. J. Antaul, B. F. Becker, *J. Phys. Chem.* **79** 2470 (1975).
61. B. M. L. Rao, R. W. Francis, H. W. Christopher, *J. Electrochem. Soc.* **124** 1490 (1977).
62. R. D. Rauh, F. S. Shuker, J. M. Marston, S. B. Brummer, *J. Inorg. Nucl. Chem.* **78** 1791 (1977).
63. J. O. Besenhard, *J. Electroanal. Chem.* **78** 189 (1977).
64. J. O. Besenhard, *J. Electroanal. Chem.* **94** 77 (1978).

65. J. Wang, I. P. King, R. A. Huggins, *Solid State Ionics* **20** 185 (1986).
66. A. Anami, S. Crouch-Baker, R. A. Huggins, *J. Electrochem. Soc.* **135** 2103 (1988).
67. A. Anami, R. A. Huggins, *J. Power Sources* **38** 351 (1992).
68. A. Anami, R. A. Huggins, *J. Power Sources* **38** 363 (1992).
69. Y. Toyoguchi, S. Nankai, J. Yamamura, T. Matsui, T. Iijima, *The 24th Batery Sympojium in Japan, Extended Abstracts, Osaka* p. 205 (1983).
70. Y. Toyoguchi, S. Nankai, J. Yamamura, T. Matsui, T. Iijima, *Third International Meeting on Lithium Batteries, Extended Abstracts, Kyoto* p. 113 (1986).
71. T. Nagaura, K. Tozawa, *Progress in Batteries and Solar Cells,* Vol. **9**, p. 209 (1990).
72. T. Nagaura, *Progress in Batteries and Solar Cells,* Vol. **10**, p. 218 (1991).
73. M. Nishijima, Y. Takeda, N. Imanishi, O. Yamamoto, *J. Solid State Chem.* **113** 205 (1994) 205
74. M. Nishijima, N. Tadokoro, Y. Takeda, N. Imanishi, O. Yamamoto, *J. Electrochem. Soc.* **141** 2966 (1994).
75. M. Nishijima, T. Kagohashi, N. Imanishi, Y. Takeda, O. Yamamoto, S. Kondo, J. *Solid State Ionics* **83** 107 (1996).
76. T. Shodai, S. Okada, S. Tobishima, J. Yamaki, *Solid State Ionics* **86–88** 785 (1996).
77. Y. Idota, US Patent, 5478671 (1995).
78. Nippon Denki Shinbun (Japanese newspaper), March 11 (1996).
79. Y. Idota, T.Kubota, A. Matsufuji, Y. Maekawa, T. Miyasaka, *Science* **276** 1395 (1997).
80. O. Mao, R. A. Dunlap, J. R. Dahn, *J. Electrochem. Soc.* **146** 405 (1999).
81. O. Mao, J. R. Dahn, *J. Electrochem. Soc.* **146** 423 (1999).
82. K. D. Kepler, J. T. Vaughey, M. M. Thackeray, *J. Power Sources* **81–82** 383 (1999).
83. L. Y. Beaulieu, D. Larcher, O. Mao, A. E. George, J. R. Dahn, *J. Electrochem. Soc.* **147** 1703 (2000).
84. L. Y. Beaulieu, D. Larcher, R. A. Dunlap, J. R. Dahn, *J. Electrochem. Soc.* **147** 3206 (2000).
85. L. Y. Beaulieu, J. R. Dahn, *J. Electrochem. Soc.* **147** 3237 (2000).
86. Y. Xia, T. Sakai, T. Fujieda, M. Wada, H. Yoshinaga, *J. Electrochem. Soc.* **148** A471 (2001).

87. H. Li, L. Shi, W. Lu, X. J. Huang, L. Chen, *J. Electrochem. Soc.* **148** A915 (2001).

88. C. Wang, A. J. Appleby, F. E. Little, *J. Power Sources* **93** 174 (2001).

89. L. J. Zhang, X. B. Zhao, X. B. Jiang, C. P. Lv, G. S. Cao, *J. Power Sources* **94** 92 (2001).

90. G. S. Cao, X. B. Zhao, T. Li, C. P. Lu, *J. Power Sources* **94** 102 (2001).

91. M. Wachtler, J. O. Besenhard, M. Winter, *J. Power Sources* **94** 189 (2001).

92. X. B. Zhao, G. S. Cao, C. P. Lv, L. J. Zhang, S. H. Hu, T. J. Zhu, B. C. Zhou, *J. Alloys Compd.* **315** 265 (2001).

93. D. G. Kim, H. Kim, H.-J. Sohn, T. Kang, *J. Power Sources* **104** 221 (2002).

94. R. Benedek, M. M. Thackeray, *J. Power Sources* **110** 406 (2002).

95. H.-Y. Lee, S.-W. Jang, S.-M. Lee, S.J. Lee, H.-K. Baik, *J. Power Sources* **112** 8 (2002).

96. J.-M. Tarascon, M. Morcrette, L. Dupont, Y. Chabre, C. Payen, D. Larcher, V. Pralong, *J. Electrochem. Soc.* **150** A732 (2003).

97. L. Y. Beaulieu, K. C. Hewitt, R. L. Turner, A. Bonakdarpour, A. A. Abdo, L. Chritensen, K. W. Eberman, L. J. Krause, J. R. Dahn, *J. Electrochem. Soc.* **150** A149 (2003).

98. S. D. Beattie, T. Hatchard, A. Bonakdarpour, K. C. Hewitt, J. R. Dahn, *J. Electrochem. Soc.* **150** A701 (2003).

99. L. Y. Beaulieu, T. D. Hatchard, A. Bonakdarpour, M. D. Fleischauer, J. R. Dahn, *J. Electrochem. Soc.* **150** A1457 (2003).

100. J. T. Vaughey, L. M. L. Fransson, H. A. Swinger, K. Edstroem, M. M. Thackeray, *J. Power Sources* **119–121** 64 (2003) 64

101. J. Wolfenstine, *J. Power Sources* **124** 241 (2003).

102. J. Yin, M. Wada, S. Yoshida, K. Ishihara, S. Tanase, T. Sakai, *J. Electrochem. Soc.* **150** A1129 (2003).

103. H. Mukaido, T. Osaka, P. Reale, S. Panero, B. Scrosati, M. Wachtler, *J. Power Sources* **132** 225 (2004).

104. Q. F. Dong, C. Z. Wu, M. G. Jin, Z. C. Huang, M. S. Zheng, J. K. You, Z. G. Lin, *Solid State Ionics* **167** 49 (2004).

105. G. A. Roberts, E. J. Cairns, J. A. Reimer, *J. Electrochem. Soc.* **151** A151 (2004).

106. J. Yin, M. Wada, S. Tanase, T. Sakai, *J. Electrochem. Soc.* **151** A583 (2004).

107. J. Yin, M. Wada, S. Tanase, T. Sakai, *J. Electrochem. Soc.* **151** A867 (2004).
108. Y.-U. Kim, C.-K. Kim, H.-J. Sohn, T. Kang, *J. Electrochem. Soc.* **151** A933 (2004).
109. N. Tamura, R. Ohshita, M. Fujimoto, S. Fujitani, M. Kamino, I. Yonezu, *J. power Sources* **107** 48 (2002).
110. S. D. Beattie, J. R. Dahn, *J. Electrochem. Soc.* **150** A894 (2003).
111. N. Tamura, R. Ohshita, M. Fujimoto, M. Kamino, S. Fujitani, *J. Electrochem. Soc.* **150** A679 (2003).
112. L. Wang, S. Kitamura, T. Sonoda, K. Obata, S. Tanase, T. Sakai, *J. Electrochem. Soc.* **150** A1346 (2003).
113. S.-J. Lee, H.-Y. Lee, S.-H. Jeong, H.-K. Baik, S.-M. Lee, *J. Power Sources* **111** 345 (2002).
114. Y.-L. Kim, S.-J. Lee, H.-K. Baik, S.-M. Lee, *J. Power Sources* **119** 106 (2003).
115. K. Sayama, H. Yagi, Y. Kato, S. Matsuta, H. Tarui, A N. Tamura, S. Fujitani, *The 11th Intern. Meeting on Lithium Batteries, Monterey, CA, June 23–28,* 2002, Abs. 52.
116. T. Yoshida, T. Fujihara, H. Fujimoto, R. Ohshita, M. Kamino, S. Fujitani, *The 11th Intern. Meeting on Lithium Batteries, Monterey, CA, June 23–28,* 2002, Abs. 48.
117. T. Takamura, S. Ohara, M. Uehara, J. Suzuki, K. Sekine, *J. Power Sources* **129** 96 (2004).
118. J. Graetz, C. C. Ahn, R. Yazami, B. Fultz, *J. Electrochem. Soc.* **151** A698 (2004).
119. H. Dong, X. P. Ai, H. X. Yang, *Electrochem. Commun.* **5** 952 (2003).
120. I.-S. Kim, G. E. Blomgren, P. N. Kumta, *Electrochem. Solid-State Lett.* **7** A44 (2004).
121. G. X. Wang, J. Yao, H. K. Liu, *Electrochem. Solid-State Lett.* **7** A250 (2004).
122. X. B. Zhao, G. S. Cao, *Electrochim. Acta* **46** 891 (2001).
123. M. Winter, J. O. Besenhard, *Electrochim. Acta* **45** 31(1999).
124. D. Larcher, L. Y. Beaulieu, O. Mao, A. E. George, J. R. Dahn, *J. Electrochem. Soc.* **147** 1703 (2000).
125. F. J. Fernandes-Madrigal, P. Lavala, C. Perez-Vicente, J.L. Tirado, *J. Electroanal. Chem.* **501** 205 (2001).

126. R. Alcantara, F. J. Fernandes-Madrigal, P. Lavala, C. Perez-Vicente, J. L. Tirado, J. C. Jumas, *J. Mater. Chem.* **9** 2517 (1999).

127. L. M. L. Fransson, J. T. Vaughey, R. Benedek, K. Edstroem, M. M. Thackeray, *Electrochem. Commun.* **3** 317 (2001).

128. O. Mao, R. L. Turner, I. A. Coutney, B. D. Fredericksen, M. I. Buckett, L. J. Krause, J. R. Dahn, *Electrochem. Solid-State Lett.* **2** 3 (1999).

129. M. M. Thackeray, J. T. Vaughey, A. J. Kahaian, K. D. Kepler, R. Benedek, *Electrochem. Commun.* **1** 111 (1999).

130. L. Fang, B. V. R. Chowdari, *J. Power Sources* **97/98** 181 (2001).

131. W. J. Weydanz, M. Wohlfahrt-Mehrens, R. A. Huggns, *J. Power Sources* **81/82** 237 (1999).

132. G. X. Wang, L. Sun, D. H. Bradhurst, S. Zhong, S. X. Dou, H. K. Liu, *J. Alloys Compd.* **306** 249 (2000).

133. G. X. Wang, L. Sun, D. H. Bradhurst, S. Zhong, S. X. Dou, H. K. Liu, *J. Power Sources* **88** 278 (2000).

134. H. Kim, J. Choi, H. J. Sohn, T. Kang, *J. Electrochem. Soc.* **146** 4401 (2001).

135. M. Wachtler, M. Winter, J. O. Besenhard, *J. Power Sources* **105** 151 (2002).

136. M. M. Thackeray, J. T. Vaughey, C. S. Johnson, A. J. Kropf, R. Benedek, L. M. L. Fransson, K. Edstrom, *J. Power Sources* **113** 124 (2003).

137. A. Saito, T. Horiba, Y. Aono, *The 36th Battery Symposium in Japan,* Abs. No.2 B07 (1995).

138. S. Matsuno, T. Kohno, N, Takami, T. Sawa, *Electrochem. Solid-State Lett.* **8(4)** A234 (2005).

139. J. Yang, M. Wachtler, M. Winter, J. O. Besenhard, *Electrochem. Solid-State Lett.* **2** 161 (1999).

140. X. H. Gleiter, *Prog. Mater. Sci.* **33** 223 (1989).

141. I.-S. Kim, G. E. Blomgren, P. N. Kumta, *J. Power Source* **130** 275 (2004).

142. X-W. Zhang, C. Wang, A. J. Appleby, F. E. Little, *J. Power Sources* **109** 136 (2002).

143. H. Li, L. Shi, X. Huang, L. Chen, Y. Liang, *J. Electrochem. Solis-State Lett.* **2** 547 (1999).

144. S.-W. Song, K. A. Striebel, X. Song, E. J. Cairns, *J. Power Sources* **119–121** 110 (2003).

Chapter 5

Nanometer Anode Materials for Li-Ion Batteries

Xuejie Huang and Hong Li

Institute of Physics, Chinese Academy of Sciences, P.O. Box 603, Beijing 100080, China

xjhuang@iphy.ac.cn

Use of nanometer materials in anode for Li-ion batteries on the one hand presents new opportunities for energy density, high rate of charge and discharge, and low cost. On the other hand, the electrochemical agglomeration, surface side reactions, and irreversible trapping by abundant defects are more significant than bulk materials, which decrease expected advantages of nanometer materials. Investigations on size effects revealed in the storage of lithium through tin oxide (SnO_2, SnO), alloy (Si, SnSb), conversion (Cr_2O_3, MnO), and interfacial charging (LiF/Ti) are introduced. Nano/micro core/shell or dispersed composite and surface modification can be used to enhance the stability of nanometer materials.

Nanomaterials for Lithium-Ion Batteries: Fundamentals and Applications

Edited by Rachid Yazami

ISBN 978-981-4316-40-8 (Hardcover), 978-981-4364-23-2 (eBook)

www.panstanford.com

5.1 Introduction

Since researchers at Sony Energytech developed the first commercial Li-ion batteries in late 1980s,[1–3] a notable effort has been undertaken to improve the battery materials for last two decades.

The anode of commercial Li-ion batteries has been dominated by graphite carbons which can form the compound LiC_6, with theoretical capacity of 372 $mA\text{-}h{\cdot}g^{-1}$. It is a severe storage limitation for high-energy batteries compared with the 3800 $mA\text{-}h{\cdot}g^{-1}$ for pure lithium. The volumetric capacity (848 $mA\text{-}h{\cdot}cm^3$) of graphite is also rather low due to its low density. Extensive research is being performed for the development of alternative anode materials that would give higher energy density for the Li-ion battery systems.

Anode material for Li-ion batteries should

(1) Have electrochemical potential values of lithium intercalation/de-intercalation (insertion/extraction) that are closely above lithium metal (0 V *vs.* Li^+/Li)
(2) Be able to store larger amount of lithium reversibly
(3) Have small variation of the electrode structure with lithium storage
(4) Have high electronic and ionic conductivities
(5) Be stable with the electrolyte in a wide range of electrochemical window
(6) Should be abundant in earth and nontoxic
(7) Should be economical for the manufacturer

A series of alloys and compounds of lithium with other metals and metalloids have been considered in early time as lithium source anode materials due to their high gravimetric capacities.[4–6] Table 5.1 shows the theoretical capacity of some element anode materials which can form alloy with lithium.

Experimental work on anodes using chemical elements which form alloys with lithium was started in early 1960s. For high temperature molten salt thermal cell, Li–Al,[7–10] Li–Si,[10–14] Li–Sn,[14–16] Li–Sb,[17,18] Li–Bi,[19] Li–Pb, Li–Cd, and Li–Mg alloys have been investigated. Among them, Seefurth and Sharma reported the formation of Li-Si alloys in formation of $Li_{12}Si_7$, $Li_{14}Si_6$, $Li_{13}Si_4$, and $Li_{22}Si_5$ at high temperature cells operating in the 400–500°C range 13. It indicates that each silicon atom can accommodate 4.4 lithium atoms leading to formation of $Li_{22}Si_5$ alloy, that is, a specific insertion capacity of 4200 mA-h·g^{-1}, the highest among the above alloying elements. In the early 1970s, several groups demonstrated that alloying reactions of metallic lithium with metallic or semi-metallic elements are electrochemically feasible at room temperature in a non-aqueous based electrolyte system,[20–22] Lithium can alloy with Al, Si, Sn, Pb, In, Bi, Sb, Ag, Mg, Zn, Pt, Cd, Au, As, Ge.[23–29]

Table 5.1 Theoretical capacity of element anode material which can form alloy with lithium

Element	Density (g·cm^3)	Alloy	Specific capacity (mA-h·g^{-1})	Specific capacity (mA-h·cm^3)
Li	0.53	Li	3860	2061
Si	2.33	$Li_{4.4}Si$	4191	9765
Sn	7.3	$Li_{4.4}Sn$	992	7241
Al	2.7	LiAl	993	2681
Sb	6.68	Li_3Sb	659	4402
Ge	5.32	$Li_{4.4}Ge$	1621	8626
Pb	11.4	$Li_{4.4}Pb$	568	6478
Bi	9.8	Li_3Bi	384	3763
Zn	7.14	LiZn	409	2920
In	7.31	Li_3In	699	5110

There are two main problems with the use of alloy-based anode. The first is poor cycle life and the second is slow diffusion rate. The main difficulties for using alloy-based materials are their dramatic volume expansion and contraction during Li insertion and extraction, that is, Sn atoms in a Sn particle that must undergo a drastic "dilution" by a factor of more than 5 (4.4 Li + Sn $\rightarrow$ $Li_{4.4}Sn$: 440% rise in the number of atoms). The volume expansion is as high as 310%.[30–31] This leads to the pulverization of the electrode materials and poor cycling performance. A report on 1997 by Idota *et al.* changed viewpoint on alloy reactions. It was found that amorphous tin composite oxide (TCO) materials showed good cyclic performance.[32] The achievement is later recognized to be attributed to the special microstructure in which active Sn nanograins are dispersed in an inert Li_2O matrix at the nanometer scale after lithium insertion into SnO.[33] Besides TCO, it was reported on 1996 by Yang *et al.* that superfine $Sn(SnSb)_{0.14}$ (200 nm) showed improved cyclic performance compared to larger particles.

These findings attract much attention on small-sized materials. In spite of this positive-size effect, it was gradually found that the stand-alone nanosized materials tend to agglomerate to form large particles upon cycling due to their large surface area and high surface energy.[35–37] The electrode fades also due to the bad electronic conducting network since the alloy particles are covered by electronic insulating solid electrolyte interphase (SEI) layer.[36,38,39] Even worse, the SEI is broken due to the large volume variation in cycling, lithium is consumed for new SEI formation for every cycle. If the lithium consumption is 1% per cycle, the capacity retention of a battery will be less than 80% after 20 cycles.

Four approaches have been taken to solve these problems.

(1) Designing a composite material, in which an electrochemically active phase is homogeneously dispersed within an electrochemically inactive matrix,[40–44] similar as the lithiated products of TCO.[33] The structural integrity of the composite electrode can be retained during the alloying and de-alloying processes. In addition, the inactive matrix as well as the active phase must possess high electronic conductivity and high Li-ion diffusivity, respectively, in order to achieve high-rate capability.

(2) Forming an intermetallic compound with active and inactive metals, in which an inert metal (e.g., Cu, Co, Ni, etc.) as a buffering matrix with an active metal (e.g., Si, Sn, Sb, etc.). In case of an optimized system like SnSb, the matrix also accepts lithium for storage, but then at different potentials. Hence, one component reacts with lithium while the other one acts as a buffer at the same time. As soon as the first active component is fully lithiated, the second one accepts lithium and then the first one acts as the buffer.[34,45–50]

(3) Dispersing active phases in a polymer-based elastic matrix. The alloys were dispersed in finely divided form throughout the conducting polymer matrix in a configuration that minimized alloy degradation processes associated

with exfoliation, surface passivation, and inefficient Li diffusion.[51]

(4) Designing an alloy/C composite material, in which an electrochemically active phase is homogeneously dispersed within in carbonaceous matrix. The carbonaceous materials are light and possess high electronic conductivity and high Li-ion diffusivity. It is presently a well-studied system.[52–60]

Due to the continuous demand on higher energy density battery, the research on nanosized materials for Li-ion batteries was stimulated since 1996 for searching high capacity materials. In this chapter, research on the nanosized anode materials are summarized following the line of tin oxide, tin and silicon alloys, and transitional metal oxides.

5.2 Tin Oxides

In 1995, Idota *et al.* in Fuji Film Co. claimed a class of amorphous tin-based composite oxides (TCO) as anode-active materials.[61] They showed 500–600 $mA\text{-}h\cdot g^{-1}$ reversible capacities and excellent cyclic performance.[32] TCO has a basic formula represented by SnM_xO_y, where M is a group of glass-forming metallic elements whose total stoichiometric number is equal to or more than that of tin ($x \geq 1$) and is typically comprised of a mixture of B(III), P(V), and Al(III).

Tin oxide is an active center for Li storage and other glass-forming elements, which make up an oxide network. Lithium-7 nuclear magnetic resonance measurements evidenced the high ionic state of lithium retained in lithiation state. Idota *et al.* believed that there was a new Li-storage mechanism. This finding attracts great attentions. It is believed that the nanocrystalline nature of tin precludes the formation of bulk phases of Li–Sn alloys, and the ratio Li:Sn is therefore varied continuously, minimizing the deleterious influence of the large volume expansion–related strain. Liu *et al.* studied nanosized SnO_2 powder calcined at different temperatures in air.[62] Higher temperature treatment results in stronger and sharper diffraction peaks, indicating an increase of crystallinity. The relationship between the logarithm of the crystallite size and the reciprocal heat-treatment temperature is given in Fig. 5.1.

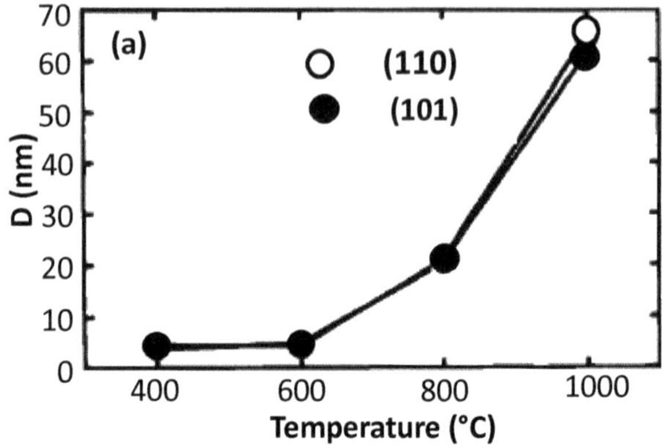

Figure 5.1 Variation of crystallite size as a function of treatment temperatures.

Figure 5.2 shows the first discharge and charge curves of Li/SnO_2 cells with SnO_2 heated at various temperatures. In the discharge curves, a plateau at about 1.0 V can be observed for each sample. For the untreated sample (dry gel), the plateau is very short. With the increase of treatment temperature the plateaus are elongated, and plateaus with a capacity of 300 mA-h·g^{-1} appear for SnO_2 samples treated at 800°C and 1000°C. Comparing with the XRD results, it is believed that formation of the plateau is closely related to crystallinity.

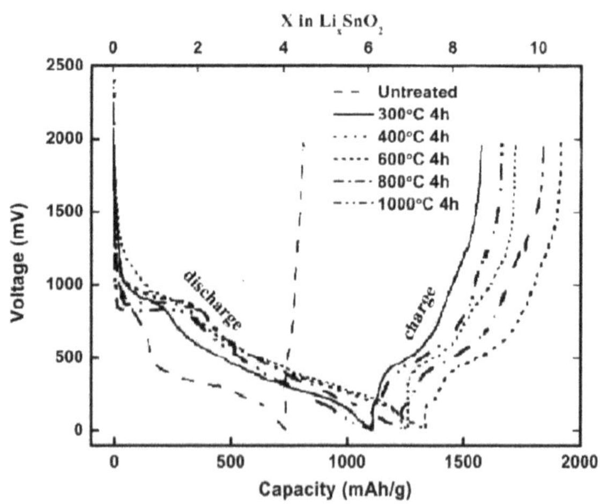

Figure 5.2 The first discharge–charge curves of Li/SnO_2 cells in which SnO_2 was heat treated at various temperatures.

Based on *in situ* XRD results obtained by Dahn's group and Raman spectroscopy performed in this group,[62,63] it is clear that lithium storage in tin-based oxides occurs in a two-step mechanism. Taking SnO as an example, the reaction equations can be written as:

$$2Li + SnO \rightarrow Li_2O + Sn \tag{5.1}$$

$$4.4Li + Sn \leftrightarrow Li_{22}Sn_5 \tag{5.2}$$

As shown in Fig. 5.3,[64] it is hard to explain the origin of five reversible redox peaks in cyclic voltammogram of SnO anode by other mechanisms rather than a series of lithium-alloy reactions.

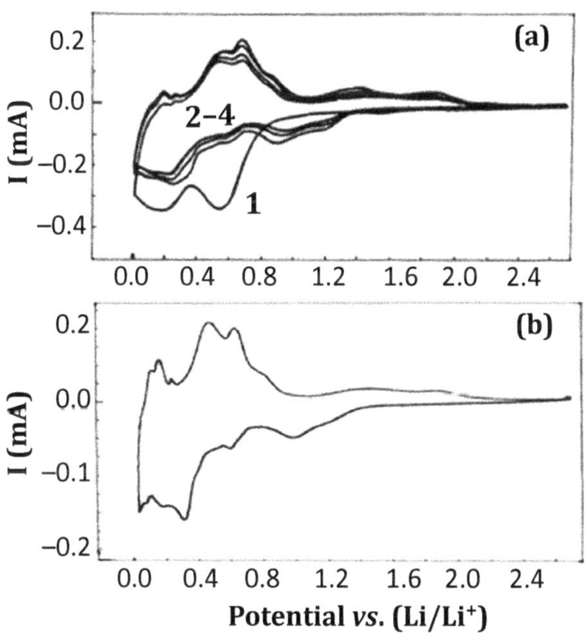

Figure 5.3 Cyclic voltammogram of SnO electrode in a lithium cell. (a) 1st to 4th cycles, 0.02 mV·sec^{-1}, (b) 5th cycle, 0.01 mV·sec^{-1}.

Sn anode was known to have very poor cyclic performance since the formation of $Li_{22}Sn_5$ leads to significant volume expansion and tends to pulverize during cycling. Therefore, Courtney *et al.* proposed that Li_2O matrix acts as a glue to retard the aggregation of tin atoms into large coherent regions in the case of tin oxides.[63] A direct image of fully lithiated nano-SnO anode was given in 1998 by us for the first time.[33] As shown in Fig. 5.4, nano-SnO maintains its particle shape after full lithiation (1551 mA-h·g^{-1}) but the interior part was converted into a nanocomposite where Li–Sn crystallite (2–20 nm) dispersed within amorphous Li_2O matrix. These results show a clear picture that enhanced cyclic performance of alloy reaction in the case of tin-based oxides could be benefited from two factors: nanosized alloy grains and volume variation can be buffered by another phase.

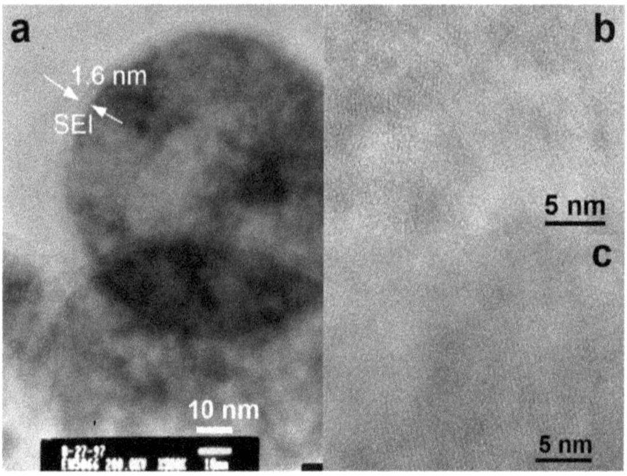

Figure 5.4 High-resolution TEM of fully lithiated nano-SnO (7.8 Li): (a) low magnification; (b, c) different regions inside the particle.

The large volume fraction of the nanocrystalline tin islands enclosed by the amorphous oxide network was sufficient to form continuous channels and thereby maintain the necessary electrical conductivity within the electrode. Forming a nanocomposite becomes an important strategy for designing high capacity anode materials later.

Besenhard found that small particle size Li-alloy anode (tin-based metallic host matrices) would improve the cycling performance.[34] The improvements are not only related to the small absolute changes in the size of the individual particles, but also to the fact that the more reactive particles are allowed to expand in a soft and ductile structure of unreacted material.

A series of SnO samples (BM0x) with different size particles are prepared by ball-milling. Figure 5.5 shows the XRD results of BM0x. It is obvious that the particle size of BM0x decreases with increasing ball-milling time. Figure 5.6 presents the discharge and charge curves of BM0x. No carbon black was added for preparing the working electrode. It is clear that anode material with smaller

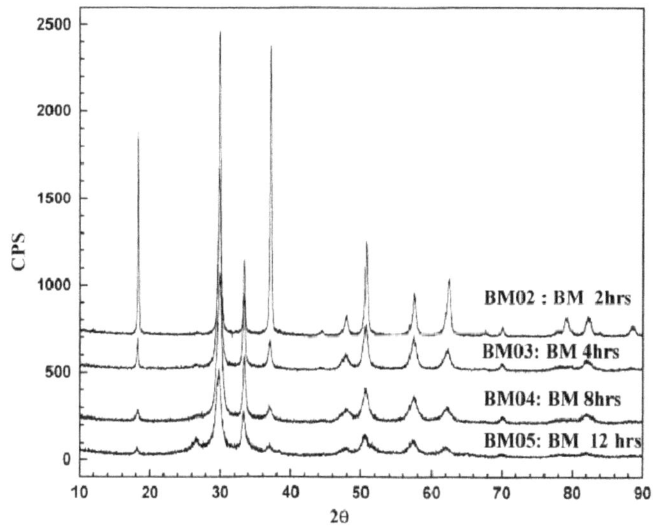

Figure 5.5 X-ray diffraction patterns of ball-milling SnO: BM0x.

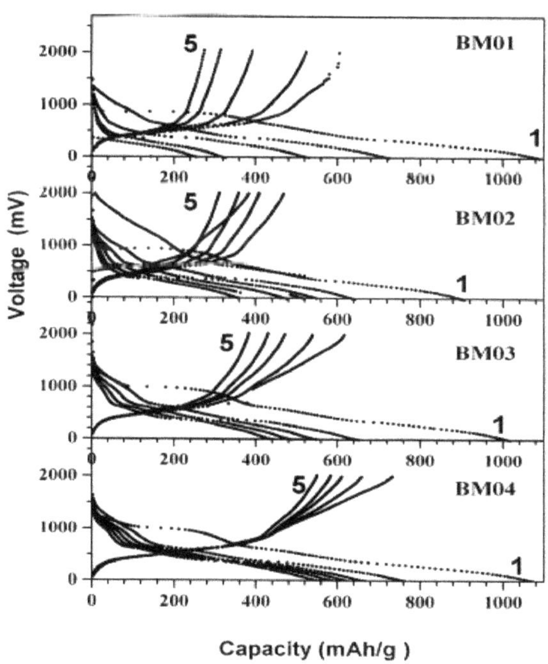

Figure 5.6 The discharge and charge curves of ball-milling SnO (BM0x) in lithium rechargeable batteries. BM0x/1 M $LiPF_6$, EC-DEC(1:1)/Li. The number in the figure is the cycle number.

particle size improves cycling performance. Since the crystal structure of SnO will be destroyed in spite of original particle size, this phenomenon may imply that SnO with small particle size can allow the expansion caused by the alloy reactions.

5.3 Nanosized Alloy Anodes

As mentioned above, electrochemical alloying reaction of lithium with metals has been widely studied since 1970s.[11–26] Many metals and alloys can store large quantity of lithium by the formation of alloys ($Li_{4.4}Si$, corresponding to Li-storage electrochemical capacity of 4200 $mA\text{-}h\cdot g^{-1}$, $Li_{4.4}Ge$: 1600 $mA\text{-}h\cdot g^{-1}$, LiAl and $Li_{4.4}Sn$: 990 $mA\text{-}h\cdot g^{-1}$, Li_3Sb: $mA\text{-}h\cdot g^{-1}$). The main difficulties for using alloy-based materials are their dramatic volume expansion and contraction during Li insertion and extraction. This leads to the pulverization of the electrode materials and poor cyclic performance. As mentioned above, investigations on tin oxide materials and report on superfine $Sn(SnSb)_{0.14}$ (200 nm) material enlighten us that nanosized alloy may overcome the disadvantage of large volume variation for alloy anode during full lithium insertion/extraction. In 1999, Li *et al.* reported the electrochemical performance of silicon nanoparticles (SiNPs) and Si nanowires (SiNWs) for the first time.[65–67]

Pure nanometer scale Si powder (80 nm) was prepared by laser-induced silane gas reaction.[68] Si nanowires were synthesized by laser ablation.[69] Figure 5.7 shows their structural features. It can be seen clearly that a layer of amorphous SiOx is covered on the surface of SiNWs.

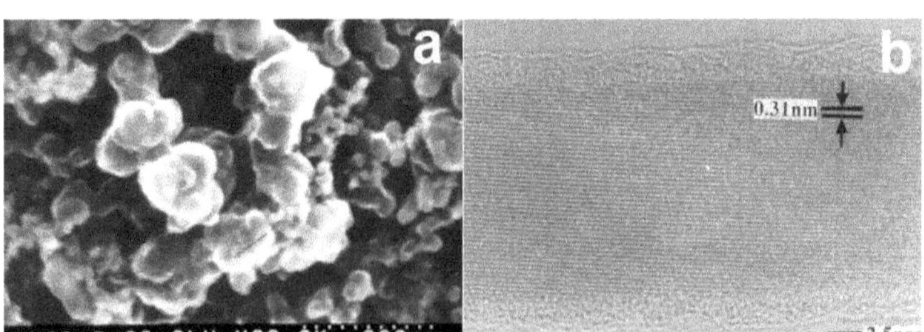

Figure 5.7 SEM images of (a) Si nanoparticles, (b) Si nanowires.

At room temperature, SiNPs and SiNWs show reversible capacities of 1700 $mA\text{-}h\cdot g^{-1}$ and 900 $mA\text{-}h\cdot g^{-1}$, respectively

(Fig. 5.8). It is found that the amount of carbon black is a key factor to improve the reversible capacity, as shown in Fig. 5.9. This means that the electronic contact between active particles is very important for SiNPs to achieve better cyclic performance. However, the same strategy is not effective for micrometer-sized Si particles (μm-Si). It was observed that serious pulverization occurs for the μm-Si electrode.[67] On the basis of this result, it was concluded that SiNPs show significant advantage in capacity retention toward large-size particles, which was consistent with our expectation.

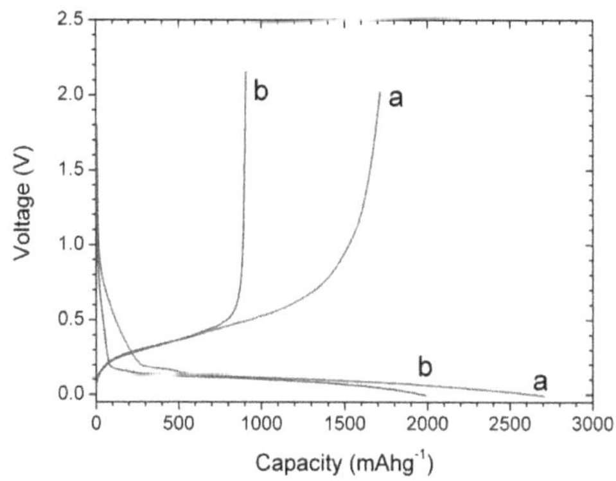

Figure 5.8 Discharge/charging curves of (a) SiNPs (weight ratio of SiNPs: carbon black = 1:1), (b) SiNWs (no conductive additive).

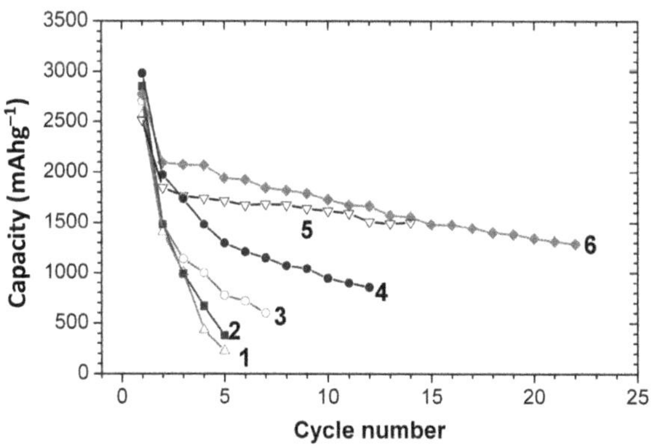

Figure 5.9 The cyclability of Si-based anodes at different conditions. The cell is Si/1 M $LiPF_6$, EC-DEC (1:1)/Li. Current density for curve 5 is 0.8 mA·cm$^{-2,}$ others are 0.1 mA·cm^{-2}. (1) μm-Si/CB (4:4), 0.0–0.8 V, (2) μm-Si/CB (4:4), 0.0–2.0 V, (3) SiNPs/CB (4:4), 0.0–2.0 V, (4) SiNPs CB (9:1), 0.0–0.8 V, (5) SiNPs/CB (4:4), 0.0–0.8 V, (6) SiNPs/CB (4:4), 0.0–0.8 V.

Previous studies revealed that Li can form various compounds with Si at elevated temperature, such as $Li_{12}Si_7$, $Li_{13}Si_4$, Li_7Si_3, $Li_{22}Si_5$. However, multiple voltage plateaus cannot be seen in the voltage profiles of SiNPs and SiNWs. Instead, only one slope can be seen, implying single two-phase transition reaction. As shown in Fig. 10a,b, it was found that electrochemical lithiation into Si at room temperature led to the formation of amorphous product. The insertion of Li into Si leads to a distorted angular bond distribution. With an increase in the doping dose, the bond distortion becomes more obvious and at last destroys the silicon structure. At room temperature, sluggish diffusion of Si inhibits the formation of ordered Li–Si alloy phase. It was observed that more ordered structure of lithiated SiNWs appeared after annealing at 400°C for 5 h under vacuum condition.[67]

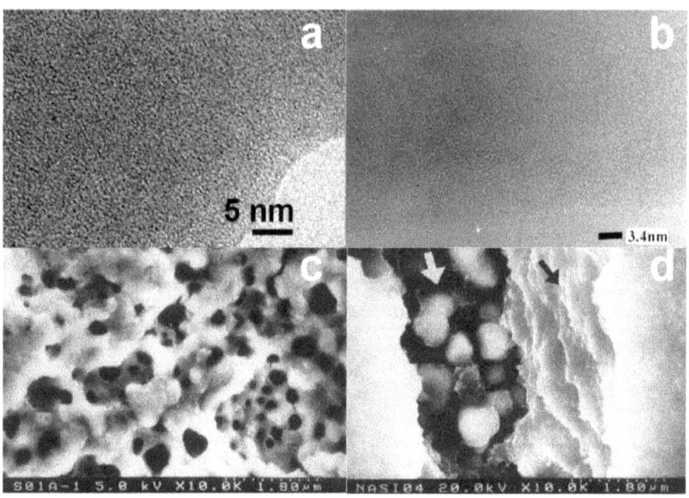

Figure 5.10 HRTEM of (a) lithiated SiNPS, (b) lithiated SiNWs, (c) lithiated SiNPs electrode (with 1:1 carbon black), (d) lithiated SiNPs electrode (no carbon black). White arrow: Cu substrate. Black arrow: Agglomerated dense Si layers.

It was also found that serious agglomeration occurs after one cycle, as shown in Fig. 5.10c,d. Nanosized materials have high surface energy. They tend to form large agglomerates. However, in most of cases, nanoparticles cannot merge together at room temperature due to slow transport kinetics of host atoms and loose contact. During electrochemical lithiation, chemical bonds among Si atom are broken. The particles are expanded, increasing their contact probability. After delithiation, most of Si atoms are in dangling bond state with high energy. They tend to bond together if no other atom can bond with them or prevent their transport, as shown in the case of SnO. Consequently, very dense agglomerates are formed after electrochemical cycling, see Fig. 5.10d. This phenomenon is called electrochemical agglomeration. It has been observed in most of alloy anodes later.

Due to the difficulty of synthesizing nanosized silicon at that time, SnSb alloy attracted our attention.

Besenhard *et al.* prepared $Sn_{0.78}Sb_{0.22}$ alloy (0.56 Sn + 0.22 SnSb) with particle size about 200 nm by electroplate and reductive precipitation.[34] Dendrite-like nanosized pure rhombohedral phase β-SnSb alloys were obtained by reductive coprecipitation method (Fig. 5.11). The morphology and particle size can be controlled by choosing solvent, temperature, and stirring conditions.[49]

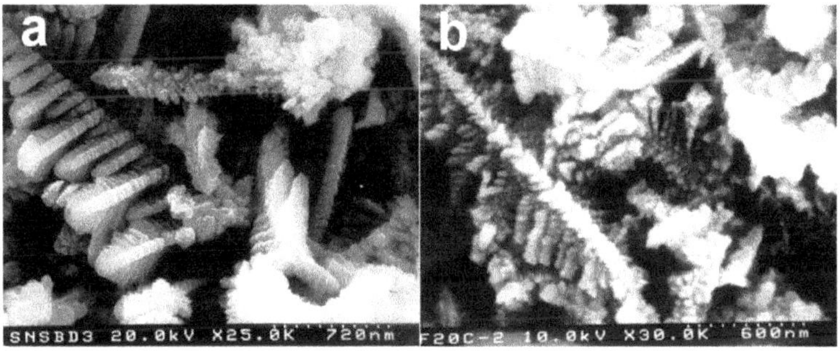

Figure 5.11 SEM images of (a) nano-SnSb prepared by reductive coprecipitation in ethane-1,2-diol solution at 0.0 ± 1.0°C under ultrasonic treatment, (b) nano-SnSb prepared by reductive coprecipitation in glyceryl alcohol solution at 0.0 ± 1.0°C under ultrasonic treatment.

The cyclability of nano-SnSb is obviously better than individual Sn and Sb anode.[36,49] Besenhard proposed that reacted SnSb may be embedded in unreacted soft and ductile Sn matrix (in their case, mole ratio of Sn:SnSb in precursor is 0.78:0.22). Therefore, SnSb alloy can be regarded as an active/active composite at atomic level.

It was interesting to find that several peaks appeared in the cyclic voltammogram of nanosized SnSb alloy anode (Fig. 5.12). It seems that Li–Sb alloy reactions occur first, followed by the Li–Sn alloy reactions. This was confirmed by *ex situ* XRD results as shown in Fig. 5.13.

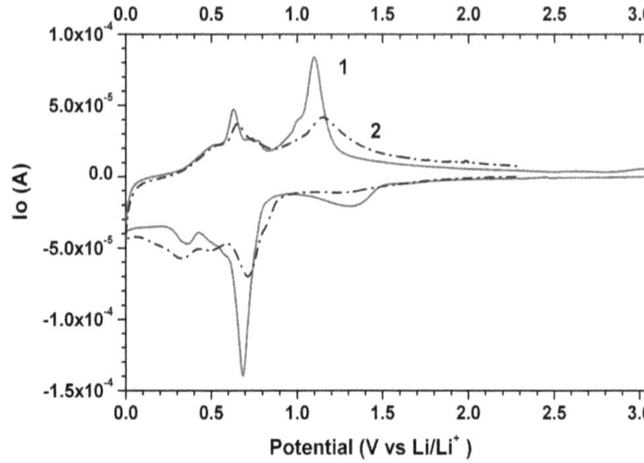

Figure 5.12 Cyclic voltammogram of nanosized SnSb alloy electrode in a lithium cell at first two cycles; r = 0.02 mV·s^{-1}.

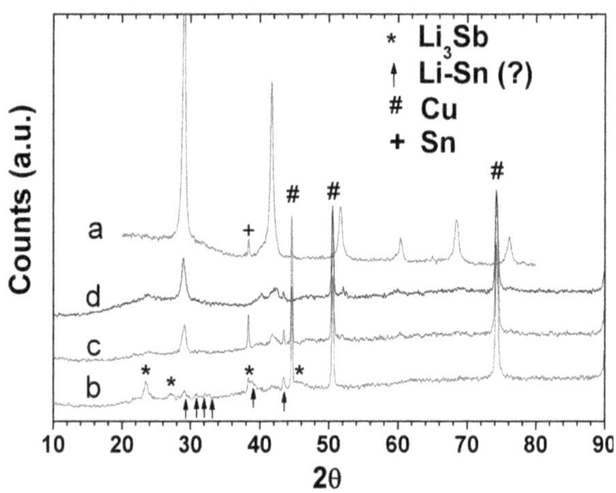

Figure 5.13 *Ex situ* XRD patterns of nano-SnSb anodes at different discharged states in lithium cells. (a) initial nano-SnSb alloy, (b) discharged to 0.0 V *vs.* Li/Li$^+$ at the first cycle, (c) after step b, then charged to 1.95 V *vs.* Li/Li$^+$, (d) after 20 cycles, stopped at 1.5 V *vs.* Li/Li$^+$.

Since initial nano-SnSb alloy is a single-phase material. This phase transition result means that insertion of lithium will first react with Sb locally. In this step, Sb atoms have to diffuse to somewhere to form Li_2Sb and Li_3Sb domains. At this step, separated Sn atoms will rearrange to form Sn domains. After Li_3Sb is formed, inserted lithium will react with Sn domains to form a series of Li–Sn alloys. Upon delithiation, reverse steps occur. It is possible to convert back into a Sn/Sb nanocomposite. However, in this case, separated Sn or Sb atoms or domains tend to rearrange to restore the original β-SnSb alloy structure. The details of the local structure evolution of this system in addition to electron distribution in this delocalized system are still not very clearly understood. It is amazing that such a complicated structure evolution is quite reversible. This result gives an example that reverse phase separation and restore could occur at room temperature driven by electrochemical reactions. It means that the diffusion of Sn, Sb, and Li at room temperature should be very fast to follow the structure variation and the structure of β-SnSb alloy phase is a thermodynamically favorable phase.

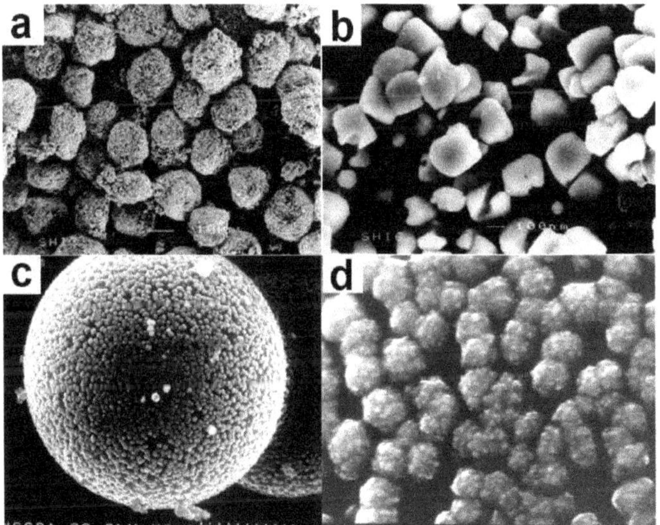

Figure 5.14 SEM images of (a, b) SnSb/nanosized MCMB and (c, d) SnSb/HCS.

Similar to the Si anode, serious electrochemical agglomeration occurs for nano-SnSb alloy. It is not good to achieve high reversible capacity and good cyclic performance since the SEI film is covered on the particles.[70] In order to solve this problem, nano-SnSb alloy was deposited on surface of mesocarbon microbead (MCMB) and hard carbon spherules (HCS), respectively,[71,72] as shown in Fig. 5.14. It was expected and confirmed that nanosized

SnSb alloy particles are pinned on the surface carbon substrate in the case of nano-SnSb/HCS.

The cyclic performance of the nano-SnSb/HCS composite is improved significantly compared to nano-SnSb, as shown in Fig. 5.15. However, it is found that the columbic efficiencies for these materials in each cycle are not very high (<99%), which is not acceptable for real Li-ion batteries. The reason for low efficiency is due to the direct exposure of nanosized alloy particle to the electrolyte. Due to significant volume variation, the SEI film on alloy particles may be not very stable during cycling, leading to formation and decomposition of the SEI in fresh exposed area appearing in each cycle. This problem could be solved by a strategy of embedding nanosized alloy particles in a stable carbon matrix or shell.[52–60] The core/shell strategy has also been used successfully for Si anodes. Consequently, high columbic efficiency (initial efficiency >85%, efficiency for each cycle >99%), high reversible capacity (>600 mA-h·g^{-1}) and excellent cyclic performance as well as high tap density (>1 g·cm^{-3}) and low surface area (<5 m^2·g^{-1}) were achieved, as shown in Fig. 5.16.

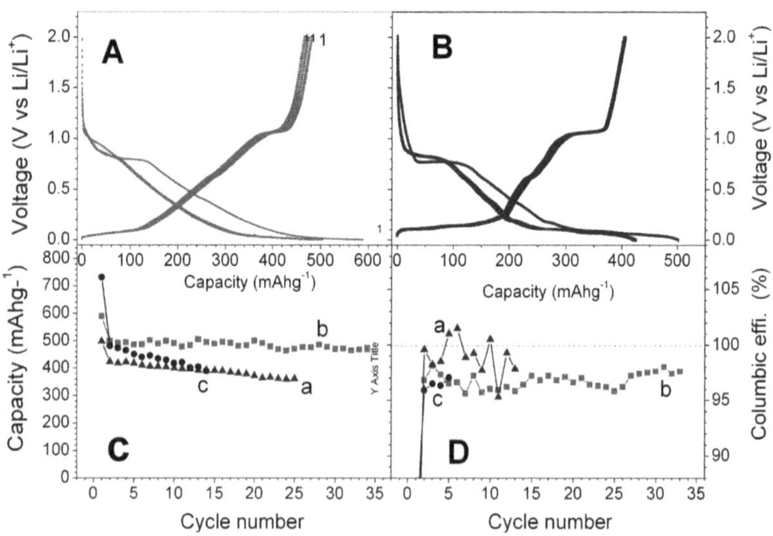

Figure 5.15 (A) Voltage profile of nano-SnSb/HCS; (B) Voltage profile of nano-SnSb/MCMB; (C) Cyclic performance of (a) nano-SnSb/MCMB, (b) nano-SnSb/HCS, and (c) nano-SnSb; and (D) Columbic efficiency of (a) nano-SnSb/MCMB, (b) nano-SnSb/HCS, and (c) nano-SnSb.

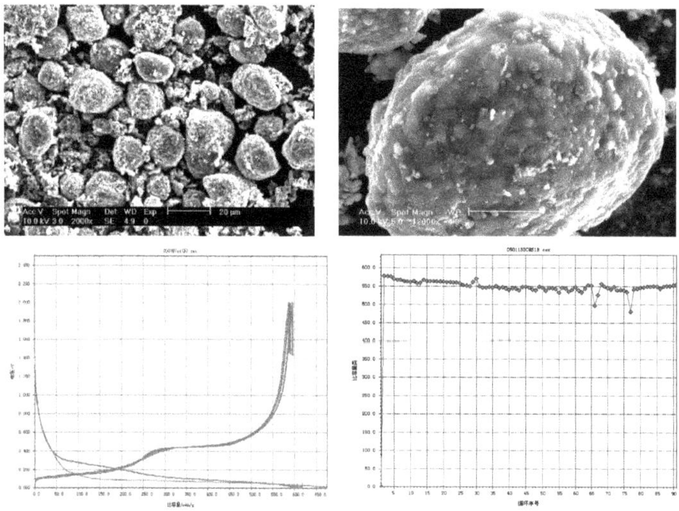

Figure 5.16 SEM images of core/shell C/nano-Si composite and electrochemical performances (0.1°C).

5.4 Transition Metal Oxides

In 2000, Poizot *et al.* for the first time reported that lithium can be stored reversibly in transition metal oxides through heterogeneous conversion reaction: Li + TMO → Li_2O + TM (1) (TM = Co, Fe, Ni, Cu).[73] Later, reversible lithium storage was also observed in transition metal fluorides as well as sulfides, nitrides, and phosphides.[74–83] It is very interesting for fundamental research that very inert LiF or Li_2O can react with transition metal TM at room temperature. This was not expected previously. It becomes clear now that enhanced electrochemical reactivity of LiF or Li_2O is mainly benefited from the special microstructure where the converted LiX and TM components show extremely small grain size (<5 nm) and intersperse with each other uniformly. Very short diffusion length and large contact area in nanocomposite are kinetically favorable for unusual reversible electrochemical behaviors of LiX/TM nanocomposite.

In view of increasing energy density of the batteries, besides high capacity, low average charging (delithiation) voltage of anode material is another important factor. Typical value of the average charging voltage at low rate is 0.3 V for graphite, 0.6 V for silicon,

0.7 V for Sn, and 1.6 V for $Li_4Ti_5O_{12}$. In addition, small polarization (voltage difference between charging and discharging curve) less than 0.2 V is necessary to achieve high-energy efficiency, which is realized for anode and cathode materials occurring intercalation or alloy mechanism. However, as for transition metal compounds occurring conversion reaction, most of materials suffer from high overpotential for both lithium insertion and extraction (about 1 V).[84] Consequently, a theoretical lithiation capacity could be approached only when the thermodynamic equilibrium voltage (also called electromotive force, emf) of the material for the conversion reaction is higher than 1 V. Cr_2O_3 and MnO are two transition metal oxides which have higher lithiation capacity but relatively lower emf values (1.085 V and 1.032 V *vs.* Li/Li^+, respectively).[84] They are more suitable as anode materials for Li-ion batteries.

Figure 5.17 SEM images of (a) commercial Cr_2O_3, (b) nanosized Cr_2O_3 obtained through hydrothermal method, (c) 2 mol% Mg-doped Cr_2O_3, and (d) 2 mol% Ni-doped Cr_2O_3.

Rhombohedral Cr_2O_3 (85-0869) has a theoretical density of 5.235 $g\cdot cm^{-3}$ and shows a dark green color. Commercial Cr_2O_3 powder material shows a particle size of 200–500 nm and the electronic conductivity is about 1.78×10^{-7} $S\cdot cm^{-1}$. It has been known that material with high electronic conductivity could achieve high columbic efficiency for conversion reaction, as shown in the case of RuO_2.[81] Mg and Ni doping can improve the conductivity of Cr_2O_3. Therefore, Mg and Ni doped Cr_2O_3 were also prepared. Their electronic conductivities were improved to 3.11×10^{-4} $S \cdot cm^{-1}$ and 9.57×10^{-5} $S \cdot cm^{-1}$, respectively after 2 mol% doping.[85]

SEM images of (a) commercial Cr_2O_3, (b) nanosized Cr_2O_3 obtained through hydrothermal method, (c) 2 mol% Mg-doped Cr_2O_3, and (d) 2 mol% Ni-doped Cr_2O_3 are shown in Fig. 5.17.

The initial discharge capacities of Cr_2O_3 materials are 1200–1400 $mA \cdot h \cdot g^{-1}$ and mainly contributed from a plateau region at 0.15 V.[86] This capacity is larger than the theoretical value of 1058 $mA \cdot h \cdot g^{-1}$ for Cr_2O_3 (6 Li needed for a full reduction of Cr_2O_3 into $3Li_2O/2Cr$ composite). The initial charge capacities are 700–800 $mA \cdot h \cdot g^{-1}$ in the voltage range from 0 V to 3.0 V and the voltage profile shows mainly two sloped regions. As discussed in recent papers for such behavior,[70,87] the sloped region at low voltage region could be assigned to extra Li storage at interfacial regions of LiX/M nanocomposite. In the case of Cr_2O_3 system, it contributes a Li-extraction capacity of about 250 $mA \cdot h \cdot g^{-1}$ as shown in Fig. 5.18. The sloped region above the emf value for this reaction is related to the phase transformation from Li_2O/Cr to Cr_2O_3, as confirmed by the XPS results. The charging capacities are about 500 $mA \cdot h \cdot g^{-1}$, much less than the theoretical capacity for a complete Li-extraction reaction. It indicates that almost half of inserted lithium is trapped irreversibly. The irreversibility is not decreased for doped Cr_2O_3.

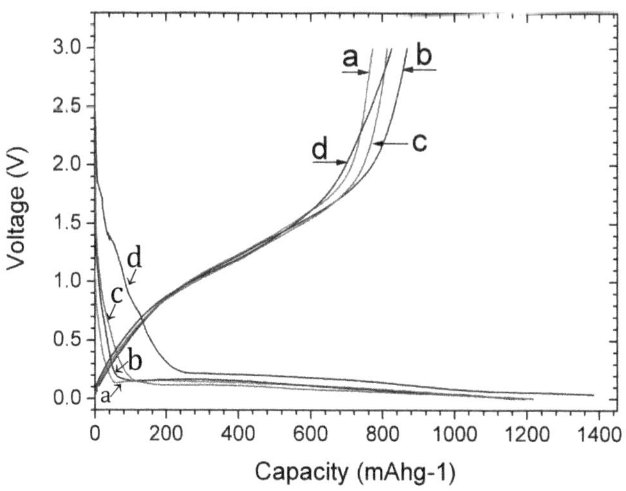

Figure 5.18 Voltage profiles of (a) commercial Cr_2O_3, (b) nanosized Cr_2O_3 obtained through hydrothermal method, (c) 2 mol% Mg-doped Cr_2O_3, and (d) 2 mol% Ni-doped electrodes.

The average charging voltage for Cr_2O_3 is about 1.2 V, much lower than most of the other transition metal compounds (TMX, in brief) (>1.8 V). This is a significant advantage for Cr_2O_3 as anode material compared with other transition metal oxides and fluorides, beneficial from its low emf value of 1.058 V *vs.* Li/Li^+.

After full insertion, all particles are covered by a layer of amorphous species with a thickness of 20–90 nm (Fig. 5.19a). This layer has been observed in many lithiated TMX anodes and represents the so-called SEI. In zoomed image, in Fig. 5.19b, obvious disintegration of Cr_2O_3 grains is observed. Some tiny grains less than 5 nm are dispersed in amorphous regions. Such a microstructure is similar to most of other TMO anodes observed. After full delithiation, the interior part converts back to more ordered Cr_2O_3 phase (Fig. 5.19d) and the SEI thickness is decreased to less than 5 nm; some particles are even naked. As observed previously, the SEI on the TMO anodes are electrochemically decomposable. Since the thick SEI is covered on all particles, it is believed that the reversible formation/decomposition of the SEI film is not favorable for the cyclic performance, it leads to the peeling off of active materials from the current collector.

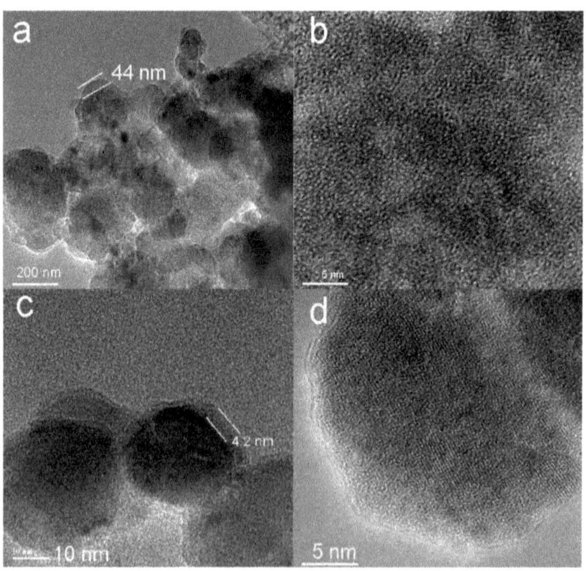

Figure 5.19 HRTEM of nano-Cr_2O_3 particle at (a and b) in fully lithiated state, and (c and d) fully delithiated state.

The SEI film is regarded as a passivating layer, which does not decompose during charging and is around several nanometers thick in cases of graphite carbon anodes. The formation of the SEI will be terminated when the electron cannot tunnel through this film. Why the SEI film is so thick on TMX anodes and what are possible components are two important issues. Recently, Gachot *et al.* studied a chromium-based oxide (named CBO) grown on stainless steel mesh tissues as anode in a half cell with 1 M $LiPF_6$ dissolved in EC/DMC (1/1 w/w) as electrolyte.[88] A layer of 20–50 nm amorphous film was observed on fully lithiated CBO electrode. After analyzing the supernatant in a borosilicate micro-fiber (millipore) film separator recovered from Li/CBO cells, the existence of MeOLi, CH_3OCO_2Li, and different PEO oligomers series were identified clearly based on high-resolution mass spectrometry (ESI-HRMS) comparison analysis on a series of reference compounds. In order to determine the SEI components on the electrode instead of the separator and the electrochemically decomposable components, a technique combining thermo-gravimetry and mass spectrometry (TG-MS) was applied, which was first used by Zhao *et al.* to analyze the SEI film on lithiated graphite electrode.[89] As shown in Fig. 5.20a, the Cr_2O_3 electrode in a fully lithiated state shows a continuous weight loss up to 8.97% at 500°C, while the electrodes charged at 1.1 V, 2.0 V, and 3.0 V show a 5.36%, 1.7%, and 1.51% weight loss at 500°C, respectively.[90] Since the main inorganic species in the SEI film, that is, Li_2CO_3 and LiF and lithiated products of Cr, Li_2O, and unreacted Cr_2O_3 cannot be decomposed below 500°C, the thermal decomposable components in the electrode are mainly organic or polymer species in the SEI film. Accordingly, the electrode at fully lithiated state should have the largest quantity of organic and/or polymer-like species. The results shown in Fig. 5.20a mean that these thermal decomposable species in the fully lithiated Cr_2O_3 electrode are decomposed electrochemically upon charging. The weight ratio of the SEI film in the Cr_2O_3 electrode is decreased at least 7.5 wt% after charging to 3.0 V. This consists of significant thinning of the SEI film on Cr_2O_3 anode after charging is observed by TEM (Fig. 5.19).

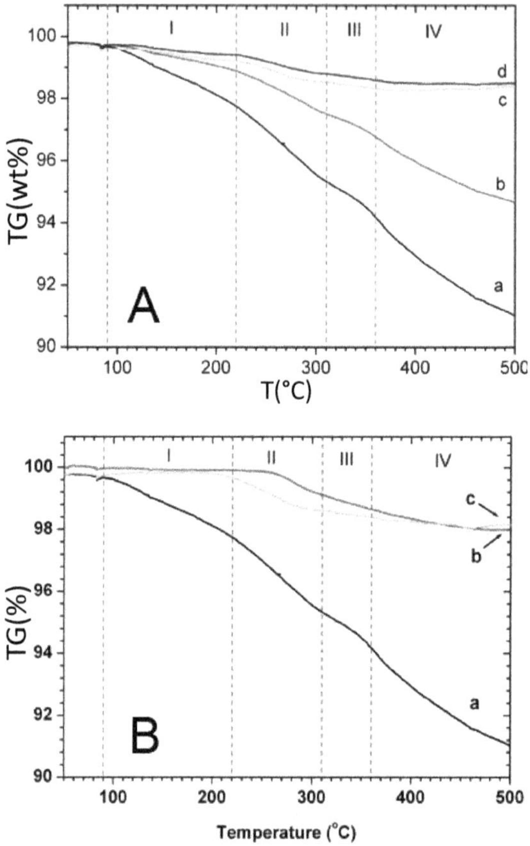

Figure 5.20 (A) TG curves of Cr_2O_3 anode discharged to (a) 0.005 V and charged to 0.005 V, then charged to (b) 1.1 V, (c) 2.0 V and (d) 3.0 V *vs.* Li/Li$^+$, B) TG curves for (a) Cr_2O_3, (b) MCMB, (c) HCS electrodes discharged to 0.005 V *vs.* Li/Li$^+$ in the first cycle.

Based on mass spectroscopy analysis, it was found that oligomer and polymer with ethylene oxide units are electrochemically decomposable species during charging and $ROCO_2Li$ is relatively electrochemically stable. The formation of large amount of these species could be related to strong catalytic effect from nanosized transition metal nanograins.

The TG curves of MCMB and HCS are similar to the delithiated Cr_2O_3 samples (c) and (d). This means that much less thermal decomposable species are formed on carbon electrode, consistent with the TG result Zhao *et al.* that the SEI films on graphite are thin.[89] In addition, it was found that $ROCO_2Li$ could be the main thermal decomposable components in the SEI layers on MCMB and HCS electrodes.

The cyclic performance of Cr_2O_3 electrode is very poor.[86] Cracks were observed in the cycled electrode. The main reasons for poor cyclic performance were loss of electronic contact of active particles and poor transport properties for electrons and ions within active material. The first issue could be due to large volume variation in the electrode layer, which could be solved by decreasing particle size, coating, and avoiding the formation of unstable SEI film. The solution for the second issue could be decreasing the particle size in addition to improving intrinsic electronic conductivity via hetero-atom doping and keeping good contact with conductive additive. Several material modifications aiming to solve above problems have been performed.[86,91] including decreasing particle size, doping with heterogeneous atoms, preparing carbon/ Cr_2O_3 composite with different microstructures.

Figure 5.21 shows the cyclic performance of several Cr_2O_3-based anodes. For samples (a) to (d), the surface of Cr_2O_3 particles is exposed to the electrolyte phase. Therefore, such a microstructure cannot avoid the formation of the unstable SEI film on Cr_2O_3 particles. Carbon coating is very effective to improve its cyclic performance, as shown in the case of samples (e) and (h). Carbon nanotubes grown on the Cr_2O_3 particle to form a composite in which each Cr_2O_3 particle is fully covered by coiled CNTs is also effective to improve the cyclic performance. It is reasonable that the volume variation of Cr_2O_3 particles should not have large influence on the contact between coiled CNT and Cr_2O_3 particles. Since CNT is a good electronic conductor, this structure is certainly good for keeping good electronic contact of Cr_2O_3 particle. However, such a structure cannot avoid the formation of SEI on exposed area of Cr_2O_3. The sample showing the best cyclic performance is the sample (f). Cr_2O_3 particles were first mixed with carbon black, then the composite particles were coated with a layer of carbon. After that, particles were treated by CVD to cover all exposed area. Such a composite has advantage for maintaining electronic contact and avoid the formation of unstable SEI film during cycling, as evidenced partially by TEM investigation.

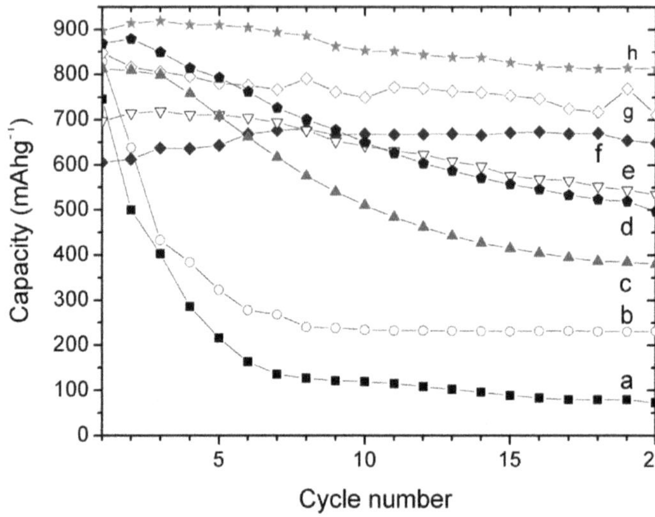

Figure 5.21 Cyclic performance of Cr_2O_3-based anodes. (a) pristine Cr_2O_3, (b) Cr_2O_3 with 10 wt% binder, (c) nanosized Cr_2O_3, (d) 2 wt% Mg-doped Cr_2O_3, (e) carbon-coated Cr_2O_3, (f) carbon-coated on Cr_2O_3/carbon black mixture, followed by further CVD treatment, (g) CNTs grown on Cr_2O_3 nanocomposite, and (h) carbon-coated Mg-doped Cr_2O_3.

For anode materials based on conversion reaction, there are three main disadvantages, such as low columbic efficiency, high average charging voltage, and high polarization.

The polarizations of different anodes have been compared by the ΔV curves. The ΔV curve is obtained by subtracting the discharging curve at the second cycle from the charging curve at the first cycle after normalization. It is the sum of the overpotentials of both charging and discharging at galvanostatic mode. ΔV curve method was used to roughly compare the polarization of different materials.[92] As shown in Fig. 5.22, the average polarization order is $Li_4Ti_5O_{12}$ < MCMB < Si < nano-SnSb < SnO < HCS < MnO < Cr_2O_3. Obviously, lithium storage in conversion reaction shows the highest high polarization. Main origin seems related to either anion X^{n-} or M^{n+} transport properties, not determined very much on the transport of electrons and lithium ions. Among transition metal oxides, MnO shows lower polarization, high capacity (450–550 mA-h·g^{-1}, 3500–4200 mA-h·cm^{-3}) and good rate performance.[93]

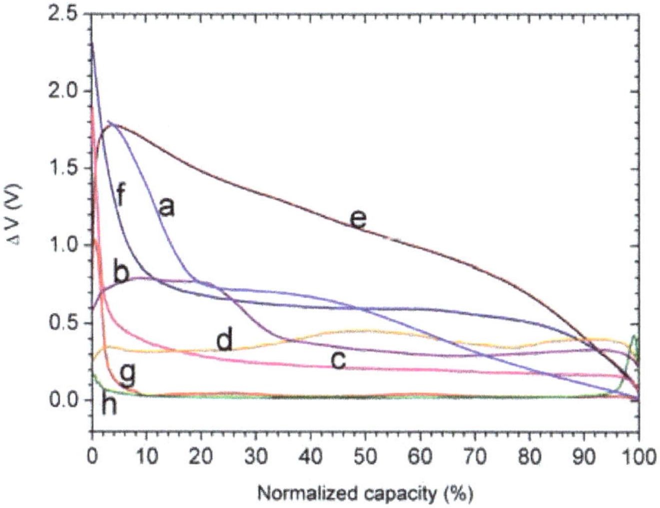

Figure 5.22 Polarization analysis (ΔV curve) of anode materials at low rate (0.05–0.2 C). (a) nano-HCS, (b) SnO, (c) Si, (d) nano-SnSb, (e) Cr_2O_3, (f) MnO, (g) MCMB, and (h) $Li_4Ti_5O_{12}$.

Actually, the only report on MnO anode material before our recent publication indicates that MnO powder electrode could be discharged and charged under a very small current density (1/300 C) by a potentiostatic intermittent titration technique (PITT) mode.[75] Pure phase MnO thin film was prepared. It was surprising to find that MnO thin film shows good rate performance and cyclic performance, as shown in Fig. 5.23.[92] Similar performances and polarizations are also observed in the cases of MnO powder materials.[94]

Recently, from the polarization perspective, MgH_2 was tested as anode material with the similar conversion reaction mechanism. It was found that the average charging voltage is about 0.5 V and ΔV is about 0.5 V at 0.05 C.[95] This finding seems to support the purpose that transport of anion is a key factor for determining the polarization. How to decrease high polarization of conversion reaction is still a great challenge for most of the TMX materials.

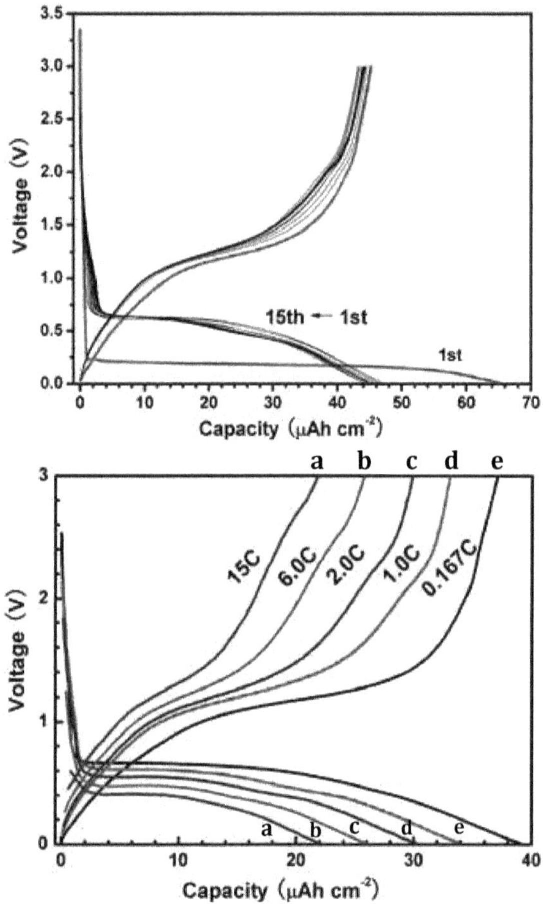

Figure 5.23 Voltage profile of MnO thin film electrode (above) and rate performance of MnO thin film electrode (below).

5.5 Conclusions and Outlook

Using nanosized and nanostructured materials presents new opportunities in rechargeable Li-ion batteries for energy density, exceptionally high rate of charge and discharge, high cycle ability, and low cost. At present, $Li_4Ti_5O_{12}$ is a good candidate for high-power Li-ion batteries and large-scale application of tin-based anode is not visible. Si is a well-recognized as a promising anode material for high-energy density Li-ion batteries. The disconnection of the active material to current collector due to large volume expansion can be alleviated by dispersing nanosized silicon in carbon matrix. However, the columbic efficiency in cycling should be improved for real applications. The loss of active lithium for each cycle is due to the thick SEI formed on the surface of Si. It happens also to other anode materials with large volume change in cycling. While the synthetic method and particle size/morphology of active component play a critical role in their electrochemical activity, electrode engineering is needed to minimize those negative effects of volume changes on cycling. Some approaches are (a) dispersing the active materials in an elastic matrix; (b) covering the surface of the active materials by a lithium permeate elastic thin film, etc.

Acknowledgment

Financial support from NSFC (50672122, 50730005, 60621061), "863" project (2006AA03Z346) and "973" project (2007CB936501) are appreciated. Contributions from Prof. L. Q. Chen, Dr. Q. Wang, Dr. L. H. Shi, Dr. X. D. Wu, Dr. J. Hu, Dr. N. Liu, and Ms. Y. Zeng to the experimental part are appreciated.

References

1. Sony Corp. [JP]. *European Patent No.* EP391281, Apr. 3 (1989).
2. T. Nagaura, K. Tozawa, in *Progress in Batteries and Solar Cells* (Eds.: JEC Press Inc. and IBA Inc.; Vol. 9, p. 209), JEC Press Inc., Brunswick, OH (1990).
3. Y. Nishi, *Chem. Rec.*, **1**, 406 (2001).
4. R. A. Huggins, *Solid State Ionics*, **113–115**, 57–67 (1998).
5. M. Winter, J. O. Besenhard, *Electrochim. Acta*, **45**, 31–50 (1999).
6. J. Wang, I. D. Raistrick, R. A. Huggins, *J. Electrochem. Soc.*, **133**, 457–460 (1986).
7. C. E. Johnson, M. S. Foster, *J. Electrochem. Soc.*, **116**, 1612 (1969).
8. C. A. Melendres, *J. Electrochem. Soc.*, **124**, 651 (1977).
9. J. R. Selman, D. K. DeNuccio, C. J. Sy, R. K. Steunenberg, *J. Electrochem. Soc.*, **124**, 1161 (1977).
10. C. A. Melendres, C. C. Sy, *J. Electrochem. Soc.*, **125**, 727 (1978).

11. S. Lai, *J. Electrochem. Soc.*, **123**, 1196 (1976).
12. C. J. Wen, B. A. Boukamp, R. A. Huggins, W. Weppner, *J. Electrochem. Soc.*, **126**, 2258 (1979).
13. R. N. Seefurth, R. A. Sharma, *J. Electrochem. Soc.*, **127**, 1101 (1980).
14. C. J. Wen, R. A. Huggins, *J. Solid State Chem.*, **37**, 271 (1981).
15. M. S. Foster, C. E. Crouthamel, S. E. Wood, *J. Phys. Chem.*, **70**, 3042 (1966).
16. C. J. Wen, R. A. Huggins, *J. Solid State Chem.*, **35**, 376 (1980).
17. C. J. Wen, R. A. Huggins, *J. Electrochem. Soc.*, **128**, 1181 (1981).
18. W. Weppner, R. A. Huggins, *J. Electrochem. Soc.*, **124**, 1569 (1977).
19. E. S. Buzzelli, in Proceedings of Fall Meeting of the Electrochemical Society, Miami Beach FL, Oct. 8–13, 1972.
20. W. Weppner, R. A. Huggins, *J. Solid State Chem.*, **22**, 297 (1977).
21. C. S. Tedmon, W. C. Hagel, *J. Electrochem. Soc.*, **115**, 151 (1968).
22. A. N. Dey, *J. Electrochem. Soc.*, **118**, 1547 (1971).
23. J. Wang, I. D. Raistrick, R. A. Huggins, *J. Electrochem. Soc.*, **133**, 457 (1986).
24. A. Anani, S. Crouch-Baker, R. A. Huggins, *J. Electrochem. Soc.*, **134**, 3098 (1987).
25. R. A. Huggins, *J. Power Sources*, **22**, 341 (1988).
26. R. A. Huggins, *J. Power Sources*, **26**, 109 (1989).
27. I. Hauke, S. Machill, D. Rahner, K. Wiesener, *J. Power Sources*, **43–44**, 421 (1993).
28. S. V. Sazhin, A. V. Gorodskii, M. Yu. Khimchenko, *J. Power Sources*, **47**, 57, (1994).
29. E. M. Shembel, I. M. Maksyuta, L. I. Neduzho, A. I. Beloshokov, A. F. Maumenko, V. V. Rozhkov, *J. Power Sources*, **54**, 416 (1995).
30. L. Y. Beaulieu, K. W. Eberman, R. L. Turner, L. J. Krause, J. R. Dahn, *Electrochem. Solid-State Lett.*, **4**, A137 (2001).
31. T. Brousse, R. Retoux, U. Herterich, D. M. Schleich, *J. Electrochem. Soc.*, **145**, 1 (1998).
32. Y. Idota, T. Kubota, A. Matsufuji, Y. Maekawa, T. Miyasaka., *Science*, **276**, 1395 (1997).
33. H. Li, X. J. Huang, L. Q. Chen, *Electrochem. Solid-State Lett.*, **1**, 241 (1998).

34. J. Yang, M. Winter, J. O. Besenhard, *Solid State Ionics*, **90**, 281–287 (1996).

35. H. Li, X. J. Huang, L. Q. Chen, G. W. Zhou, Z. Zhang, D. P. Yu, Y. J. Mo, *Solid State Ionics*, **135**, 181 (2000).

36. H. Li, L. H. Shi, W. Lu, X. J. Huang, L. Q. Chen, *J. Electrochem. Soc.*, **148**, A915 (2001).

37. H. Li, L. H. Shi, Q. Wang, L. Q. Chen, X. J. Huang, *Solid State Ionics*, **148**, 247 (2002)

38. M. Winter, J. O. Besenhard, *Electrochim. Acta*, **45**, 31 (1999).

39. I. S. Kim, P. N. Kumta, G. E. Blomgren, *Electrochem. Solid-State Lett.*, **3**, 493 (2000).

40. O. Mao, R. L. Turner, I. A. Courtney, B. D. Fredericksen, M. I. Buckett, L. J. Krause, J. R. Dahn, *Electrochem. Solid-State Lett.*, **2**, 3 (1999).

41. I. S. Kim, G. E. Blomgren, P. N. Kumta, *J. Power Sources*, **130**, 275 (2004).

42. I. S. Kim, P. N. Kumta, G. E. Blomgren, *Electrochem. Solid-State Lett.*, **3**, 493 (2000).

43. I. S. Kim, G. E. Blomgren, P. N. Kumta, *Electrochem. Solid-State Lett.*, **6**, A157 (2003).

44. M. K. Datta, P. N. Kumta, *J. Power Sources*, **158**, 557 (2006).

45. J. Yang, Y. Takeda, N. Imanishi, T. Ichikawa, O. Yamamoto, *Solid State Ionics*, **135**, 175 (2000).

46. J. Yang, M. Wachtler, M. Winter, J. O. Besenhard, *Electrochem. Solid-State Lett.*, **2**, 161 (1999).

47. J. Yang, Y. Takeda, N. Imanishi, T. Ichikawa, O. Yamamoto, *J. Power Sources*, **79**, 220 (1999).

48. J. Yang, Y. Takeda, Q. Li, N. Imanishi, O. Yamamoto, *J. Power Sources*, **90**, 64 (2000).

49. H. Li, G. Y. Zhu, X. J. Huang, L. Q. Chen, *J. Mater. Chem.*, **10**, 693 (2000).

50. I. Rom, M. Wachtler, I. Papst, M. Schmied, J. O. Besenhard, F. Hofer, M. Winter, *Solid State Ionics*, **143**, 329 (2001).

51. M. Maxfield, T. R. Jow, S. Gould, *J. Electrochem. Soc.*, **135**, 299 (1988).

52. R. L. Turner, D. J. Mcclure, L. J. Krause, M. M. Buckett, J. R. Dahn, O. Mao. *U.S. Patent No.* 6,203,944 (2001).

53. S. Kawakami, M. Asao. *U.S. Patent No.* 6,949,312 (2005).

54. S. Mizutani, H. Inoue. *U.S. Patent No.* 0208378 (2005).

55. S. Mizutani, Y. Kudo. *U.S. Patent No.* 0121348 (2006).
56. A. D. W. Todd, R. E. Mar, J. R. Dahn, *J. Electrochem. Soc.*, **154**, A597 (2007).
57. K. T. Lee, Y. S. Jung, S. M. Oh, *J. Am. Chem. Soc.*, **125**, 5652 (2003).
58. W. M. Zhang, J. S. Hu, Y. G. Guo, S. F. Zheng, L. S. Zhong, W. G. Song, L. J. Wan, *Adv. Mater.*, **20**, 1160 (2008).
59. H. Li, J. Hu, X. J. Huang, L. Q. Chen, *Chinese Patent No.* ZL 1328805c (2004).
60. H. Li, X. J. Huang, *Chinese Patent No.* ZL 100422112c (2005).
61. Y. Idota, M. Nishima, Y. Miyaki, T. Kubota, T. Miyasaki. *European Patent No.* 0651450a1 (1995).
62. W. F. Liu, X. J. Huang, Z. X. Wang, H. Li, L. Q. Chen, *J. Electrochem. Soc.* **145**, 5 (1998).
63. I. A. Courtney, J. R. Dahn, *J. Electrochem. Soc.*, **144**, 2045 (1997).
64. H. Li, X. J. Huang, L. Q. Chen, *Solid State Ionics*, **123**, 189 (1999).
65. H. Li, X. J. Huang, L. Q. Chen, Z. G. Wu, Y. Liang, *Electrochem. Solid-State Lett.*, **2**, 547 (1999).
66. G. W. Zhou, H. Li, H. P. Sun, D. P. Yu, Y. Q. Wang, X. J. Huang, L. Q. Chen, Z. Zhang, *Appl. Phys. Lett.*, **75**, 2447 (1999).
67. H. Li, X. J. Huang, L. Q. Chen, G. W. Zhou, Z. Zhang, D. P. Yu, Y. J. Mo, *Solid State Ionics*, **135**, 181 (2000).
68. Y. L. Li, Y. Liang, K. S. Xiao, F. Zheng, Z. Q. Hu, *Acta Metall. Sin.*, **31**, B21 (1995).
69. D. P. Yu, C. S. Lee, I. Bello, X. S. Sun, Y. H. Tang, G. W. Zhou, Z. G. Bai, Z. Zhang, S. Q. Feng, *Solid State Commun.*, **105**, 403 (1998).
70. H. Li, L. H. Shi, W. Lu, X. J. Huang, L. Q. Chen, *J. Electrochem. Soc.*, **148**, A915 (2001).
71. L. H. Shi, H. Li, Z. X. Wang, X. J. Huang, L. Q. Chen, *J. Mater. Chem.*, **11**, 1502 (2001).
72. H. Li, Q. Wang, L. H. Shi, L. Q. Chen, X. J. Huang, *Chem. Mater.*, **14**, 103 (2002).
73. P. Poizot, S. Laruelle, S. Grugeon, L. Dupont, J. M. Tarascon, *Nature*, **407**, 496 (2000).
74. A. Débart, L. Dupont, P. Poizot, J. B. Leriche, J. M. Tarascon, *J. Electrochem. Soc.*, **148**, A1266–A1274 (2001).
75. P. Poizot, S. Laruelle, S. Grugeon, L. Dupont, J. M. Tarascon, *J. Electrochem. Soc.*, **149**, A1212–A1217 (2002).

76. N. Pereira, L. C. Klein, G. G. Amatucci, *J. Electrochem. Soc.*, **149**, A262–A271 (2002).
77. D. C. C. Silva, O. Crosnier, G. Ouvrard, J. Greedan, A. Safa-Sefat, L. F. Nazar, *Electrochem. Solid-State Lett.*, **6**, A162–A165 (2003).
78. H. Li, G. Richter, J. Maier, *Adv. Mater.*, **15**, 736–739 (2003).
79. F. Badway, N. Pereira, F. Cosandey, G. G. Amatucci, *J. Electrochem. Soc.*, **150**, A1209–A1218 (2003).
80. F. Badway, F. Cosandey, N. Pereira, G. G. Amatucci, *J. Electrochem. Soc.*, **150**, A1318–A1327 (2003).
81. P. Balaya, H. Li, L. Kienle, J. Maier, *Adv. Funct. Mater.*, **13**, 621–625 (2003).
82. Y. Wang, Z. W. Fu, X. L. Yue, Q. Z. Qin, *J. Electrochem. Soc.*, **151**, E162–E167 (2004).
83. S. Grugeon, S. Laruelle, L. Dupont, F. Chevallier, P. L. Taberna, P. Simon, L. Gireaud, S. Lascaud, E. Vidal, B. Yrieix, J. M. Tarascon, *Chem. Mater.*, **17**, 5041–5047 (2005).
84. H. Li, P. Balaya, J. Maier, *J. Electrochem. Soc.*, **151**, A1878–A1885 (2004).
85. J. Hu, H. Li, X. J. Huang, L. Q. Chen, *Solid State Ionics*, **177**, 2791 (2006).
86. J. Hu, H. Li, X. J. Huang, *Electrochem. Solid-State Lett.*, **8**, A66 (2005).
87. J. Jamnik, J. Maier, *Phys. Chem. Chem. Phys.*, **5**, 5215–5220 (2003).
88. G. Gachot, S. Grugeon, M. Armand, S. Pilard, P. Guenot, J-M. Tarascon, S. Laruelle. *J. Power Sources*, **178**, 409 (2008).
89. L. W. Zhao, I. Watanabe, T. Doi, S. Okada, J. I. Yamaki. *J. Power Sources*, **161**, 1275 (2006).
90. Y. Zeng, L. F. Li, H. Li, X. J. Huang, L. Q. Chen, *Ionics*, **15**, 91 (2009).
91. J. Hu, H. Li, X. J. Huang, L. Q. Chen, *Solid State Ionics*, **177**, 2791 (2006).
92. J. P. Sun, K. Tang, X. Q. Yu, J. Hu, H. Li, X. J. Huang, *Solid State Ionics*, **179**, 2390 (2008).
93. X. Q. Yu, Y. He, J. P. Sun, K. Tang, H. Li, L. Q. Chen, X. J. Huang, *Electrochem. Commun.*, **11**, 791 (2009)
94. K. F. Zhong, X. Xia, B. Zhang, H. Li, Z. X. Wang, L. Q. Chen, *J. Power Sources*, **195**, 3300 (2010).
95. Y. Oumellal, A. Rougier, G. A. Nazri, J. M. Tarascon, L. Aymard, *Nat. Mater.*, **7**, 916 (2008).

Chapter 6

Lithium Reaction with Metal Nanofilms

Rachid Yazami
School of Materials Science and Engineering, Nanyang Technological University, 50 Nanyang Avenue, N. 4-1 Singapore 639798
rachid@ntu.edu.sg

Nanofilms of metals alloying and not alloying with lithium were prepared by physical vapor deposition and used as active electrode materials for electrochemical lithiation/delithiation at ambient temperatures. A significant lithium storage capacity was achieved even with metals that were known to not alloy with lithium, which we ascribed to lithium being stored in the vicinity of metal–oxygen bonds formed on the nanoparticles surfaces and in the grain boundaries. In metals forming alloys with lithium, a much larger capacity was achieved owing to successive alloys formation. Galvanostatic charge and discharge profiles showed successive voltage plateaus matching with current peaks in the corresponding cyclic voltammograms. The expected crystal lattice expansion upon the metal alloying with lithium, which causes capacity decay upon

Nanomaterials for Lithium-Ion Batteries: Fundamentals and Applications
Edited by Rachid Yazami

ISBN 978-981-4316-40-8 (Hardcover), 978-981-4364-23-2 (eBook)
www.panstanford.com

premature cycling was mitigated by the nanosize character of the metal film domains, preventing them from decrepitating.

Our proposed mechanism of lithium reversible storage in both alloying and non-alloying metal nanofilms on surface metal–oxygen bonds and in grain boundaries should be sustained by further evidences by local probe methods and first principles of computational approaches.

6.1 Introduction

Lithium-ion batteries (LIBs) have become the most popular autonomous electric power source for portable electronics, smart grid energy storage, and electric vehicles applications, owing to superior energy storage performances.[1] In fact LIB chemistry is rich and versatile; therefore it offers great opportunities for performance improvements in terms of energy density (Wh/kg and Wh/l), power density (W/kg and W/l), cycle life, thermal stability, safety, and cost as compared with other secondary battery chemistries. Tremendous basic research and R&D activity has been continuously pursued during the last two decades, which has resulted, for example, in more than doubling the energy density since LIBs were commercialized in 1991, owing to improved electrode and electrolyte materials formulation and to electrode and cell fabrication optimized engineering. Current research efforts in LIB are mostly directed toward discoveries of new anode, cathode, and electrolyte materials through both theoretical[2–8] and experimental approaches and on electrode size reduction to the nanometer scale, which is expected to enhance the lithium (Li) storage capacity and the charge and discharge rate.[9–63]

In fact, owing to higher surface/volume ratio, nanostructured electrode materials provide additional sites for Li storage compared with micron-sized particles of similar chemical composition and crystal structure. Moreover, the Li diffusion length is many orders of magnitude shorter in nanomaterials than in micron-sized ones, which translates into an even shorter time to fill and empty the host material available sites with Li, time t being proportional to the diffusion length l squared ($t \sim \frac{l^2}{D}$, D = Li chemical diffusion coefficient). This nanostructure feature in principle allows batteries to be charged and discharged at much faster rates, thereby, improving the battery power density. Nanosize also allows for the confinement of the electrode crystal structure defects and reduces internal stress induced by electrode material volume change upon lithiation/delithiation. Accordingly, cycle life should be enhanced in nanostructured materials.

On the drawback side, a high surface/volume ratio in nanostructured materials can affect the electrode stability toward the electrolyte, which is in general thermodynamically unstable at both high-voltage cathodes and low-voltage anodes. Such an electrode/electrolyte-promoted reactivity is detrimental to self-discharge rates, thermal stability, and cycle life. For each material, a trade-off regarding particle size should be achieved so that the advantages, such as higher energy and power density and extended cycle life, of using nanomaterials should not be canceled by drawbacks, such as enhanced self-discharge and thermal instability.

In this chapter, the process of reaction of metals in their nanofilm form factor with Li in half-cells at ambient temperatures is shown. Two types of metals are considered in this study—those that do not form alloys with Li, such as Cr, Mn, Co, Cu, and Fe, and those that form alloys with Li, such as Au, In, Sb, Ag, and Zn,—in an attempt to understand the effect of size on Li storage mechanism. In fact, the author and his coworkers were intrigued on finding that metals not alloying with Li show significantly high Li storage capability and reasonable cycle life when used as nanosized films. They also found that metals with Li alloying capability have excess storage capacity compared with the theoretical one. These results will be discussed based on surface chemistry and on defects confinement.

6.2 Experimental Procedure

Metal nanofilm electrodes were prepared by physical evaporation directly on the internal surfaces of a positive can of 2016 stainless coin cells. Weighted metal chunks and cans were preliminarily rinsed with ethanol and dried in a vacuum oven at 150°C for a couple of hours. Then they were rapidly transferred into the evaporation chamber. Metal chunks were loaded into a tungsten basket and the cans were put into a multicell holder beneath the tungsten basket. The cans were then covered with a polyethylene film with circular holes of 16 mm diameter to allow for circular nanofilms to be deposited on the internal faces of the cans. After evacuating the evaporation chamber under secondary vacuum of about 3×10^{-6} torr, the tungsten basket was resistively heated at temperature slightly above the metal sublimation point. The evaporation of the metal was achieved only in a few seconds. Cans were then weighted with the same high-precision balance (~5 μg accuracy) before evaporation to determine the film mass. Nanofilm thickness was computed from the metal density assuming no pores were formed. Cans with deposited metals were then dried at 120°C in vacuum for several hours and transferred into a glove box filled with argon for half-cells assembly. The latter consisted of metallic Li disc of 0.75 mm thickness, a Celgard microporous separator, and 1 M $LiPF_6$ in EC(1):DMC(1) electrolyte. A stainless steel spacer was use on the Li side to ensure good half-cell internal pressure once crimped. Galvanostatic regime and cyclic voltammetry were used for electrochemical tests at ambient temperatures.

6.3 Results

6.3.1 Metals Not Alloying with Lithium

(a) Chromium

A typical galvanostatic charge and discharge profile of a Cr-based nanofilm electrode (58 nm thick, 100 μg in mass) under 120 μA (12 C-rate, i.e., full charge and discharge cycle in 12 min) is shown in Fig. 6.1. An unusual increase in voltage yielding a peak can be seen only during the first discharge. This behavior is reproducible over many tested cells. As shown in Fig. 6.2, the first discharge capacity is about 285 mAh/g, which corresponds to $Li_{0.55}Cr$ theoretical formulae. The charge and discharge capacity is then stabilized around 100 mAh/g, corresponding to ~$Li_{0.2}Cr$ end of discharge composition. It is important to note that the lithiation/delithiation cycle efficiency is close to 100% after 30 cycles. Cyclic voltammogram (CV) achieved at 1 mV/s rate is displayed in Fig. 6.3. No oxidation and reduction current peaks are found in the CV profile, which is in agreement with the absence of plateaus in the galvanostatic cycle profiles of Fig. 6.2. This indicates that no alloying has taken place between Li and the Cr nanofilm. Moreover, the Cr-based electrode voltage was maintained positive relative to metallic Li. Accordingly no under-potential Li metal deposition is expected to take place on the Cr nanofilm surface. A similar behavior was observed with all other non-Li alloying metals in this study, that is, no voltage plateaus were observed and no Li-metal plating occurred. Therefore, in the next sections only galvanostatic charge/discharge cycling will be discussed, and no CV will be shown.

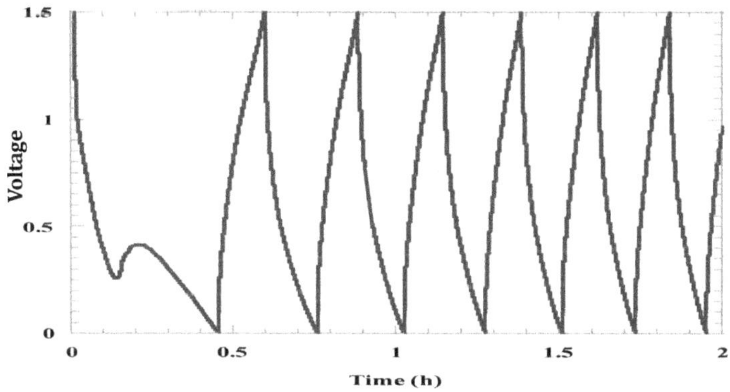

Figure 6.1 Galvanostatic cycle profile of a Cr-based nanofilm.

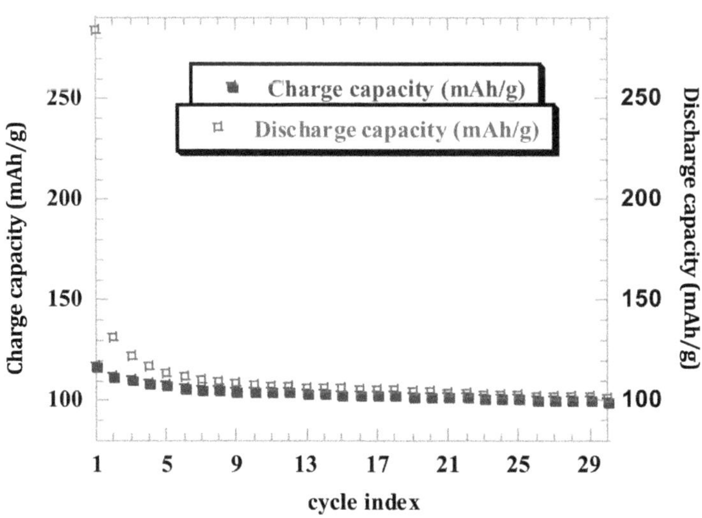

Figure 6.2 Cycle capacity of the Cr-based nanofilm of Fig. 6.1.

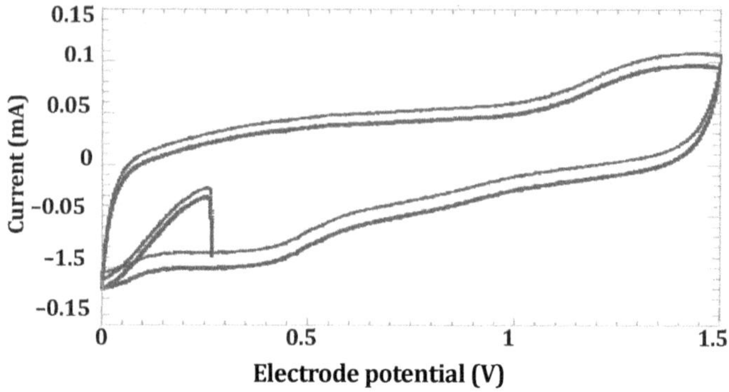

Figure 6.3 Cyclic voltammogram of a Cr-based nanofilm.

(b) Manganese

An Mn-based film of 96 nm thickness and 169 mg mass was used. Figure 6.4 shows the galvanostatic cycle profile under 5 μA current; equivalent to ~C/3 rate. A short plateau can be seen during the first discharge only. The cycle capacity and equivalent Li cycle composition x in Li_xMn are displayed in Fig. 6.5a and 6.5b, respectively. The initial discharge capacity is about 315 mAh/g corresponding to $Li_{0.65}Mn$ and the first charge is only 85 mAh/g, then the cycle capacity stabilizes around 85 mAh/g (or $Li_{0.17}Mn$) and shows high columbic efficiency. In the Mn nanofilm, most of the Li storage capacity occurs between 0.85 V and 5 mV, whereas delithiation covers all the voltage range up to 1.5 V.

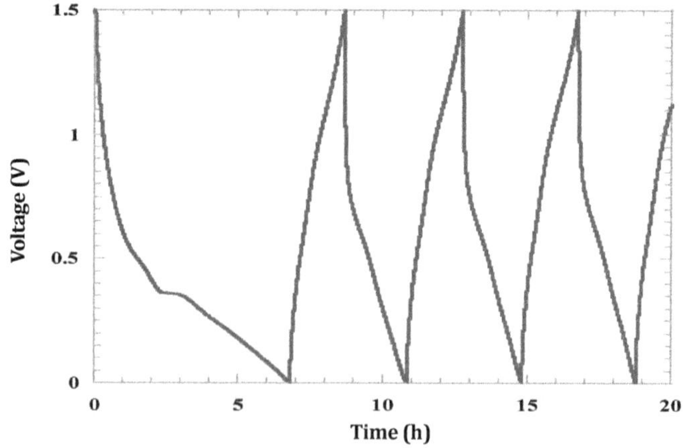

Figure 6.4 Galvanostatic cycle profile of a Mn-based nanofilm.

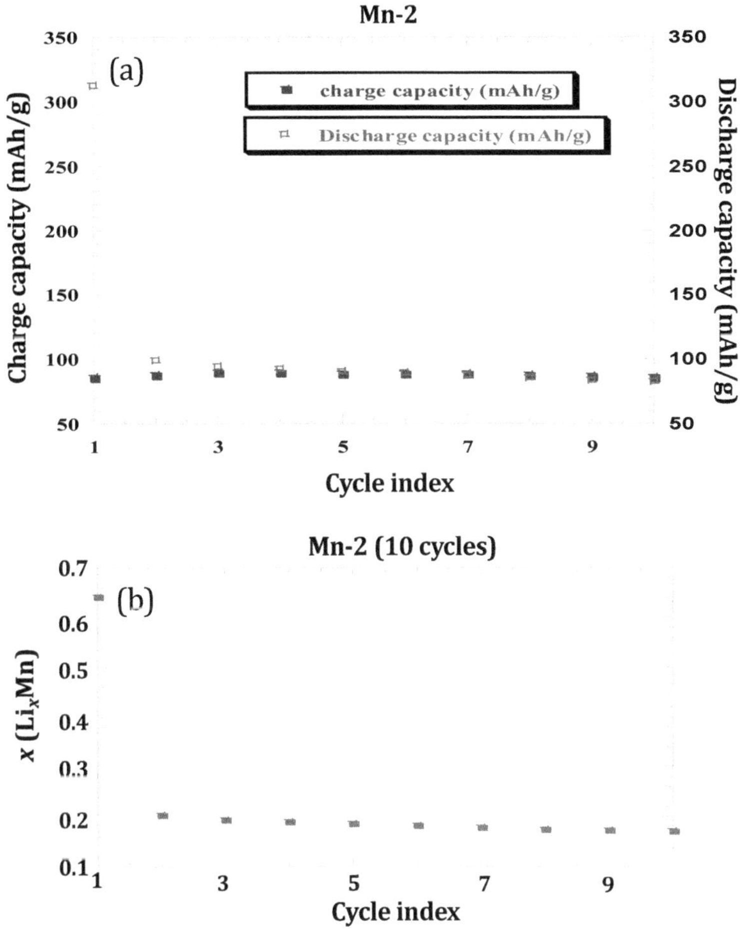

Figure 6.5 Cycle capacity (a) mAh/g and (b) x in Li_xMn of the Mn-based nanofilm of Fig. 6.4.

(c) Cobalt

The Co nanofilm used here is 53 nm thick and 113 μg in mass. The charge and discharge profile under 2 μA (C/5 rate) is displayed in Fig. 6.6, and the associated cycle capacity in mAh/g and in Li/Co composition at the end of discharge are presented in Fig. 6.7a and 6.7b, respectively. The initial capacity is about 400 mAh/g (or $Li_{0.87}Co$), which dropped monotonously to 85 mAh/g (or ~$Li_{0.17}Co$) after 30 cycles. The cycle efficiency, however, is lower than in the Cr and the Mn cases, reaching 85% after 30 cycles.

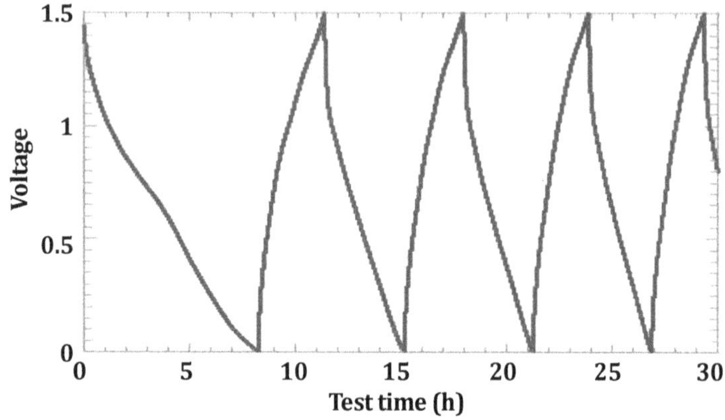

Figure 6.6 Galvanostatic cycle profile of a Co-based nanofilm.

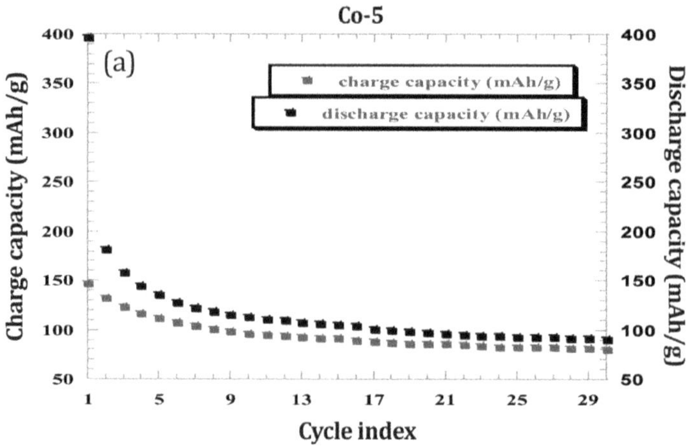

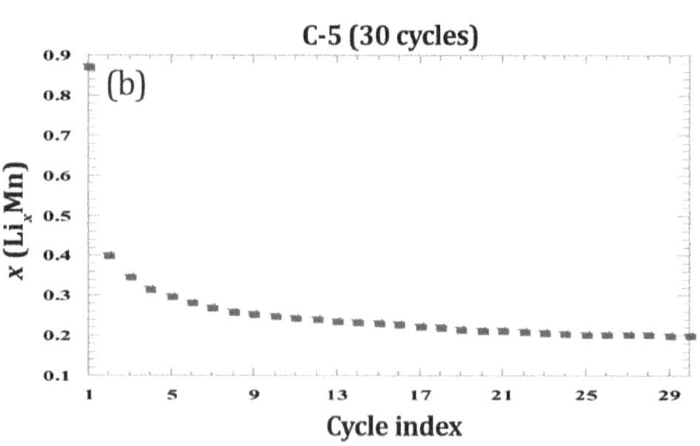

Figure 6.7 Cycle capacity (a) mAh/g and (b) x in Li_xMn of the Co-based nanofilm of Fig. 6.6.

(d) Copper

Figure 6.8 shows the charge and discharge profile of a copper-based nanofilm electrode of 30 nm thickness and 65 μg mass under 3 μA current intensity (~1.5 C-rate). Cycle capacity expressed in mAh/g and in Li/Cu theoretical ratio is depicted in Figs. 6.9a and 6.9b, respectively. Unlike in Cr, Mn, and Co metals case, surprisingly the initial discharge capacity of the copper nanofilm is rather low (39 mAh/g), which increases to 75 mAh/g in the second cycle and then stabilizes around 70 mAh/g ($Li_{0.17}Cu$) after 8 cycles, with a cycle efficiency of 85%.

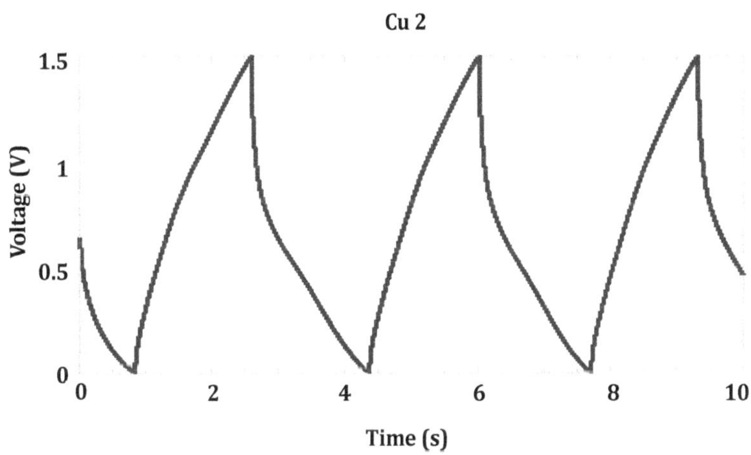

Figure 6.8 Galvanostatic cycle profile of a Cu-based nanofilm.

(e) Cycle profile comparison

In order to draw a graphic comparison between different transition metals during lithiation/delithiation cycle, metal nanofilms raging between 50 and 60 nm in thickness were prepared and were cycled at similar rates ranging between C/4 and C/3. Figure 6.10 shows the charge and discharge profiles during the second cycle of Cr, Mn, Fe, Co, and Cu, and Fig. 6.11 shows their respective cycle capacity up to 30 cycles. Mn and Cr yield the highest cycle capacity, which stabilizes around 150 mAh/g at the 30th cycle followed by Co, Fe, and Cu in decreasing capacity order.

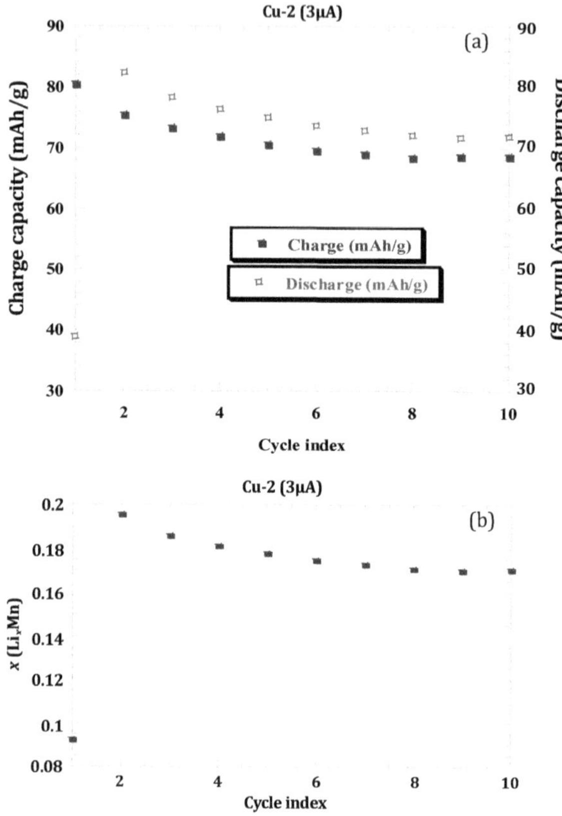

Figure 6.9 Cycle capacity (a) mAh/g and (b) x in Li_xCu of the Cu-based nanofilm of Fig. 6.8.

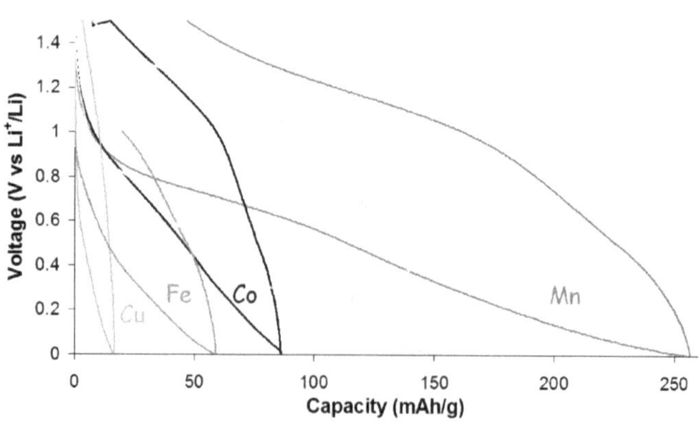

Figure 6.10 Galvanostatic cycle profile of a Cu, Fe, Co, Cr, and Mn-based nanofilms.

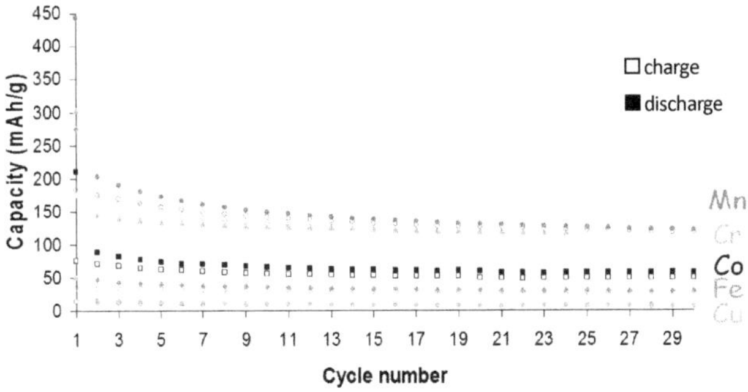

Figure 6.11 Cycle capacity of the Cu, Fe, Co, Cr, and Mn-based nanofilms of Fig. 6.10.

6.3.2 Metals with Lithium-Alloying Capability

(a) Gold

The phase diagram of the Au–Li binary system is shown in Fig. 6.12. Alloys of Au_5Li_4, AuLi, $AuLi_3$, and Au_4Li_{15} compositions are expected to form.

Figure 6.13a displays the voltage profile of an Au-based nanofilm electrode of 27 nm thickness and 124 µg mass under 5 mA (C/11 rate). During the first lithiation/delithiation cycle, three flat voltage plateaus are observed denoting alloying. On the following cycles only two flat plateaus remained together and a shoulder at high voltage during delithiation, which almost vanished after four cycles. Figure 6.13b magnifies the first and second cycle using Li composition x in Li_xAu as the x-axis. During the first lithiation, about 5 Li/Au atom were introduced in four steps. In the first step, a pseudo-plateau of about 0.5 Li/Au appeared, followed by a first plateau of 1 Li/Au range ($0.5 < x < 1.5$), a second plateau of 1.5 Li/Au range ($1.5 < x < 3$), and a third plateau of 2 Li/Au range ($3 < x < 5$). During delithiation, a first plateau of about 1.5 Li/Au range ($5 > x > 3.5$) appeared followed by a second plateau of 1 Li/Au range ($3.5 > x > 2.5$), a third plateau of ~Li/Au range ($2.5 > x > 1.5$), and a pseudo-plateau of about 0.8 Li/Au range ($1.5 > x > 0.7$). In the following cycle in Fig. 6.13b, a pseudo-plateau of about 0.25 Li/Au followed by two plateaus can be seen during discharge ranging 1.25 Li/Au and 1.75 Li/Au, respectively. During the following charging, a first plateau of about 1.35 Li/Au range followed by a short semi-plateau (in the 0.3–0.4 V area) of about 0.25 Li/Au length, a plateau of about 0.7 Li/Au and a less well-defined semi-plateau of 0.6 Li/Au total length.

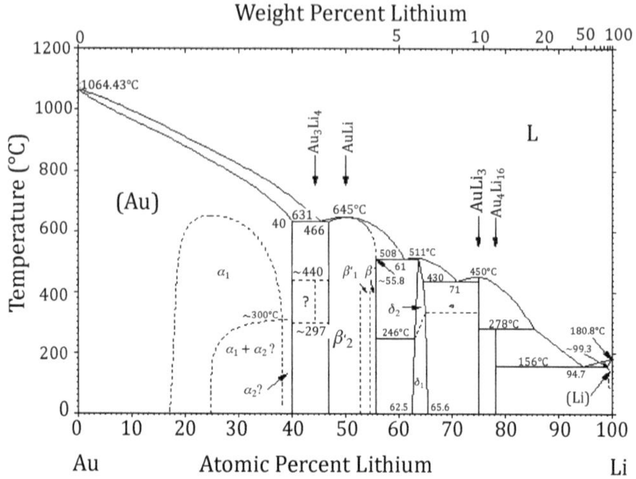

Figure 6.12 Phase diagram of the Au–Li binary system.

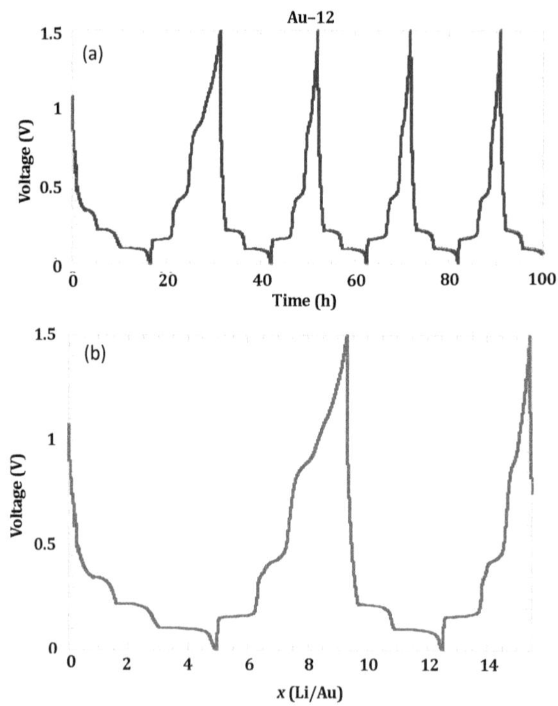

Figure 6.13 (a) Galvanostatic cycle profile of an Au-based nanofilm; (b) Charge and discharge profiles versus x during the first two cycles of an Au-nanofilm.

The cycle capacity in mAh/g and in Li/Au atomic ratio composition is displayed in Figs. 6.14a and 6.14b, respectively. It shows a drop in capacity after 5 cycles from 300 mAh/g ($\Delta x = 3$) to about 30 mAh/g ($\Delta x = 0.3$) after the 9th cycle.

Figure 6.15 shows the X-ray diffraction patterns (using the Cu *Kα* radiation) of the stainless steel holder before (Fig. 6.15a) and after (Fig. 6.15b) the Au nanofilm deposition and after the first lithiation (Fig. 6.15c). The peaks circled in Fig. 6.15b are attributed to the disappearance of Au metal after lithiation and no appearance of new peaks as a result.

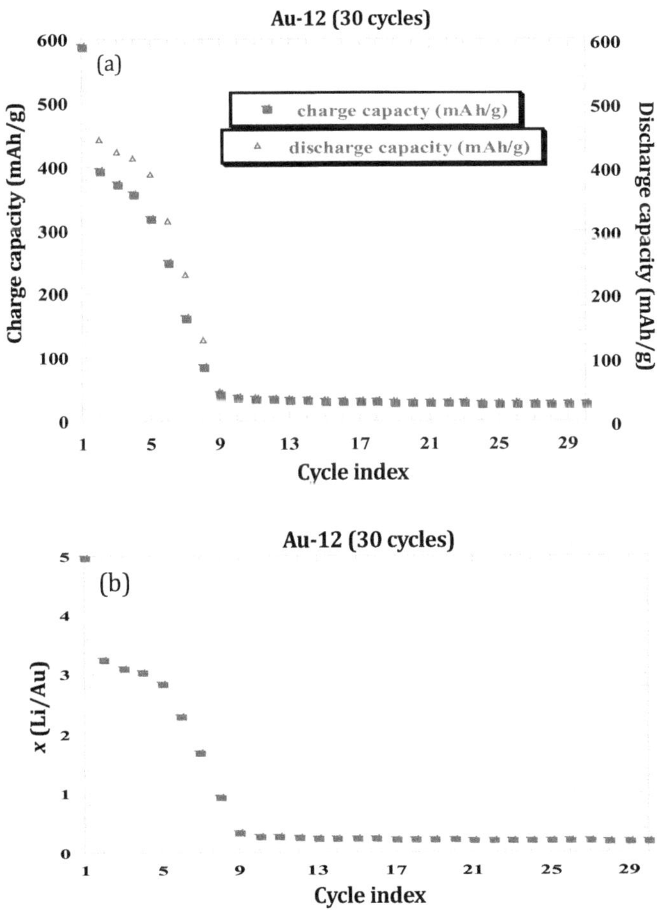

Figure 6.14 Cycle capacity (a) mAh/g and (b) x in Li_xAu of the Au-based nanofilm of Fig. 6.13.

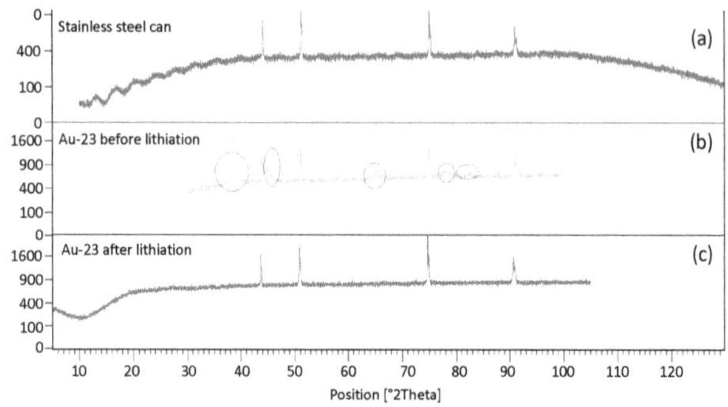

Figure 6.15 X-ray diffractograms of the stainless steel holder (a), the Au nanofilm before (b) and after (c) lithiation.

CVs achieved under powers of 10 decreasing sweeping rates of 10 mV/s, 1 mV/s, 0.1 mV/s, and 0.01 mV/s are presented in Figs. 6.16a–d, respectively. As the sweeping rate decreases, more and more well-defined lithiation/delithiation peaks appear with maximum four anodic (delithiation) peaks and two cathodic peaks. The profile of current peaks and their number highlights the kinetics effect on the alloying process, which also includes the galvanostatic cycling rate and the nanofilm thickness.

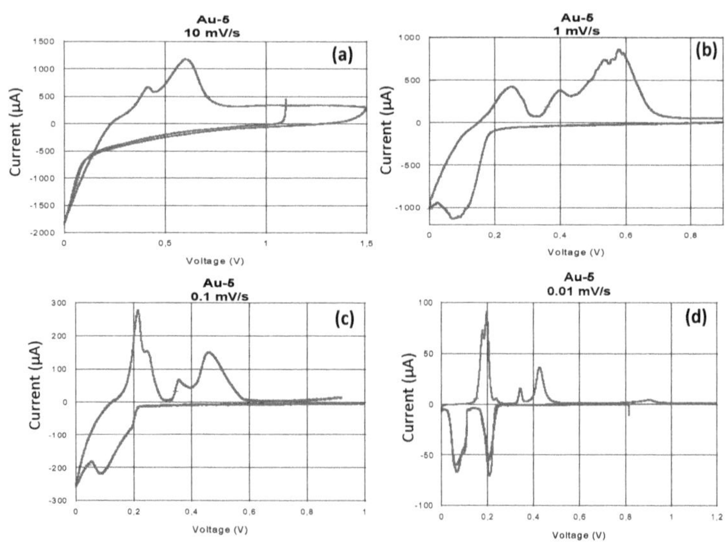

Figure 6.16 CVs of an Au-based nanofilm at different voltage sweeping rates: (a) 10 mV/s, (b) 1 mV/s, (c) 0.1 mV/s, and (d) 0.01 mV/s.

Figure 6.17 highlights the correlation existing between the CV peaks and the galvanostatic charge plateaus.

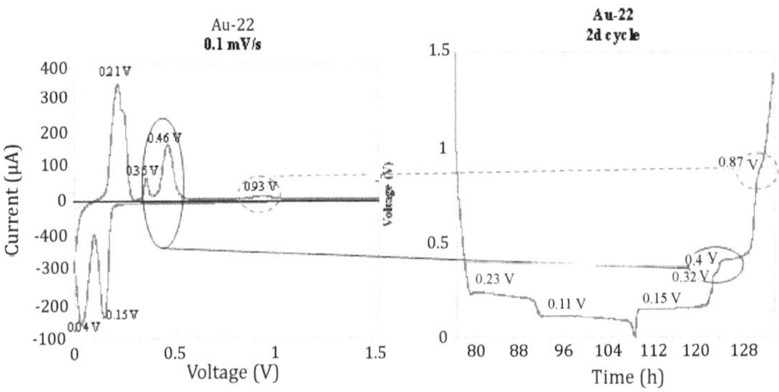

Figure 6.17 CV and associated charge and discharge profile of an Au-based nanofilm.

(b) Indium

The phase diagram of the In–Li binary system is presented in Fig. 6.18. Four alloys form with the following compositions: In_4Li_5, In_2Li_3, $InLi_2$, and In_3Li_{13}.

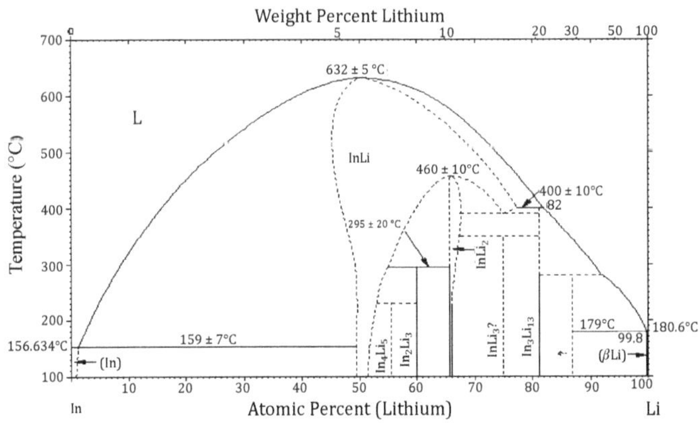

Figure 6.18 Phase diagram of the In–Li binary system.

Figure 6.19 shows the galvanostatic charge and discharge profile achieved on an In nanofilm of 7 nm thickness and 12 μg mass under 12 μA (1.5 C-rate). Figures 6.20a and 6.20b show the cycle capacity in mAh/g and in Li composition, respectively. The initial discharge capacity is a high 1200 mAh/g corresponding to 5.28 Li per In atom.

Then the discharge capacity decreased to 1067 mAh/g (i.e., 4.65 Li/In) after 14 cycles. The CV profile at 0.1 mV/s rate in Fig. 6.20c shows three lithiation peaks at 0.42 V, 0.21 V, and 0.06 V, respectively. It also shows two delithiation peaks at 0.3 V and 0.48 V, respectively.

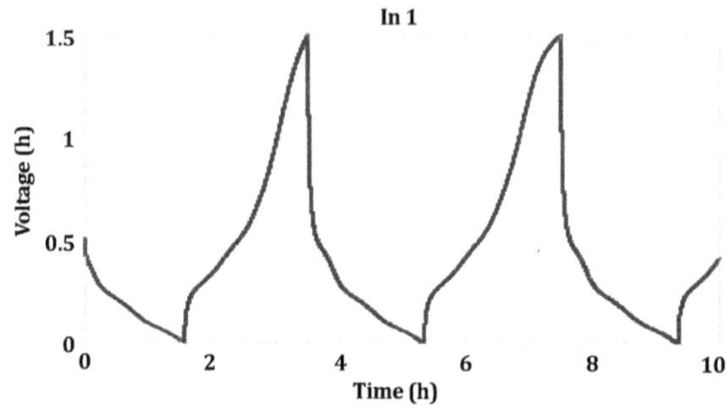

Figure 6.19 Galvanostatic cycle profile of an In-based nanofilm electrode.

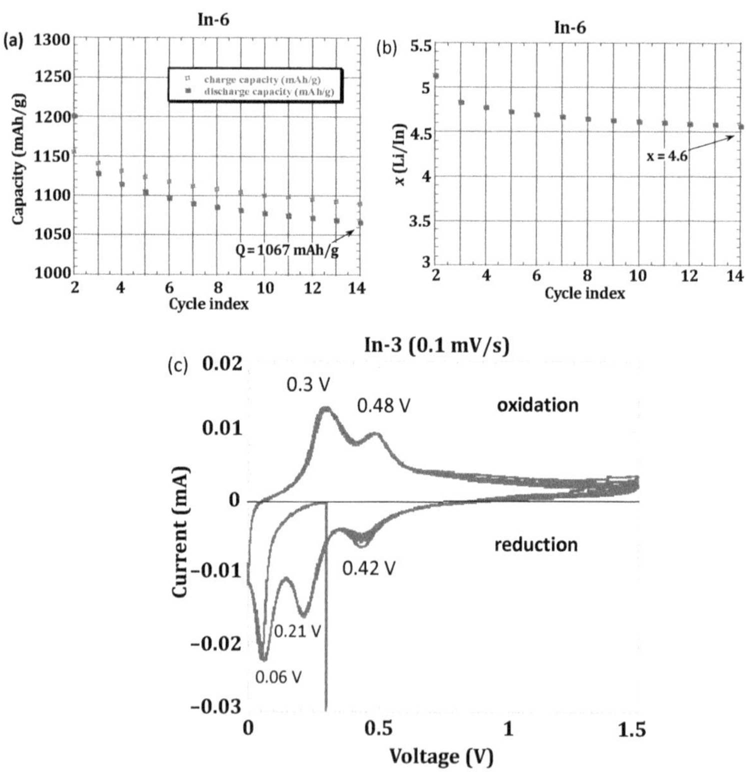

Figure 6.20 Cycle capacity (a) mAh/g, (b) x in Li_xIn of the Au-based nanofilm of Fig. 6.19, and (c) cyclic voltammogram of Li/In cell at 0.1 mV/s voltage sweeping rate.

(c) Antimony and its alloys

Figure 6.21 shows the charge and discharge profiles achieved with Sb, Sb–Cu, Sb–Au, and Sb–Co nanofilms. While Sb alone shows a single plateau, two plateaus are present in the Sb–Cu system and three in the Sb–Co and the Sb–Au systems. The additional plateau(s) in the Sb alloys, which appear at lower voltages than the one of Sb alone, come from contribution of the transition metal in agreement with results reported in the previous section.

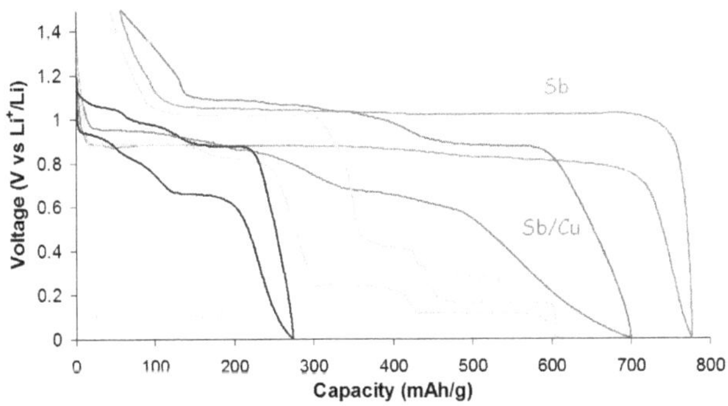

Figure 6.21 Galvanostatic cycle profile of tin and some metal–tin alloy-based nanofilms.

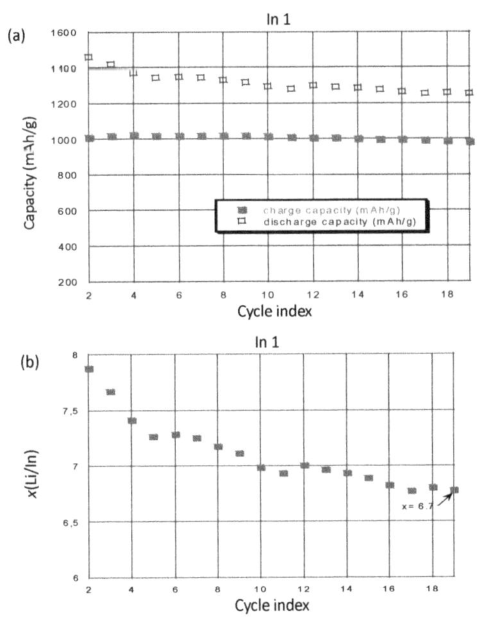

Figure 6.22 Cycle capacity (a) mAh/g and (b) x in Li_xSb of the Au-based nanofilm of Fig. 6.21.

(d) Silver and zinc

As shown in the Li–Ag phase diagram in Fig. 6.23, Li and silver are very miscible and form a series of alloys including the following increasing Li content order: LiAg, Li_9Ag_4, $Li_{10}Ag_3$, and $Li_{12}Ag$. Similarly Li forms many alloys with zinc including $LiZn_4$, Li_2Zn_5, $LiZn_2$, and Li_2Zn_3 as shown in their phase diagram depicted in Fig. 6.24.

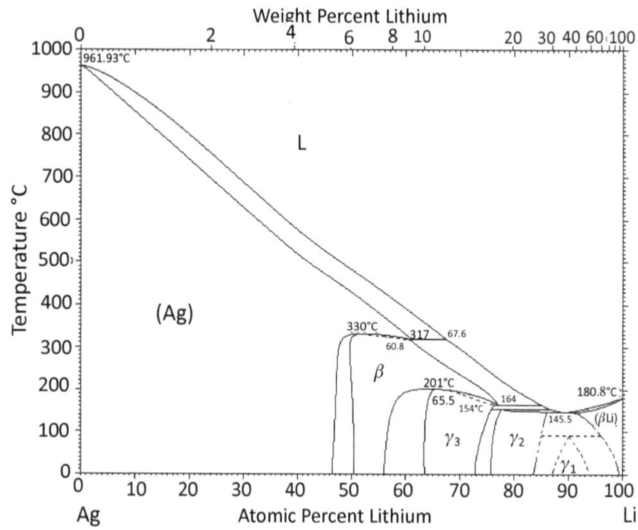

Figure 6.23 Phase diagram of the Ag–Li binary system.

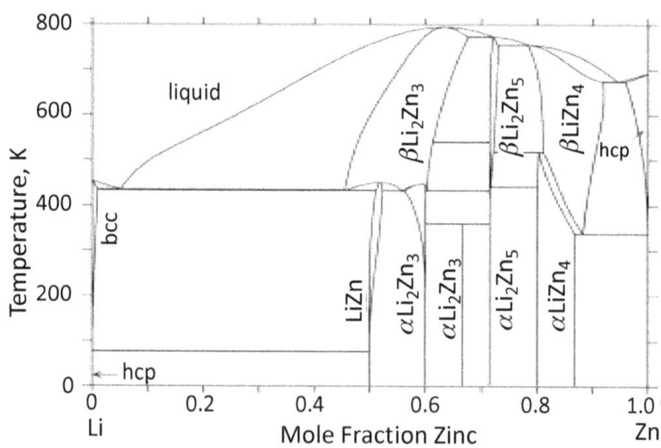

Figure 6.24 Phase diagram of the Zn–Li binary system.

A comparative cycle capacity of Ag, Zn, and Au nanofilm electrodes is displayed in Fig. 6.25. The initial discharge capacity decreases in the order Ag > Zn > Au. However, upon cycling, Ag's capacity deceased sharply followed by that of Zn. Au sustained a rather high capacity of 320 mAh/g after 16 cycles but the capacity dropped to a low of 10 mAh/g after 26 cycles.

(e) Blank test

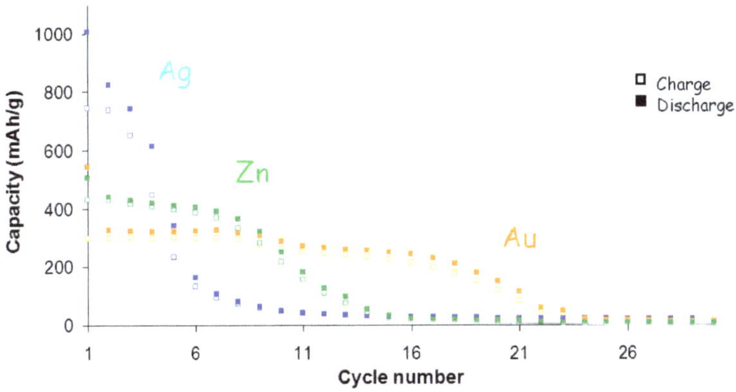

Figure 6.25 Cycle capacity of Ag, Zn, and Au nanofilm electrodes.

In order to evaluate the contribution to the capacity that may come from the stainless steel can holder, a blank test was carried out without any deposited nanofilm. The achieved cycle capacity is below 5 μAh, which is negligible compared with the actual nanofilms anodes. Accordingly the results shown is this study come mainly from the metal nanofilms capacity.

Discussion

In this study we have disclosed different electrochemical behaviors of metal nanofilms depending on the metal nature, thickness, charge and discharge rates and alloying or non-alloying capability with Li at the ambient temperatures.

Our results raise two main questions:

(1) What is the mechanism of reversible Li storage in metals not alloying with Li?

(2) What is the origin of excess capacity during the first cycle discharge in metals forming alloys with Li compared with their respective phase diagram?

Nanostructured materials are known to provide higher surface to bulk ratios compared with micron-sized particle materials and larger concentration of grain boundaries. Surface atoms and grain boundaries offer energetically and kinetically favorable sites for Li storage.[64–67] Moreover, during physical evaporation, some metal–oxygen physical or chemical bonds may form due to non-zero oxygen partial pressure in the evaporator under secondary vacuum.[68–73] It is therefore our argument that in both Li alloying and non-alloying metal nanofilms, Li can be stored at the vicinity of surface atoms, grain boundaries, and metal–oxygen bonds. However, local characterization probes that have such high-resolution transmission electron microscopy in combination with elemental analyses tools would be useful to support our assumptions. Figure 6.26 illustrates the grains boundary structure, which offers additional capacity and accommodates internal strains.

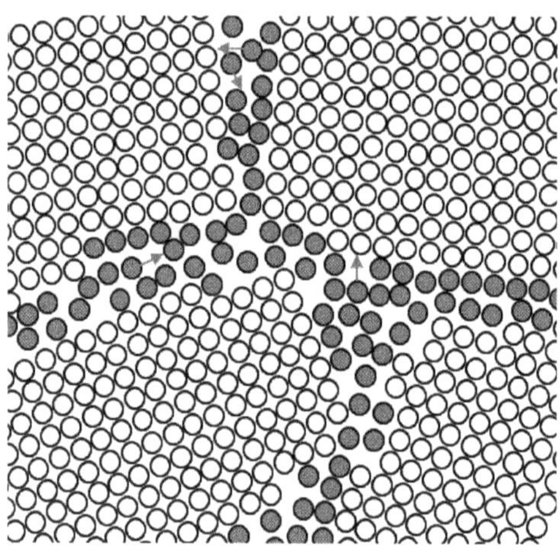

Figure 6.26 Schematics of grains boundary structure in metal nanofilms.

In fact metal–oxygen bonds play a double role in the Li electrochemical storage processes: (1) they contribute to the irreversible capacity losses during the first lithiation/delithiation cycle due to the formation of Li oxides or peroxide and (2) they stabilize Li–M system especially for those metals that do not alloy with Li.

In another set of experiments not shown here, we found that the Li storage performances of metal nanofilms strongly depend on the nanofilm thickness; thicker films yielding lower storage capability in mAh/g. This highlights kinetic limitations due to Li diffusion throughout the film thickness.

It is also important to note that the nanofilm electrode voltage remained positive relative to metallic Li, which excludes any possible Li metal plating and stripping on/from the nanofilm surface.

The lack of voltage plateaus in the charge and discharge profiles of non-Li alloying metals confirms that no alloy is formed in the nanofilms that are in agreement with the Li–M binary system thermodynamics. Therefore cycle capacity in theses metals come from the above mentioned mechanisms of surface chemistry, including oxygen bound to the metal and from the grain boundaries.

The voltage peak that appears during the first discharge of the Cr nanofilm shown in Fig. 6.1a should be due to excess oxygen on the electrode surface, which is irreversibly reduced to Li_2O and Li_2O_2 during the first lithiation, thus enhancing the electrode electrical conductivity. Once the surface oxygen is reduced, Li is reversibly stored on the metal surface atoms and on the grain boundaries, which stabilize the cycle capacity to ~85 mAh/g after 30 cycles as shown in Fig. 6.2a. A similar, although less-pronounced, phenomenon is observed in the Mn-based nanoelectrode as shown in Fig. 6.1. A short plateau forms during the first discharge, which probably comes from the surface oxygen reduction. More generally other mechanisms than the metal–oxygen reduction may account for the irreversible capacity loss during the first cycle. Li may be "trapped" in the electrode nanopores and/or in high-energy sites in the grain boundaries.[74–77] It is also interesting that nanofilms based on Cr, Mn, Co, and Cu yield reversible capacity of the same order of magnitude in the 70–85 mAh/g range. This suggests that the surface chemistry and the grain boundaries are of similar structure after the metal–oxygen bond has been reduced during the first cycles.

In Li-alloying metals, voltage plateaus are observed during lithiation/delithiation, which goes along with phase transitions predicted in the Li–M phase diagram. Table 6.1 shows theoretical and experimental capacity according to successive phase transition in Li-alloying metals together with difference in total capacity between the experimental, achieved at the end of the first discharge, and the calculated one.

Table 6.1 Metal–Li alloys and corresponding theoretical and achieved capacity during phase-to-phase transitions

Metal	Alloys composition	Capacity (mAh/g) Theoretical	Experimental (first discharge)	difference
Zn	$LiZn_4$	102.5		
	Li_2Zn_5	61		
	$LiZn_2$	40.9		
	LiZn	205		
	(Total)	(410)	(500)	90
Ag	LiAg	248		
	Li_9Ag_4	310		
	$Li_{10}Ag_3$	269		
	$Li_{12}Ag$	2151		
	(Total)	(2981)	(1000)	-1981
In	Li_5In_4	292		
	Li_3In_2	58		
	Li_3In	642.5		
	$Li_{13}In_3$	369.3		
	(Total)	(1011.8)	(1200)	188
Au	Li_2Au_5	59.4	68	
	LiAu	76.7	136	
	Li_3Au	331	204	
	$Li_{15}Au_4$	179	272	
	(Total)	(510)	(680)	170

It can be seen that the capacity difference ranges between 90 and 188 mAh/g for all metals except for Ag, where it is as high as 1981 mAh/g This excess capacity is in the same order of magnitude as in Li non-alloying metals, which supports a similar Li storage mechanism in metals in Table 6.1, that is, in grain boundaries, surface atoms, and metal–oxygen bonds in addition to Li alloying.

6.4 Conclusion

Nanofilms of Li alloying and non-alloying metals have been prepared and used in Li half-cells to study lithiation/delithiation reactions processes at ambient temperatures. We found a significantly high Li storage capacity during the first cycles in metals known not to alloy with Li,[78] which we ascribed to Li being stored on the surface of the metal nanodomains, including grain boundaries, surface atoms, and surface metal–oxygen bonds. The same surface chemistry mechanism applies to metals forming alloys with Li in excess to the capacity arising from alloy formation as shown in Table 6.1. In the other metals, voltage plateaus and current peaks appeared in the galvanostatic and CV profiles, respectively, as evidence of alloying. This contrasts with non-alloying metals for which no voltage plateaus and current peaks were observed and accordingly yielded lower storage capacity.

The Li storage in metal nanofilms is controlled by the film thickness and by the charge and discharge current rates in the galvanostatic mode and by the voltage-sweeping rate when CV is used. Therefore, for an accurate comparison of the kinetics and thermodynamics of Li storage in metal nanofilms, these parameters should be taken into account.

Stable and reasonably high cycle capacity is achieved in metal nanofilms owing to lower crystal structure stresses upon unit cell volume expansion induced by Li alloying and/or stabilization of the Li–metal and Li–oxygen bonds formed on the nanodomain surfaces, favored by a high surface-to-volume ratio in metal nanofilms.

The irreversible capacity during the first cycles may be associated with the reduction of the metal–oxygen bonds but also to the growth of a solid electrolyte interphase (SEI) due to low operation voltages on the metal nanofilms.

Further investigations are needed to better understand the mechanism(s) of Li storage in nanostructured metals, including theoretical quantum chemistry first principles, methods, and local probes tools such as electron transmission microscopy, X-ray photoelectron spectroscopy, and electron energy loss spectrometry.

Acknowledgments

The work presented in this chapter was performed at the Department of Materials Science, Division of Engineering and Applied Science of the California Institute of Technology, California. The author would like to thank Prof. Brent Fultz for allowing access to his facility and for fruitful discussions and the collaborators who contributed to the experimental part of this work: Dr. Anne Dailly, Miss Kenza Lahlou, and Miss Aurélie Serrès.

References

1. M. Armand, J-M. Tarascon, *Nature*, **45**, 652–657 (2008).
2. A. Jain, G. Hautier, C. J. Moore, S. P. Ong, C. C. Fischer, T. Mueller, K. A. Persson, G. Ceder, *Comp. Mater. Sci.*, **50**, 2295–2310 (2011).
3. S. P. Ong, V. L. Chevrier, G. Hautier, A. Jain, C. Moore, S. Kim, X. H. Ma, G. Ceder, *Energy & Environ. Sci.*, **4**, 3680–3688 (2011).
4. X. H. Zhu, N. Chen, F. Lian, Y. P. Song, Y. Li, *Chin. Sci. Bull.*, **56**, 3229–3232 (2011).
5. G. H. Zhong, Y. L. Li, P. Yan, Z. Liu, M. H. Xie, H. Q. Lin, *J. Phys. Chem. C*, **114**, 3693–3700 (2011).
6. C. Combelles, M-L., Doublet, *Ionics*, **14**, 279–283 (2008).
7. Y. S. Meng, Y. W. Wu, B. J. Hwang, Y. Li, G. Ceder, *J. Electrochem. Soc.*, **151**, A1134–A1140 (2004).
8. K. R. Kganyago, P. E. Ngoepe, Catlow Cra, *Phys. Rev. B*, **67**, 104103 (2003).
9. L. W. Ji, X. W. Zhang, *Energy & Environ. Sci.*, **3**, 124–129 (2010).
10. J. P. Sullivan, J. Huang, M. J. Shaw, A. Subramanian, N. Hudak, Y. Zhan, J. Lou, Proceedings SPIE—The International Soc. Optical Engineering, Conference on Energy Harvesting Storage: Materials, Devices, and Applications, FL, Orlando, April 05–06 (2010).
11. I. Bilecka, M. Niederberger, *Nanoscale*, **2**, 358–1374 (2010).
12. H. W. Liu, H. M. Yang, J. L. Li, *Electrochimica Acta*, **55**, 1626–1629 (2010).
13. G. Yang, H. M. Ji, H. D. Liu, K. F. Huo, J. J. Fu, P. K. Chu, *J. NanoSci. Nanotechnol.*, **10**, 980–986 (2010).
14. G. Neumann, A. Wursig, *Physica Status Solidi-Rapid Res. Lett.*, **4**, A21–A23 (2010).
15. B. L. Ellis, K. T. Lee, L. F. Nazar, *Chem. Mater.*, **22**, 691–714 (2010).
16. C. Liu, F. Li, L. P. Ma, H. M. Cheng, *Adv. Mat.*, **22**, E28 (2010).
17. J. Zhang, Eerdemutu, C. X. Yang, J. Y. Feng, X. W. Di, Z. L. Liu, G. Xu, *J. NanoSci. Nanotechnol.*, **10**, 2092–2098 (2010).
18. A. R. Kamali, D. J. Fray, *J. New Mat. Electr. Sys.*, 13147–13160 (2010).

19. A. Magasinski, P. Dixon, B. Hertzberg, A. Kvit, J. Ayala, G. Yushin, *Nature Mater.*, 9353–358 (2010).

20. S. P. Kuksenko, *Russ. J. Appl. Chem.*, **83**, 641–647 (2010).

21. Munoz-D Rojas, Casas-M Cabanas, E. Baudrin, *Solid State Ionics*, **181**, 536–544 (2010).

22. L. W. Ji, Z. Lin, M. Alcoutlabi, X. W. Zhang, *Energy & Environmental Sci.*, **4**, 2682–2699 (2011).

23. S. C. Zhang, Y. L. Xing, T. Jiang, Z. J. Du, F. Li, L. He, W. B. Liu, *J. Power Sources*, **196**, 6915–6919 (2011).

24. D. Kushnir, B. A. Sanden, *J. Clean. Prod.*, **19**, 1405–1416 (2011).

25. J. C. Guo, Y. H. Xu, C. S. Wang, *Nano Lett.*, **11**, 4288–4294 (2011).

26. R. Sanjines, M. D. Abad, C. Vaju, R. Smajda, M. Mionic, A. Magrez, *Surf. Coat. Technol.*, **206**, 727–733 (2011).

27. H. Q. Li, H. S. Zhou, *Chem. Commun.*, **48**, 1201–1217 (2011).

28. P. G. Bruce, S. A. Freunberger, L. J. Hardwick, J. M. Tarascon, *Nature Mater.*, **11**, 19–29 (2011).

29. K. Y. Xie, Z. G. Lu, H. T. Huang, W. Lu, Y. Q. Lai, J. Li, L. M. Zhou, Y. X. Liu, *J. Mater. Chem.*, **22**, 5560–5567 (2011).

30. C. Joyce, L. Trahey, S. A. Bauer, F. Dogan, J. T. Vaughey, *J. Electrochem. Soc.*, **159**, A909–A914 (2012).

31. L. Dimesso, C. Förster, W. Jaegermann, J. P. Khanderi, H. Tempel, A. Popp, J. Engstler, J. J. Schneider, A. Sarapulova, D. Mikhailova, L. A. Schmitt, S. Oswald, H. Ehrenberg, *Chem. Soc. Rev.*, **41**, 5068–5080 (2012).

32. W. X. Zhang, H. Zeng, Z. H. Yang, Q. Wang, *J. Solid State Chem.*, **186**, 58–63 (2012).

33. K. Guerin, M. Dubois, A. Houdayer, A. Hamwi, *J. Fluorine Chem.*, **134**, 11–17 (2012).

34. T. L. Kulova, A. M. Skundin, *Russ. J. Electrochem.*, **48**, 330–335 (2012).

35. L. C. Qian, Y. Xia, W. K. Zhang, H. Huang, Y. P. Gan, H. J. Zeng, X. Y. Tao, *Micropor. Mesopor. Mat.*, **152**, 128–133 (2012).

36. L. M. Dai, D. W. Chang, J. B. Baek, W. Lu, *Small*, **8**, 1130–1166 (2012).

37. W. S. Chang, C. M. Park, J. H. Kim, Y. U. Kim, G. Jeong, H. J. Sohn, *Energy & Environ. Sci.*, **5**, 6895–6899 (2012).

38. H. L. Wang, Y. Y. Liang, M. Gong, Y. G. Li, W. Chang, T. Mefd, J. G. Zhou, J. Wang, T. Regier, F. Wei, H. J. Dai, *Nature Commun.*, **3**, 917 (2012).

39. R. S. Kalubarme, C. J. Park, *Chem. Lett.*, **41**, 612–613 (2012).

40. X. L. Wang, H. Y. Chen, J. M. Bai, W. Q. Han, *J. Phys. Chem. Lett.*, **3**, 1488–1492 (2012).

41. M. Hagen, S. Dorfler, H. Althues, J. Tubke, M. J. Hoffmann, S. Kaskel, K. Pinkwart, *J. Power Sources*, **213**, 239–248 (2012).

42. F. Sauvage, V. Bodenez, J. M. Tarascon, K. R. Poeppelmeier, *J. Am. Chem. Soc.*, **132**, 6778–6782 (2010).

43. B. Delobel, D. Larcher, J. F. Blach, L. Croguennec, M. Menetrier, E. Simon, E. Baudrin, *Solid State Ionics*, **181**, 623–630 (2010).

44. R. Yang, J. Zheng, J. Huang, X. Z. Zhang, J. L. Qu, X. G. Li, *Electrochem. Commun.*, **12**, 784–787 (2010).

45. J. C. He, H. L. Zhao, M. W. Wang, X. D. Jia, *Mater. Sci. Eng. B-Solid.*, **171**, 35–39 (2010).

46. K. Amine, I. Belharouak, Z. H. Chen, T. Tran, H. Yumoto, N. Ota, S. T. Myung, Y. K. Sun, *Adv. Mater.*, **223**, 052–3057 (2010).

47. M. Au, T. Adams, *J. Mater. Res.*, **25**, 1649–1655 (2010).

48. D. W. Liu, G. Z. Cao, *Energy Environ. Sci.*, **3**, 1218–1237 (2010).

49. J. Y., Duan, B. W., Wei, R. J., Yang, J. Z., Pan, *Functional Mat. Lett.*, **3**, 193–195 (2010).

50. J. Cabana, L. Monconduit, D. Larcher, M. R. Palacin, *Adv. Mater.* **22**, E170–E192 (2010).

51. R. F. Li, S. Q. Wu, Y. Yang, Z. Z. Zhu, *J. Phys. Chem. C*, **114**, 16813–16817 (2010).

52. C. J. Cui, G. M. Wu, H. Y. Yang, S. F. She, J. Shen, B. Zhou, Z. H. Zhang, *Electrochimica Acta*, **55**, 8870–8875 (2010).

53. X. Y. Wang, Z. Y. Wen, Y. Liu, *Electrochimica Acta*, **56**, 1512–1517 (2011).

54. P. Balaya, K. Saravanan, S. Hariharan, V. Ramar, H. S. Lee, M. Kuezma, S. Devaraj, D. H. Nagaraju, Ananthanarayanan, K. C. W. Mason, Proceedings, SPIE, Conference on Energy Harvesting Storage—Applications II, Oslo, FL, April 25–28, 2011.

55. L. J. Zhang, P. Hu, X. Y. Zhao, R. L. Tian, R. Q. Zou, D. G. Xia, *J. Mat. Chem.*, **45**, 18289–18283 (2011).

56. W. W. Zhou, J. P. Liu, T. Chen, K. S. Tan, X. T. Jia, Z. Q. Luo, C. X. Cong, H. P. Yang, C. M. Li, T. Yu, *Phys. Chem. Chem. Phys.*, **13**, 14462–14465 (2011).

57. X. W. Zhang, L. W. Ji, O. Toprakci, Y. Z. Liang, M. Alcoutlabi, *Polym. Rev.*, **51**, 239–264 (2011).

58. F. Huang, X. Z. Lu, Y. Liu, J. Liu, R. Wu, *Mater. Sci. Technol.*, **27**, 29–34 (2011).

59. Y. N. Zhou, Z. W. Fu, *Prog. Chem.*, **23**, 336–348 (2011).

60. M. Baibarac, I. Baltog, S. Lefrant, *Curr. Org. Chem.*, **15**, 1160–1197 (2011).

61. Y. N. Zhou, W. J. Li, H. J. Chen, C. Liu, L. Zhang, Z. W. Fu, *Electrochem. Commun.*, **13**, 546–549 (2011).

62. Z. X. Dong, S. J. Kennedy, Y. Q. Wu, *J. Power Sources*, **196**, 4486–4904 (2011).

63. X. W. Guo, X. P. Fang, Y. Mao, Z. X. Wang, F. Wu, L. Q. Chen, *J. Phys. Chem. C*, **115**, 3803–3808 (2011).

64. J. Maier, *Phys. Chem. Chem. Phys.*, **11**, 3011–3022 (2009).

65. X. Q. Yu, J. P. Sun, K. Tang, H. Li, X. J. Huang, L. Dupont, J. Maier, *Phys. Chem. Chem. Phys.*, **11**, 9497–9503 (2009).

66. P. Balaya, *Energy Environ. Sci.*, **1**, 645–654 (2009).

67. M. Schildenberger, M. R. Prins, Y. C. Bonetti, *J. Phys. Chem. B*, **104**, 3250–3260 (2000).

68. T. Schalow, M. Brandt, M. Laurin, S. Schauermann, S. Guilmond, H. Kuhlenbeck, J. Libuda, H. J. Freund, *Surf. Sci.*, **600**, 2528–2542 (2006).

69. B. D. Sosnowchik, J. Y. Ha, L. Lin, 3rd IEEE International Conference on Nano/Micro Engineered Molecular System (Vols 1–3, pp. 1005–1010), January 06–09, 2008, Sanya, Peoples Republic of China.

70. S. F. Lee, Y. P. Chang, L. Y. Lee, *New Carbon Mater.*, **23**, 302–308 (2008).

71. U. Diebold, S. C. Li, M. Schmid, *Annu. Rev. Phys. Chem.*, **61**, 129–148 (2010).

72. G. X. Zhang, S. H. Sun, M. Bostetter, S. Poulin, E. Sacher, *J. Colloid Interface Sci.*, **350**, 16–21 (2010).

73. P. Rao, S. Kumar, G. L. N. Reddy, S. Veena, S. Kalavathi, J. V. Ramana, V. S. Raju, Nuclear instruments and methods, *Phy. Res. Section B—Beam Interactions with Mater. Atoms*, **268**, 3395–3402 (2010).

74. J. W. Zheng, S. C. Zhang, W. B. Liu, Y. L. Xing, Z. J. Du, *J. New Mater. Electrochem. Sys.*, **14**, 213–216 (2011).

75. X. H. Liu, S. Huang, S. T. Picraux, J. Li, T. Zhu, J. Y. Huang, *Nano Lett.*, **11**, 3991–3997 (2011).

76. S. C. Zhang, Y. L. Xing, T. Jiang, Z. J. Du, F. Li, L. He, W. B. Liu, *J. Power Sources*, **196**, 6915–6919 (2011).

77. G. S. Attard, J. M. Elliott, P. N. Bartlett, A. Whitehead, J. R. Owen, Macromolecular Symposia IUPAC 8th International Symposium on Macromolecule–Metal Complexes (MMC-8), Tokyo, Japan September 05–09, 1999.

78. Rajendra Kumar, *J. Mat. Sci.*, **7**, 1426–1433 (1972).

Chapter 7

High-Rate Li-Ion Intercalation in Nanocrystalline Cathode Materials for High-Power Li-Ion Batteries

Masashi Okubo[a] **and Itaru Honma**[b]

[a]*National Institute of Advanced Industrial Science and Technology (AIST), Umezono 1-1-1, Tsukuba, Ibaraki 305-8568, Japan*

[b]*Institute of Multidisciplinary Research for Advanced Materials (IMRAM), Tohoku University, Sendai, Miyagi 980-8577, Japan*

m-okubo@aist.go.jp

Discovering functionalities of nanomaterials that are superior to those of bulk materials has been a challenge in fundamental science and technology. In particular, increasing industrial needs for higher-power rechargeable batteries has encouraged advanced research on nanosized electrode materials. In this chapter, we first introduce the relationship between the high-charge/discharge-rate capability and particle size of the electrode materials based on the diffusion equation. Then, we describe size-controlled synthesis of cathode materials by means of the hydrothermal method, which is indispensable for estimating size dependency. Finally, we discuss the nanosize effect and its influence on high-charge/discharge-rate capability of cathode materials, as well as the related structural, physical, and electrochemical properties, of cathode materials is discussed.

Nanomaterials for Lithium-Ion Batteries: Fundamentals and Applications
Edited by Rachid Yazami

ISBN 978-981-4316-40-8 (Hardcover), 978-981-4364-23-2 (eBook)
www.panstanford.com

7.1 Introduction

Li-ion batteries are widely used for consumer electronic devices such as cellular phones and portable computers. Although their high-voltage, portability, and excellent cyclability have contributed to their commercial success, new applications such as hybrid electric vehicles require a high-charge/discharge-rate capability.

Commercial Li-ion batteries usually consist of a lithiated carbon anode and Li-ion intercalative transition-metal-oxide cathode. The high-charge/discharge-rate capability is generally dominated by Li-ion diffusion in the cathode.[1] Therefore, using nanosized cathode materials is an effective way of achieving better capability, since these materials possess a shorter Li-ion diffusion length d due to their small particle size. In this context, various nanosized cathodes with short d have been intensively investigated.[2–10]

However, it is expected that the presence of a large electrolyte–electrode heterointerface in nanosized cathode materials might also alter their structural, physical, and electrochemical properties. For example, the importance of the space charge region in nanosized ionic or mixed conductors has been stressed recently.[11,12] Furthermore, the impact of crystallite size on phase morphology and stoichiometry due to the surface energy has also been recognized very recently.[13,14] To clarify the effect of nanosize on cathode materials for practical applications, precise size-controlled synthesis and systematic studies are indispensable.

In this chapter, we first introduce the correlation between the particle size and the high-rate capability based on the diffusion equation. Then, we present a systematic study of the nanosize effect on the Li-ion intercalation reaction, especially in $LiCoO_2$.

7.2 Correlation between Li-Ion Diffusion Length and High-Charge/Discharge-Rate Capability

Let us first consider the electrochemical intercalation of Li-ions into a host electrode. In this section, we ignore polarization of the electrolyte and simply assume that Li-ion diffusion in the host electrode is the rate-limiting step.

Under equilibrium conditions, the open-circuit voltage determines the Li-ion concentration C within the host electrode.

If we use $Li_{1-\alpha}CoO_2$ as the host electrode with a cylinder shape of radius d, the equivalent electric circuit and schematic illustration of the Li-ion intercalation into $Li_{1-\alpha}CoO_2$ are as shown in Scheme 7.1.[15]

Scheme 7.1 Equivalent electric circuit and illustration of Li-ion intercalation into the host cylindrical material, $Li_{1-\alpha}CoO_2$.[15]

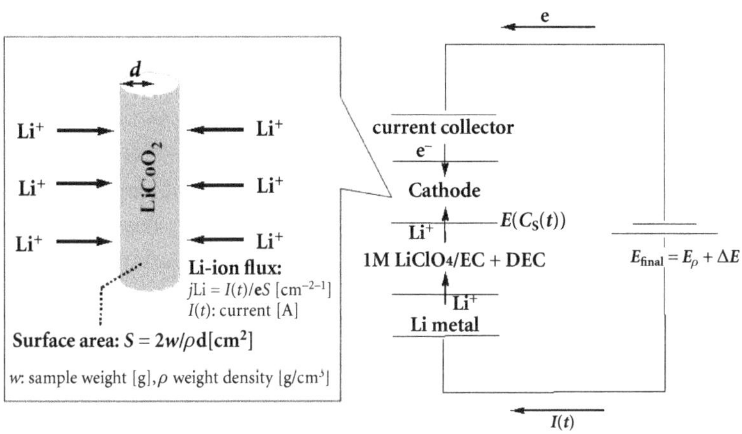

When a constant current I is forced to transverse the electrochemical cell, the diffusion equation within the host electrode

$$D\left(\frac{\partial C}{\partial r}\right)_{r=d} = \frac{I}{eS}, \quad I = \text{constant}, \tag{7.1}$$

gives the Li-ion concentration at the surface, C_s:[16]

$$C_s = C_0 + \frac{\rho\tau I}{2ew}\left\{\frac{2t}{\tau} + \frac{1}{4} - 2\sum_{n=1}^{\infty}\exp\left(-\frac{\alpha_n^2 t}{\tau}\right)\frac{1}{\alpha_n^2}\right\}, \tag{7.2}$$

where D is the chemical Li-ion diffusion coefficient, r the cylindrical coordinate, e the elementary electric charge, S the surface area, C_0 the initial Li-ion concentration, ρ the weight density of the electrode, τ the characteristic diffusion time $(= d^2/D)$, w the weight of the electrode, and α_ns the positive roots of $J_1(\alpha_n) = 0$ (J_n = the nth Bessel function).

For the condition $t/\tau > 0.125$, the solution approximates to

$$C_s \sim C_0 + \frac{\rho\tau I}{2ew}\left(\frac{2t}{\tau} + \frac{1}{4}\right), \tag{7.3}$$

which provides the current density (J, [A/g]) dependence of the capacity:

$$Q(J) = Q_0 \frac{1}{8}\tau J \tag{7.4}$$

where Q_0 is the ideal open-circuit voltage capacity. Equation 7.4 clearly suggests that the charge/discharge capacity decreases with increasing charge/discharge current density if τ (=d^2/D) is large enough.

Figure 7.1 shows the simulated Ragone plot for layered $LiCoO_2$ with various diffusion lengths, d. In the simulation, we assume a constant chemical diffusion coefficient $D = 10^{-12}$ [cm^2/s] and average charge/discharge voltage 3.9 [V *vs.* Li/Li^+] during the entire charge/discharge process, although, for real $LiCoO_2$, there is a two-phase region dominated by the moving boundaries, and the diffusion coefficient usually depends on the composition. Nevertheless, the figure shows that a shorter Li-ion diffusion length can provide higher power density; for example, electrodes with d = 100 nm can be almost fully charged/discharged even at 10 C rate (1.37 A/g).

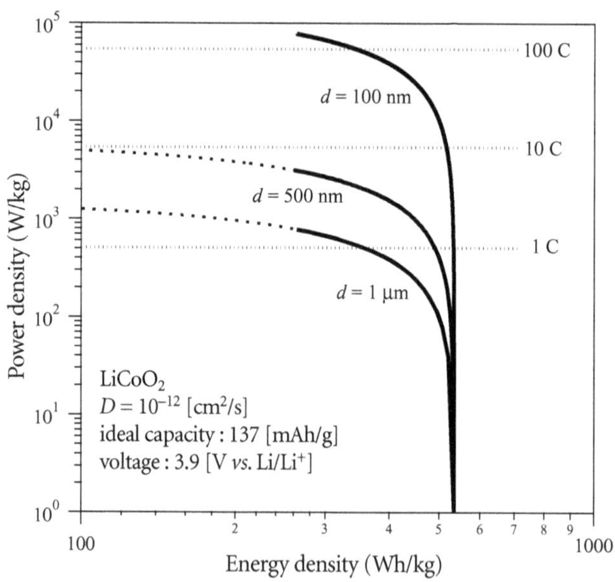

Figure 7.1 Simulated Ragone plot for $LiCoO_2$ with various diffusion lengths, d (100 nm, 500 nm, and 1 μm).

The validity of Eq. 7.4 was confirmed by high-discharge-rate experiments with commercial bulk $LiCoO_2$. Figure 7.2 shows discharge curves for bulk $LiCoO_2$ at various discharge rates ($C/2$ to 100 C). Note that the initial charged state, $Li_{0.5}CoO_2$, was achieved by slow charging at $C/2$ rate, and the discharge process was carried out at various rates. The IR drop was corrected appropriately to become the same initial voltage, since the initial Li-ion concentration in the host corresponding to the initial voltage at any discharge rate should be the same under this slow-charge condition. As shown in the figure, the discharge capacity decreases with increasing discharge-current density. The decrease in discharge capacity results apparently from slow Li-ion diffusion in $LiCoO_2$ with long diffusion length, as predicted from Eq. 7.4.

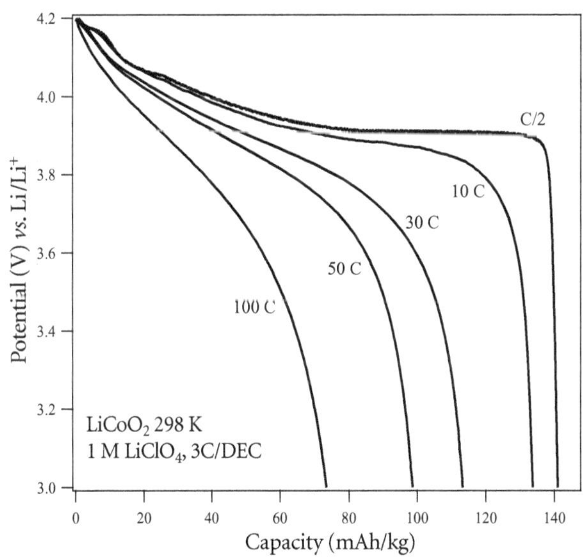

Figure 7.2 Discharge curve for bulk $LiCoO_2$ at various discharge rates.

Note that regardless of the diffusion dimensionality, the solution for the diffusion equation gives a similar current-density dependence for the capacity:[16]

$$Q(J) = Q_0 \frac{1}{\alpha} \tau J \tag{7.5}$$

where a is a constant related to the diffusion dimensionality (a = 3, 8, and 15 for the thin film, cylinder, and sphere models, respectively). This equation is valid for $Q > 0.5Q_0$.

As with Eq. 7.4, Eq. 7.5 also indicates that high-charge/discharge-rate capability strongly depends on d regardless of the diffusion dimensionality, i.e., the high-charge/discharge-rate capability decreases with increasing current density. Therefore, for any electrode material, where Li-ion diffusion is slow enough to be rate-limiting during the charge/discharge process, nanosizing is an effective way to achieve good high-charge/discharge-rate capability.

7.3 Synthesis of Nanocrystalline Electrode Materials by Hydrothermal Reaction

To date, an enormously large number of nanoelectrodes, including the nanocrystalline or nanostructured electrode materials, have been studied for the purpose of achieving good high-charge/discharge-rate capability.[17,18] While many synthetic methods have been applied for their synthesis, solution-based methods have strong advantages in controlling crystallite size and morphology. For example, a sol–gel method with templates enables the design of ordered nanostructures.[3,19–21] The reverse micelle technique also enables the control of nanostructures.[22] In this chapter, we put special emphasis on hydrothermal synthesis, which is also a powerful solution-based technique to fabricate nanoelectrodes.

Hydrothermal synthesis has been applied successfully to various nanocrystalline electrode materials such as $LiCoO_2$, $LiNiO_2$, $LiMn_2O_4$, and $LiFePO_4$. Table 7.1 summarizes selected previous reports on the hydrothermal synthesis of nanocrystalline cathode materials.

In general, all the parameters in hydrothermal synthesis are important for controlling the resulting phase, morphology, and crystallite size. Among them, we can easily control synthetic conditions such as the pH of the reaction medium, ratio of reactants, concentration of ions, presence of templating organic ions, reaction temperatures, and reaction times. For example, by varying the pH with acetic acid, the reaction of V_2O_5, TMA(OH) (TMA = $(CH_3)_4N^+$) and LiOH in the molar ratio 1:2:1 at 185°C for 72 h was reported to yield $TMA_{0.17}H_{0.1}V_2O_5$, $TMAV_8O_{20}$, $TMAV_4O_{10}$, $Li_xV_2O_4 \cdot H_2O$, $TMAV_3O_7$, and Li_3VO_4 at pH = 2, 3, 4, 5, 7, and 9, respectively.[46]

Table 7.1 Selected nanocrystalline electrode materials synthesized by the hydrothermal method

	Precursor	Reaction temperature (°C)	Reaction time (hour)	Crystallite size (nm)	Reference
$LiCoO_2$	β-CoOOH	160	12 – 60	100	23
	Co (metal)	125 ~ 170	2	300	24
	$Co(OH)_2$	130 ~ 200	48	<100	25
	$Co(NO_3)_2 \cdot 6H_2O$	120 ~ 250	0.5 ~ 24	69 ~ 164	26
	β-CoOOH	130 ~ 170	6 ~ 12	8 ~ 27	4
$LiFe_xCo_{1-x}O_2$	$(Fe_xCo_{1-x})OOH$	220 ~ 350	48	50	27
$LiMn_xCo_{1-x}O_2$	$(Mn_xCo_{1-x})_3O_4$	170	96	200	28
$LiNiO_2$	β-NiOOH	160	24	200	23,29
	γ-NiOOH	160	24	200	23,29
α $LiFeO_2$	α-$NaFeO_2$	220	7	500	30,31
$LiMn_2O_4$	γ-MnO_2	170	48 ~ 576	unknown	32
	$Mn(NO_3)_2$	90 ~ 110	8	100	33
	MnO_2	100 ~ 200	24	unknown	34
	$Mn(NO_3)_2$	400	0.25	200	35
	γ-MnO_2	200	24 ~ 168	100	36
	MnO_2	280	36	200	37
$LiFePO_4$	$FeSO_4$	120	5	3000	38
	$Fe_3(PO_4)_2 \cdot 5H_2O$	220	1	1000	39
	$FeSO_4$	150 ~ 220	3	4000	40
	$FeSO_4$	170	12	500	41
	$FeSO_4$	120	5	200	42
	$(NH_4)_2Fe(SO_4)_2 \cdot 6H_2O$	140 ~ 220	0.5 ~ 24	50 ~ 200	43
$LinPO_4$	$MnSO_4$	200	48	2000	44
TiO_2	$TiCl_4$	250	6	100	45
$Li_xV_2O_4$	V_2O_5	185	72	1000	46

For fabrication of nanocrystalline cathode materials with a short Li-ion diffusion length, reaction temperature and reaction time are apparently the most important parameters. As shown in Table 7.1, regardless of the target material, lower reaction temperature and shorter reaction time appear to give a smaller crystallite size, since they prevent the crystallite growth. However, for the purpose of complete phase transformation from precursor to resultant phase under these conditions, the rapid dissolution of the precursor into the aqueous solution is indispensable. Thus, when a nanocrystalline material with extremely small crystallite size is desired, the crystallite size of the precursor should be as small as possible.

Regarding $LiCoO_2$, the most commonly used cathode material in the commercial Li-ion batteries, many groups have reported synthesis of nanocrystalline $LiCoO_2$ by a post-templating method,[2] a sol–gel method,[10,47] a spray-drying method,[48] and a coprecipitation method.[49] As shown in Table 7.1, the hydrothermal method has also been applied successfully by some groups.[23–26] However, accurate control of crystallite size has not been achieved, and thus it has been difficult to discuss systematically the nanosize effect on structural, physical, and electrochemical properties.

As discussed above, to synthesize a nanocyrstal by the hydrothermal method, the precursor needs to be as small as possible. Therefore, nanocrystalline CoOOH was selected as a precursor for the synthesis of nanocrystalline $LiCoO_2$. The crystallite size of CoOOH can be easily controlled by synthetic conditions during the oxidation reaction of $Co(OH)_2$.[50] In fact, by controlling oxidation conditions (low temperature and high pH), we obtained nanoplate CoOOH with a thickness of 8 nm (Fig. 7.3). This extremely nanosized precursor enabled quick fabrication of the extremely nanosized $LiCoO_2$ with control of the crystallite size by varying the reaction temperature, concentration of LiOH aqueous solution, and reaction time.

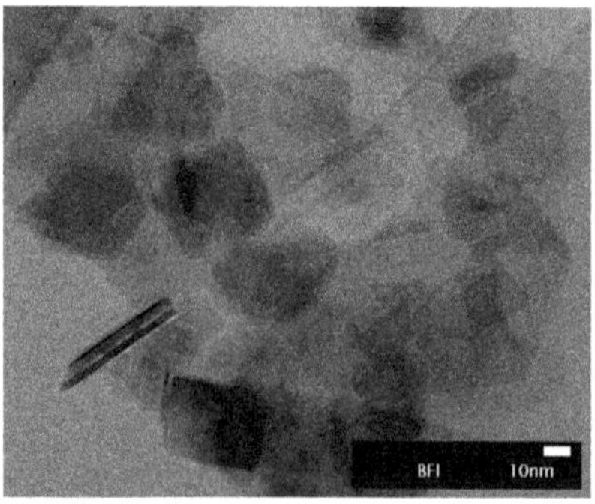

Figure 7.3 Nanoplate CoOOH with the thickness of 8 nm as a precursor for nanocrystalline $LiCoO_2$.

Figure 7.4 shows the phase diagram of the resultant as a function of reaction temperature and concentration of LiOH aqueous solution. The numbers in the diagram denote the crystallite size L of the resultant $LiCoO_2$ estimated from the linewidth of the (003) peak in the X-ray diffraction pattern. Although a detailed analysis of the X-ray diffraction pattern is presented in the following section, the hydrothermal reaction with 5 M LiOH solution produces $LiCoO_2$ after heating even at 120°C for 6 h. As shown in the figure, for lower concentration of LiOH solution, a higher reaction temperature is needed to give $LiCoO_2$. With respect to crystallite size, lower reaction temperature, higher concentration of LiOH aqueous solution, and shorter reaction time give a smaller crystallite size. This tendency is consistent with that mentioned in the previous section.

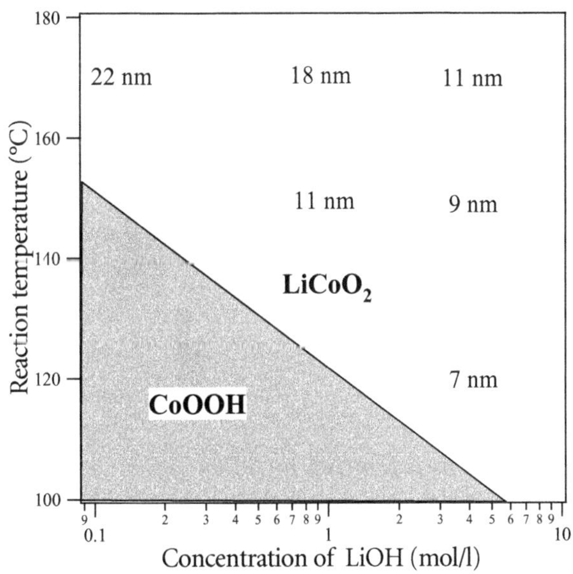

Figure 7.4 Phase diagram of the resultant from the hydrothermal reaction between CoOOH and LiOH aqueous solution.

Another feature of the hydrothermal method is the narrow distribution of particle size. Figures 7.5a,b show TEM images for nanocrystalline $LiCoO_2$ with L = 9 nm [the value estimated from the (003) peak]. The hexagonal layered platelet morphology is clearly visible. The thickness visible in Fig. 7.5b agrees well with the value of *L*. Figure 7.5c shows the SEM image for the same compound and suggests a uniform particle-size distribution. In fact, the distribution of nanoplatelet thickness counted from the TEM images for each compound (Fig. 7.5d) indicates a narrow size distribution. Therefore, nanocrystalline $LiCoO_2$ synthesized by the hydrothermal method can be a good model for a systematic study of the nanosize effect.

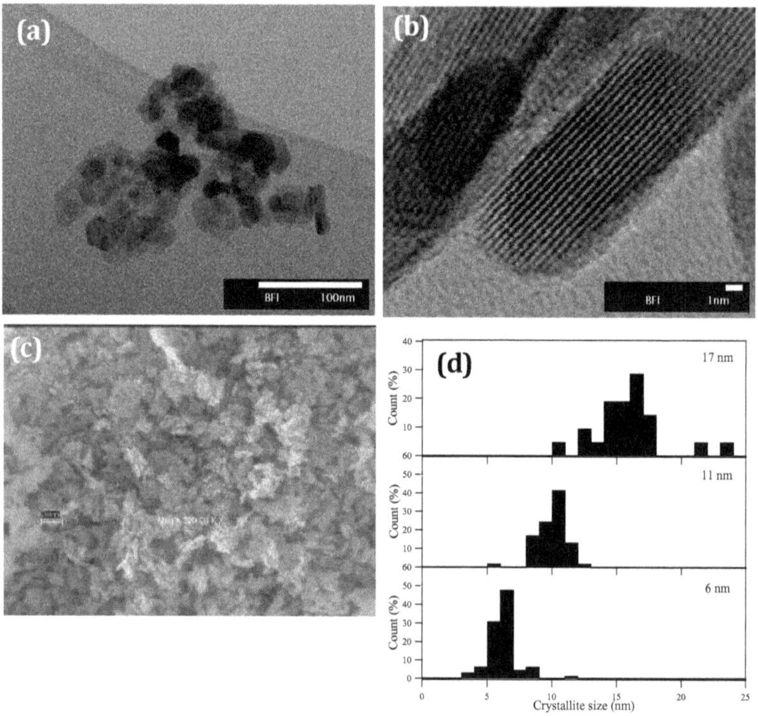

Figure 7.5 (a, b) TEM images, and (c) SEM image for nanocrystalline $LiCoO_2$ with L = 9 nm. (d) Thickness distributions of nanocrystalline $LiCoO_2$ (L = 6, 11, and 17 nm, respectively). Reprinted with permission from Ref. 4. Copyright 2007, American Chemical Society.

7.4 Nanosize Effect on Structural and Physical Properties of an Electrode

When crystallite size is restricted to the range 1–100 nm, the material's structural and physical properties can be significantly different from those of bulk materials. However, it should be emphasized that although crystallite size can have a drastic influence on the properties of a material, there generally does not exist a single simple factor but several simultaneous factors such as the confinement effect, structural defects, and elastic constraints. Therefore, attempts to identify a generalized critical size (e.g., of $LiCoO_2$ for the high-power batteries) or arrive at a single cause for all nanosize effects on electrodes are unrealistic, and attempts to clarify the most significant factors at play in a material from phenomenological size effects are more practical.

For example, nanocrystalline cerium dioxide, CeO_2, has been intensively investigated because of its potential for solid oxide fuel cell electrolytes,[51] high-performance capacitors,[52] and other applications,[53] and its nanosize effect is now well understood, at least with respect to structural and physical properties.[54–56] With respect to the structural properties, nanocrystalline CeO_2 shows lattice expansion with decreasing crystallite size, since decreasing size increases oxygen deficiency near the surface, reducing Ce^{4+} ions to Ce^{3+} ions to give CeO_{2-x}. With respect to physical properties, the phonon vibration frequencies detected by Raman spectroscopy cannot be explained completely by a simple lattice-expansion model by non-stoichiometry. Rather, to reproduce experimental results perfectly, an explanation must invoke a combination of lattice-expansion model and phonon-confinement effect.[56] In this chapter, we discuss the nanosize effect on the structural and physical properties of $LiCoO_2$ synthesized by the hydrothermal method.

The structural properties of materials are usually evaluated by the X-ray diffraction analysis. However, the X-ray diffraction relies on coherent scattering from the spatial average of a large number of unit cells. It is, therefore, insensitive to subtle structural characteristics and symmetry changes. Therefore, X-ray diffraction analysis is just a measure of the global structure of nanocrystalline materials. To probe the local structure of nanocrystalline materials, other techniques that are sensitive to a local structure are indispensable.

In an attempt to characterize the global and local structures of nanocrystalline $LiCoO_2$ synthesized by the hydrothermal method, we employed a combination of powder X-ray diffraction and Raman spectroscopy, as described in the following section. Raman spectroscopy is known to be sensitive to symmetry changes in $LiCoO_2$ as discussed below. Furthermore, the detailed Raman spectra have been obtained, even under hydrostatic pressure.[57]

Thus, Raman spectroscopy can be used to probe, at least partially, the local structure of $LiCoO_2$.

First, the resulting products were characterized by powder X-ray diffraction. Selected diffraction patterns are shown in Fig. 7.6. All of the peaks correspond to $LiCoO_2$,[58] indicating the formation of pure $LiCoO_2$. Note that the numbers in the figure denote the crystallite size L estimated from the linewidth of the (003) peak, which corresponds to the crystallite size along the stacking direction of the CoO_2 layers. The estimated value of L agrees well with the thickness of the nanoplatelet from the TEM images. Hereafter, for simplicity, we use L estimated from the (003) peak as the crystallite size of nanocrystalline $LiCoO_2$.

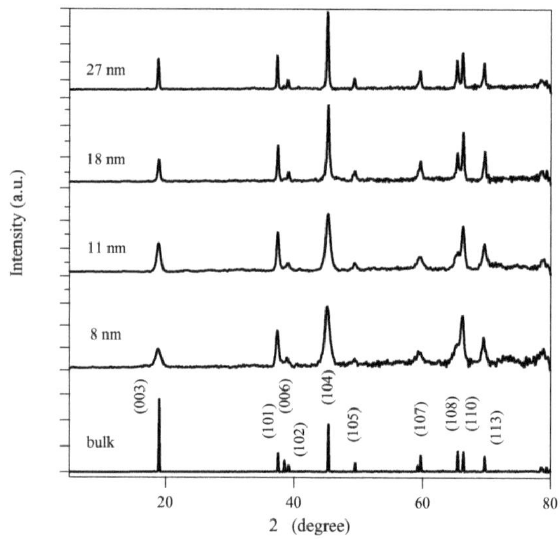

Figure 7.6 Powder X-ray diffraction patterns for nanocrystalline $LiCoO_2$ synthesized by the hydrothermal method. Reprinted with permission from Ref. 4. Copyright 2007, American Chemical Society.

Unit cell parameters for each sample were calculated from the powder X-ray diffraction patterns. Figure 7.7 shows the values of a, c, and V as a function of the crystallite size L. As shown in the figure, the lattice parameters depend on the crystallite size. Compared to bulk $LiCoO_2$ (a = 2.8122(8) Å, c = 14.025(7) Å, and V = 96.06(7) Å^3), the values of a, c, and V with a particle size of 6 nm increased to 2.820(1) Å, 14.12(2) Å, and 97.24(13) Å^3, respectively. The solid line in the figure is least-square fitted to $V(L) = V_0 + k \cdot L^{-2}$, where $V_0 = 96.3$ Å^3 and $k = 36.7$ Å^3·nm^2. This fitting result suggests lattice expansion by nanosizing $LiCoO_2$. Similar lattice expansion by nanosizing has been observed for $BaTiO_3$[59] and CeO_2.[60]

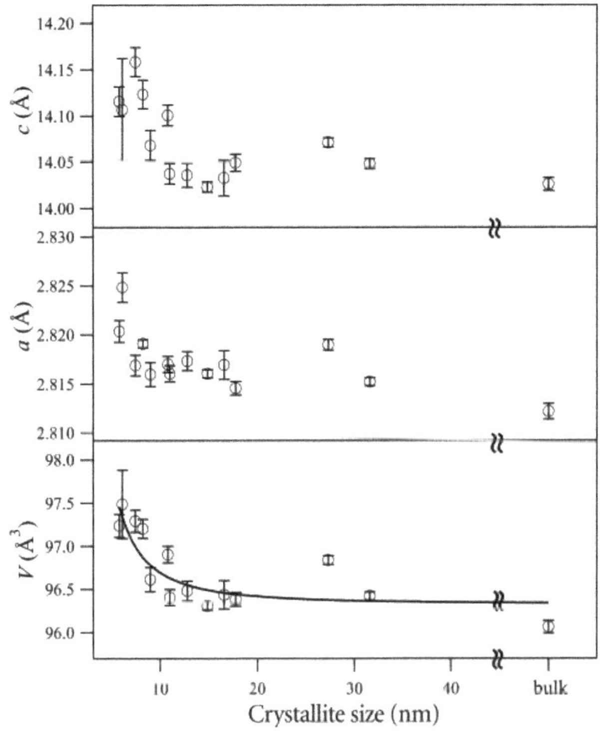

Figure 7.7 Unit cell parameters of nanocrystalline $LiCoO_2$ as a function of the crystallite size. Reprinted with permission from Ref. 4. Copyright 2007, American Chemical Society.

As mentioned in previous reports,[61-63] from a structural viewpoint, $LiCoO_2$ has many possible nonstoichiometry and structural defects. Among them, cation mixing between Co and Li ions ($[Li_{1-x}Co_x][Li_xCo_{1-x}]O_2$) has been reported to occur frequently.[64] While $LiCoO_2$ without cation mixing crystallizes in a layered α-$NaFeO_2$-type structure (hexagonal, *R*3m), $LiCoO_2$ with a large amount of the cation mixing crystallizes in a spinel-type structure (cubic, *F*d3m).[62] However, the powder X-ray diffraction patterns for the two structures are similar, and it is difficult to distinguish them at a glance. Furthermore, structural properties near the surface should differ significantly from those at the inner core, and hence the detailed structural properties can be detected not by the X-ray diffraction patterns, but by Raman spectroscopy.

Figure 7.8 shows selected Raman spectra for nanocrystalline $LiCoO_2$ with the crystallite sizes in the range of 6–27 nm. The spectrum for bulk $LiCoO_2$ is also shown for comparison. Nanocrystalline and bulk $LiCoO_2$ have two strong Raman bands at about 590 cm^{-1} and 480 cm^{-1}. From factor group analysis on the space group $R3m$, where the cation mixing does not exist, two Raman bands for the Co–O stretching A_{1g} mode and O–Co–O bending E_g mode are expected.[65] On the other hand, for the space group $Fd3m$, where a large amount of the cation mixing exists, four Raman bands of A_{1g}, E_g, and two F_{2g} modes are expected.[66,67] Therefore, the two bands observed for nanocrystalline and bulk $LiCoO_2$ should be attributed to the Co–O stretching A_{1g} mode and O–Co–O bending E_g mode, respectively, of space group $R3m$. This result strongly suggests that there does not exist a large amount of cation mixing within nanocrystalline $LiCoO_2$.

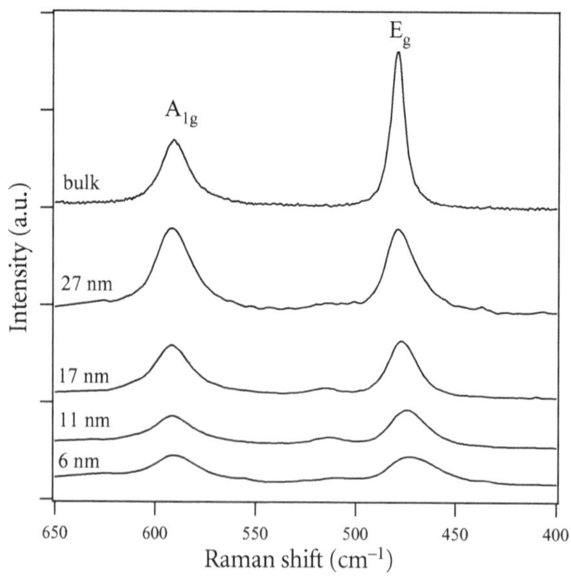

Figure 7.8 Raman spectra for nanocrystalline $LiCoO_2$ synthesized by the hydrothermal method. Reprinted with permission from Ref. 4. Copyright 2007, American Chemical Society.

Further careful comparison of the Raman spectra reveals a systematic change in the peak position of these two Raman bands. Figure 7.9 shows the relationship between Raman peak position and inverse crystallite size (L^{-1}), as determined by the (003) peak from powder X-ray diffraction. The top *x*-axis scale is the crystallite size *L*. By using the above value $V(L)$ ($=96.3 + 36.7 \cdot L^{-2}$) and the previously reported Grüneisen constants (0.79 for the A_{1g} mode and 1.15 for the E_g mode),[57] the crystallite size dependence of the Raman frequency shifts is deduced to be $\omega = 590.5 - 177.8 \cdot L^{-2}$ for the A_{1g} mode and $\omega = 480.2 - 210.0 \cdot L^{-2}$ for the E_g mode. The calculated results based on the lattice-expansion model are shown as dotted lines. Comparing calculated and experimental results, we see that the observed frequency shift in the A_{1g} mode is much less than expected from the lattice-expansion model. It is clear that the lattice expansion model is not sufficient to explain the dependence of frequency shift on crystallite size. Therefore, another contribution — the phonon confinement effect — should be considered.

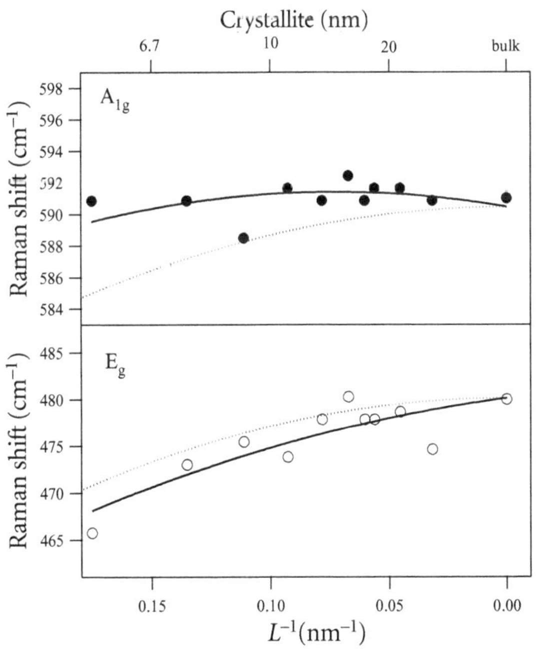

Figure 7.9 Dependence of Raman peak position on inverse crystallite size. Dotted lines and solid lines are calculated and fitted results, respectively Reprinted with permission from Ref. 4. Copyright 2007, American Chemical Society.

The Raman peak position and linewidth for a nanocrystalline material are affected by phonon confinement. In bulk-defect-free crystalline materials, only $q = 0$ phonons are Raman active. Contrary to this, in the case of a nanocrystalline material, q vectors in the range of $\Delta q = L^{-1}$ (L : crystallite size) are Raman active due to the uncertainty principle, which results in frequency shifts and broadening. The phonon-confinement model studied by Richter *et al.*[68] and extended by Campbell and Fauchet[69] generates the first-order Raman spectrum from,

$$I(\omega) = \int_{BZ} \frac{\exp\left(\frac{-q^2L^2}{8\beta}\right) d^3q}{(\omega - \omega(q))^2 + \left(\frac{\Gamma_0}{2}\right)^2} \tag{7.6}$$

where the integration is carried out over the Brillouin zone, $\omega(q)$ is the phonon dispersion, β is a confinement parameter, and Γ_0 is the natural linewidth. Equation 7.6 mathematically expresses the Raman selection-rule breaking that is induced by phonon confinement with a weighted exploration of the dispersion curves.

In the phonon-confinement effect, it is well known that there is a linear correlation between the Raman frequency shift and $L^{-\gamma}$, where γ is a scaling exponent equal to $d/2$ (d = dimensionality).[70] In fact, analyses on three-dimensional materials such as silicon[68] and gallium arsenide[71] yield a scaling factor of about 1.5, whereas analyses on two-dimensional materials such as graphite[72] and boron nitride[73] yield a scaling factor of about 1. Since $LiCoO_2$ is a two-dimensional material, the relationship between Raman frequency shifts and inverse of the crystallite size is expected to be linear. Thus, by combining the lattice-expansion model and the phonon-confinement effect, we fitted the experimental data with $\omega = 590.5 + k_1 \cdot L^{-1} - 177.8 \cdot L^{-2}$ for the A_{1g} mode and $\omega = 480.2 + k_2 \cdot L^{-1} - 210.0 \cdot L^{-2}$ for the E_g mode. The first term is the Raman frequency for the bulk crystal, the second is the contribution from the phonon-confinement effect, and the last term is the contribution from the lattice-expansion effect. The solid lines in Fig. 7.9 show least-square fitting results with $k_1 = 26.0$ nm cm^{-1} for the A_{1g} mode and $k_2 = -33.3$ nm cm^{-1} for the E_g mode. For A_{1g} mode, calculated and experimental results agree when the phonon-confinement effect is included. For the E_g mode, although the lattice expansion model is good, the agreement between the calculated and experimental results was better in this case and is consistent with the lattice expansion.

In short, the above Raman spectroscopy reveals that:

(i) There does not exist a large amount of the cation mixing between Co and Li ions within nanocrystalline $LiCoO_2$.

(ii) The lattice expansion by nanosizing that is observed in the X-ray diffraction pattern plays the most significant role in the nanosize effect on the phonon properties of $LiCoO_2$, even though another nanosize effect, the phonon-confinement effect exists.

These results clearly demonstrate that Raman spectroscopy is more powerful than X-ray diffraction analysis for detecting the local structure of nanocrystalline electrode materials. By combining X-ray diffraction analysis and Raman spectroscopy, we can clarify the detailed structural properties of nanocrystalline electrode materials.

Regarding the origin of lattice expansion in nanocrystalline $LiCoO_2$, again we cannot arrive at a single cause. At least, based on magnetic measurements and inductively coupled plasma atomic emission spectroscopy (ICP-AES), the existence of high-spin Co(II) ions in the form of $Li_{1+\delta}Co^{II}{}_{\delta}Co^{III}{}_{1-\delta}O_2$ is suggested, presumably due to excess Li-ions at the surface. Since the Co^{II}–O distance is longer than the Co^{III}–O distance due to not only the electrostatic force reduction but also to the antibonding character of the electron that occupies the e_g orbital, the existence of Co^{II} ion can explain the lattice expansion in nanocrystalline $LiCoO_2$. However, it should be mentioned that the finite lattice of nanocrystals should affect the Madelung potential near the surface, which also largely alters the structural properties. It should also be considered that elastic constraints such as surface tension can also affect structural properties.

7.5 Nanosize Effect on the Electrochemical Properties of Electrode

The above discussion described the nanosize effect on the structural and physical properties of nanocrystalline $LiCoO_2$. It is important to know whether the structural and physical nanosize effects affect the electrochemical properties. In this section, we discuss the electrochemical properties of nanocrystalline $LiCoO_2$.

If a Li^+-electron pair is intercalated into a host electrode under the potential difference E [V *vs.* Li/Li^+] with a Li metal anode, the work done within the electrochemical cell, $-eE$, must be equal to the free energy change of the electrode:

$$-eE = \frac{dG_{\text{electrode}}}{dn}. \tag{7.7}$$

Here, n is the number of intercalated Li atoms, and $dG_{\text{electrode}}/dn$ is the free energy change of the electrode during Li intercalation, usually defined as the Li chemical potential, μ_{Li}. Thus, μ_{Li} directly determines the voltage of the electrochemical cell in the equilibrium state. The Li chemical potential can be separated into contributions from the enthalpy H and entropy S of the electrode, giving

$$\mu_{\text{Li}} = \frac{dH}{dn} - T\frac{dS}{dn}, \tag{7.8}$$

where T is the temperature. This expression can be further transformed to,

$$\mu_{\text{Li}} = \mu_{\text{Li}}^0 - kT\frac{d}{dn}\{\ln W(n,N)\}, \tag{7.9}$$

where $\mu_{\text{Li}} (\equiv dH/dn)$ is the so-called Li-site energy, and W is the number of ways to arrange n Li atoms on N Li-sites.

Considering the simplest case — all the sites have same Li-site energy — n Li atoms occupy N Li-sites randomly:

$$W(n, N) = \frac{N!}{n!(N-n)!}. \tag{7.10}$$

Then the Li chemical potential is readily calculated as

$$\mu_{\text{Li}} = \mu_{\text{Li}}^0 + kT\ln\left(\frac{n}{N-n}\right). \tag{7.11}$$

This is the well-known Nernst equation. In the mean-field approximation in which the additional repulsive energy (u) is assumed for the Li-ion surrounded by fully occupied Li-sites, an extra energy nu/N is added to the above expression:

$$\mu_{\text{Li}} = \mu_{\text{Li}}^0 + \frac{nu}{N} + kT\ln\left(\frac{n}{N-n}\right). \tag{7.12}$$

Equation 7.12 is often called a mean-field expression, in which each intercalated Li feels the mean interaction field from the neighboring Li atoms.[74–78]

As indicated above, while the Li intercalation potential is affected by interaction between Li atoms or configuration entropy, it should be mainly dominated by the Li-site energy, μ_{Li}. Since $H = U + PV$ and dV/dn is negligibly small, the Li intercalation potential is largely dominated by the internal energy change dU/dn. If one Li is intercalated into the space coordinate **r**, within the crystal, the corresponding $dU(\mathbf{r})/dn$ is expressed as[79,80]

$$\frac{dU(r)}{dn} = \frac{e^2}{4\pi\varepsilon_0}\sum_j \frac{z_j}{\left|\mathbf{r}-\mathbf{R}_j\right|} - \lambda\sum_j \exp\left(-\frac{\left|\mathbf{r}-\mathbf{R}_j\right|}{\xi}\right), \tag{7.13}$$

where the first term is the Madelung energy from the jth ion at the space coordinate $\mathbf{R}_j$ with the charge number z_j, and the second term is the repulsion energy with the empirical parameters of λ and ξ. The summation is carried out over the entire volume of the material. Therefore, the finite lattice of nanocrystals should affect the Li intercalation voltage, especially for intercalation near the surface.[81] Furthermore, since lattice expansion is observed for nanocrystalline $LiCoO_2$, the Li intercalation voltage should further be affected. Likewise, Jamnik *et al.* previously expected that for small particles less than 100 nm, the electrochemical property is also affected by the surface tension as

$$\mu_{\mathrm{Li}}(L) = \mu_{\mathrm{Li}}(L=\infty) + 2\frac{\gamma(L)}{L}\nu_{\mathrm{Li}}, \tag{7.14}$$

where $\mu_{\mathrm{Li}}(L)$ is the particle-size dependence of the Li chemical potential, γ is the surface tension, and ν_{Li} is the partial molar volume of Li.[82]

In fact, the discharge curves for various nanocrystalline $LiCoO_2$ (6 nm bulk) at 1 C rate (137 mA/g), as shown in Fig. 7.10, are strongly affected by crystallite size. Capacitor-like behavior becomes more dominant with decreasing crystallite size, while the plateau region at 3.9 V decreases. Note that, since the measured electrochemical signals for nanoelectrode materials are frequently affected by the microstructure (e.g., poor electronic wiring), we mixed nanocrystalline $LiCoO_2$ with 45 wt% acetylene black to be free from such effects. Assuming the electric double-layer capacity to be 30 μF/cm^2, the electric double layer capacity for each compound can be calculated from the Brunauer–Emmett–Teller (BET)surface area as 2.6 mAh/g (17 nm), 5.4 mAh/g (11 nm), and 6.4 mAh/g (8 nm), respectively. These values are so small that we can neglect the contribution from the electric double-layer capacitor. Thus, the increasing capacitor-like behavior is ascribed to the nanosize effect on the Li intercalation potential, that is, the energy distribution of the Li-site energy μ^0_{Li} near the surface.

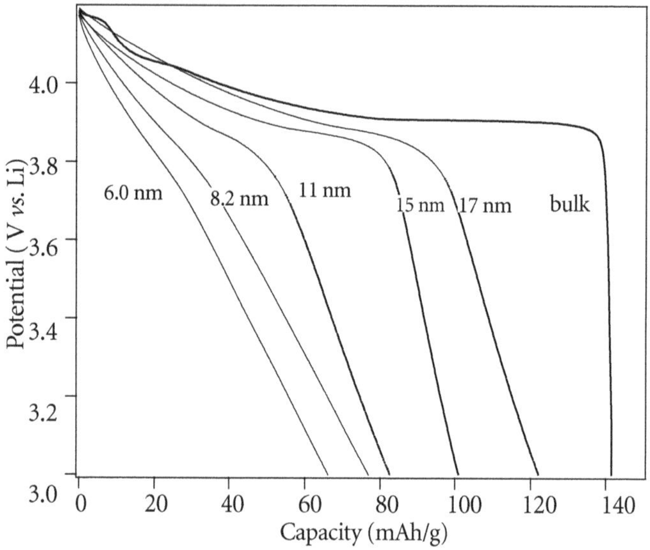

Figure 7.10 Discharge curves for nanocrystalline $LiCoO_2$ synthesized by the hydrothermal method Reprinted with permission from Ref. 4. Copyright 2007, American Chemical Society.

To estimate the energy distribution of μ^0_{Li}, we employed the lattice gas model to simulate the discharge curve. Let us consider the lattice-gas model where each site in a lattice has two states: full or empty. In the mean-field approximation, we considered the simplest case where all Li-sites have the same Li-site energy. We further consider the case where the site energy μ^0_{Li} has an energy distribution expressed by a probability density function $g(\varepsilon)$.[77]

It should be emphasized here that the Li chemical potential must be spatially uniform under the stationary state and depend only on the composition α in $Li_{1-\alpha}CoO_2$ as $\mu_{Li}(\alpha)$.[11] In contrast, the energy distribution of Li-site energy near the surface requires the spatial distribution of Li-site energy, $\mu^0_{Li}(\mathbf{r})$. Since the lattice-gas model ignores the composition dependence of μ^0_{Li} (=dH/dn) at the first approximation, the relationship between Li chemical potential and Li-site energy can be described as

$$\mu_{Li}(\alpha) = \mu^0_{Li}(\mathbf{r}) + kT \ln a_{Li}(\mathbf{r}, \alpha), \tag{7.15}$$

where $a_{Li}(\mathbf{r}, \alpha)$ is the activity of Li as a function of **r** and α. Since the activity of Li is the product of Li concentration and activity coefficient, the spatial equality of μ_{Li} should be maintained by compensating for the spatially distributed μ^0_{Li} with the gradient of Li concentration.

If the number of Li-sites with energy ε is $N(\varepsilon)$ and the occupancy $f(\varepsilon)$ of the Li-site at voltage E [V *vs.* Li/Li$^+$] obeys the Fermi distribution, the number of the occupying Li, $n(\varepsilon)$, can be expressed as

$$n(\varepsilon) = \frac{N(\varepsilon)}{1 + \exp\left(\dfrac{\varepsilon + eE}{kT}\right)}. \tag{7.16}$$

Integration of $n(\varepsilon)$ over the whole energy range must equal the number of intercalated Li in the host material, n. Thus,

$$n = \int_{-\infty}^{+\infty} \frac{N(\varepsilon)}{1 + \exp\left(\dfrac{\varepsilon + eE}{kT}\right)} d\varepsilon. \tag{7.17}$$

If we approximate the actual probability density function with rectangular distribution functions as

$$g(\varepsilon) \sim \frac{N_i}{\varepsilon_{i+1} - \varepsilon_i} \quad \text{for } \varepsilon_i < \varepsilon < \varepsilon_{i+1}, \quad \text{and} \quad i = 0, 1, \ldots, n,$$

the integration can be calculated as

$$n = N - \sum_{i=0}^{n} \frac{NikT}{\varepsilon_{i+1} - \varepsilon_i} \ln\left\{\frac{1 + \exp\left(\dfrac{\varepsilon_i + eE}{kT}\right)}{1 + \exp\left(\dfrac{\varepsilon_{i+1} + eE}{kT}\right)}\right\}, \tag{7.18}$$

where N is the total number of Li-sites ($N = N_0 + N_1 + \cdots + N_n$). Therefore, the discharge curve for $Li_{1-\alpha}CoO_2$ ($0 < \alpha < 1$) can be expressed as

$$\alpha = \frac{1}{2}\sum_{i=0}^{n}\frac{N_i}{N}\frac{kT}{\varepsilon_{i+1}-\varepsilon_i}\ln\left\{\frac{1+\exp\left(\frac{\varepsilon_i+eE}{kT}\right)}{1+\exp\left(\frac{\varepsilon_{i+1}+eE}{kT}\right)}\right\}. \tag{7.19}$$

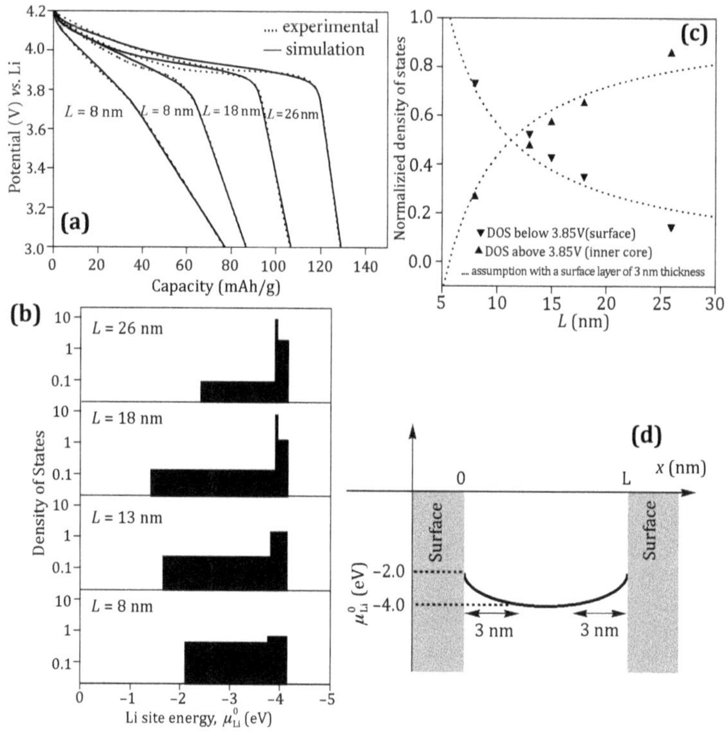

Figure 7.11 (a) Experimental discharge curves at $C/2$ rate (dotted lines) and simulated discharge curves (solid lines) based on Eq. 7.19 for nanocrystalline $LiCoO_2$ ($L = 8, 13, 18,$ and 26 nm). (b) Energy distribution of Li-site energy obtained from the simulation. (c) The crystallite size dependence of normalized density of states above −3.85 eV and below −3.85 eV. The dotted lines are the calculated results assuming a surface layer of 3 nm thickness along the stacking direction. (d) Schematic description of the spatial distribution of μ^0_{Li} along the c-direction.

The solid lines in Fig. 7.11a are the simulated discharge curves for nanocrystalline $LiCoO_2$ based on Eq. 7.19; the dotted curves are the experimental results. Figure 7.11b shows the obtained energy distribution of μ^0_{Li}. In this simulation, the distribution of μ^0_{Li} is approximated by three N_is (N_1, N_2, and N_3) for $L = 26$ and 18 nm, and two N_is (N_1 and N_2) for $L = 13$ and 8 nm. As expected, the number of Li-sites above −3.85 eV increases with decreasing L, since a decrease in L drastically increases the number of Li-sites near the surface. Figure 7.11c shows the crystallite size dependence of the normalized number of Li-sites $N(\varepsilon)$ above −3.85 eV and below −3.85 eV. The dotted lines are the calculated results, assuming a surface layer of 3 nm thickness along the c-direction (i.e., the stacking direction of the CoO_2 layer), have the Li-site energy above −3.85 eV, which agrees well with experimental results. Therefore, if we introduce the space coordinate along the c-axis, x ($0 < x < L$), the spatial distribution of μ^0_{Li} (x) could be schematically described as in Fig. 7.11d.

The spatial distribution of $\mu^0_{\text{Li}}(x)$ at the surface can be understood by the electrostatic interaction change due to the finite lattice suggested in Eq. 7.13. Other possible explanations include the previously discussed lattice expansion in the nanocrystalline $LiCoO_2$, the reduction in Co-ions near the surface due to the excess Li-ions suggested by ICP measurement, and the structural disorder near the surface such as with cation mixing between Co and Li-ions. In addition, oxygen deficiency or the hydroxyl ions could also be responsible.

It should be noted that the lattice-gas model does not consider the polarization effect during discharge or the decrease in discharge capacity due to cation mixing between Co and Li-ions. Presumably, the difference in the highest Li-site energy of each compound (e.g., −1.5 eV for $L = 18$ nm, and −2.1 eV for $L = 8$ nm) can be ascribed to such contributions as well as to the surface state for each compound. Nevertheless, the line analysis from electron energy loss spectroscopy (EELS) performed on nanocrystalline $LiCoO_2$ (to be published elsewhere[83]) has clarified that reduction of the Co-ion from Co^{3+} to Co^{2+} occurs especially along the c-direction at a depth of 3 nm. The obtained distribution of μ^0_{Li} agrees well with the EELS results, confirming that the Li-site energy distribution plays a crucial role rather than such contributions.

For practical applications of nanosized electrode materials, the nanosize effect on high-charge/discharge-rate capability is important. Therefore, high-discharge-rate measurements were carried out for nanocrystalline $LiCoO_2$ ($L = 8$ nm, 13 nm, 18 nm, and 26 nm). Figure 7.12 shows selected discharge curves for nanocrystalline $LiCoO_2$ (L = 8, 13, 18, and 26 nm) at various discharge rates from 1 C to 100 C. The averaged Li-ion diffusion length d is estimated from the TEM photos and also noted in Fig. 7.12. The discharge capacity for nanoplate $LiCoO_2$ (L = 26 nm, d = 45 nm) abruptly decreases with increasing discharge rate. The retained discharge capacity at 20 C rate is only 69% of the capacity at 1 C rate. It is clear that the incomplete Li-ion intercalation is caused by the long diffusion length of $LiCoO_2$, as discussed in the previous section. In contrast, when the nanoplate thickness decreases, the high-discharge-rate capability is drastically increased. For example, the retained discharge capacity at 100 C rate for nanocrystalline $LiCoO_2$ (L = 8 nm, d = 11 nm) reaches 91% of the 1 C rate capacity.

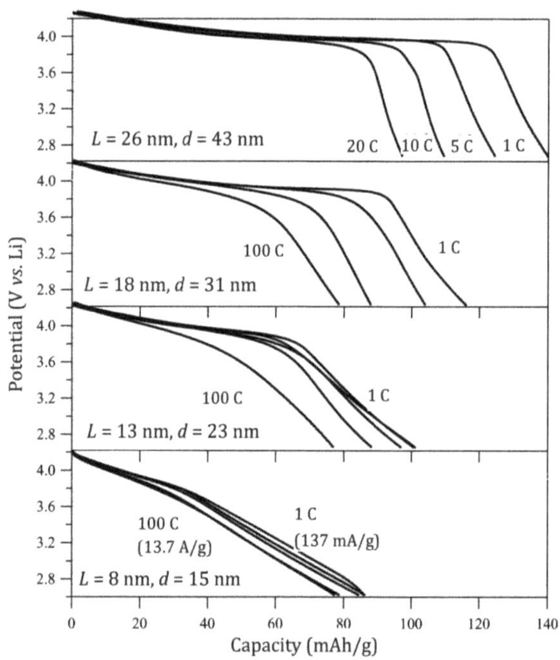

Figure 7.12 High rate discharge curves for nanocrystalline $LiCoO_2$ synthesized by the hydrothermal method (L = 8, 13, 18, and 26 nm).

The Ragone plot shown in Fig. 7.13 obtained from the high-rate discharge curves is totally different from the ordinary one (Fig. 7.1), because it contains the nanosize effects in nanocrystalline $LiCoO_2$. The Ragone plot clearly suggests that the energy density at low current density decreases with decreasing L, since the Li-ion intercalation potential is lowered by nanosizing. However, the smaller $LiCoO_2$ with shorter d could provide the higher power density, for example, 300 W-h/kg at 70 kW/kg for L = 8 nm, which should be achieved by the short Li-ion diffusion length.

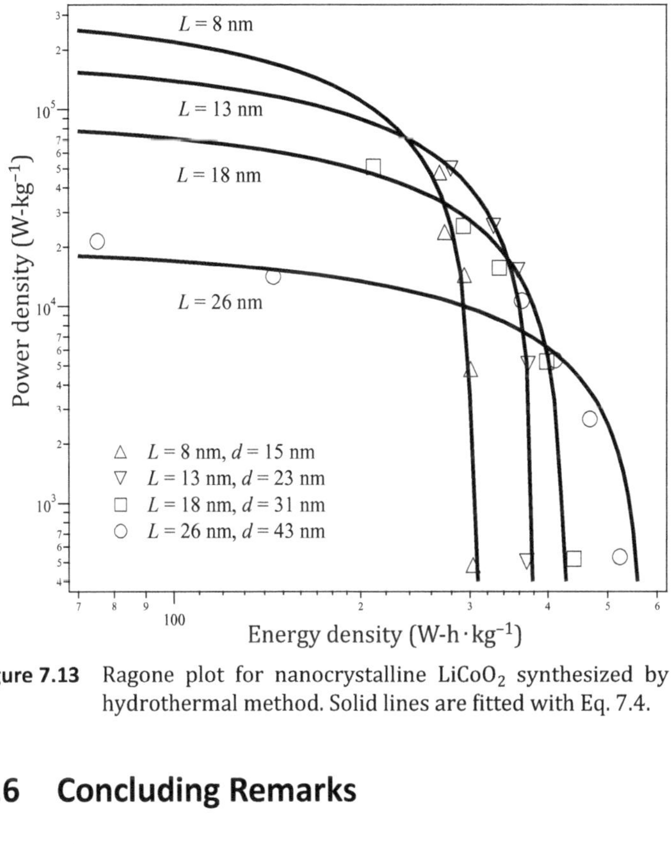

Figure 7.13 Ragone plot for nanocrystalline $LiCoO_2$ synthesized by the hydrothermal method. Solid lines are fitted with Eq. 7.4.

7.6 Concluding Remarks

In this chapter, we presented a systematic study of the nanosize effect on the Li-ion intercalation, especially in $LiCoO_2$. Decreasing crystallite size and increasing the surface area drastically affects structural, physical, and electrochemical properties. A high rate of Li-ion intercalation is achieved by shortening the Li-ion diffusion length; however, the Li-ion intercalation potential, and thus the energy density, are lowered due to the surface effect. From a practical viewpoint, for application of nanosized electrode materials to higher-power Li batteries, the appropriate particle size for a desired power density and energy density should be determined by considering not only the short Li-ion diffusion length but also the influence of the nanosize effect on electrochemical properties.

Concerning a more fundamental aspect of nanosized electrode materials, the influence of the nanosize effect on the Li-ion diffusion coefficient has not been clarified. Like the Li-ion intercalation potential, the Li-ion diffusion coefficient should be affected by the finite lattice in the nanocrystals. However, based on our potentiostatic intermittent titration technique (PITT) for nanocrystalline $LiCoO_2$,[83] the Li-ion diffusion in the solid electrode is no longer the rate-limiting step under the PITT condition due to the extremely short Li-ion diffusion length. Thus, the Li-ion diffusion coefficient could not be obtained by this conventional technique. Further study involving novel experimental techniques to determine the Li-ion diffusion coefficient is required to gain a better understanding of the Li-ion dynamics in nanosized electrode materials, especially near the surface.

Furthermore, in the above discussion, we considered Li-ion intercalation under the assumption that an Li^+-electron pair behaves as one particle. However, as mentioned in the introduction, the importance of the space charge region in the nanosized ionic or mixed conductors has been stressed recently. In the space charge region, the Li-ion and electron separate from each other and redistribute themselves individually so as to achieve minimum free energy.[11] For example, when excess Li-ions at the surface of nanocrystalline $LiCoO_2$ are assumed to be the positive-charged surface, the concentration of negative-charged species such as the electron rises whereas the concentration of positive-charged species such as the Li ion falls, which appears to explain the reduction in Co ions near the surface. To understand the stationary state of nanosized electrode materials completely, further study is indispensable including study of electronic structural changes and Li ion concentration changes near the surface.

List of symbols

d	Li-ion diffusion length
β	composition in $Li_{1-\alpha}CoO_2$
C	Li-ion concentration
I	current
D	chemical Li-ion diffusion coefficient

r	cylindrical coordinate
e	elementary electric charge
S	surface area
C_0	initial Li-ion concentration
ρ	weight density of electrode
τ	characteristic diffusion time ($=d^2/D$)
w	weight of electrode
J_n	nth Bessel function
t	time
J	current density
Q	capacity
Q_0	ideal capacity
L	crystallite size estimated from the (003) peak in the powder X-ray diffraction pattern for $LiCoO_2$
a	unit cell parameter
c	unit cell parameter
V	unit cell volume
q	wave vector
I	Raman intensity
ω	phonon dispersion
β	confinement parameter
Γ_0	natural Raman linewidth
γ	Raman scaling exponent ($=d/2$)
E	potential of the electrochemical cell
G	free energy
n	the number of intercalated Li
μ_{Li}	Li chemical potential
H	Enthalpy
S	Entropy
T	Temperature
μ_{Li}^0	Li-site energy
W	the number of ways to arrange n Li on N Li-sites
N	the number of Li-site
k	the Boltzmann constant
u	repulsive energy between Li atoms
U	internal energy

(*Continued*)

(*Continued*)

P	pressure
V	volume
$\mathbf{r}$	space coordinate
ε_0	permittivity of free space
$\mathbf{R}_j$	space coordinate of the *j*th ion
z_j	charge number of the *j*th ion
λ, ξ	empirical parameters for repulsion energy
a_{Li}	activity of Li
ε	energy
$N(\varepsilon)$	the number of the Li-site
$f(\varepsilon)$	occupancy of Li-site
$g(\varepsilon)$	density function of Li-site
N_i	the number of Li-site for $\varepsilon_i < \varepsilon < \varepsilon_{i+1}$

Acknowledgments

This work is a contribution from the National Institute of Advanced Industrial Science and Technology (AIST), Japan, and was financially supported by the New Energy and Industrial Technology Development Organization, (NEDO), Japan. One of the authors was financially supported from grant-in-aid for scientific research from the Ministry of Education, Culture, Sports, Science and Technology. M.O. was also financially supported from Tokuyama Science Foundation and Mitsubishi Chemical Corporation Fund. We thank Dr. J.-D. Kim and Mr. T. Sasaki (National Institute for Materials Science) for the TEM observation, Prof. N. Kojima and Dr. M. Enomoto for the ICP and magnetic measurements, Dr. H. S. Zhou and Dr. E. Hosono for the SEM and BET measurements. The basic concept, analysis of the experimental data, and theoretical consideration of this work was mainly directed by Emeritus Professor Tetsuichi Kudo, University of Tokyo, Japan.

References

1. T. Kudo, K. Fueki, *Solid State Ionics*, (Kodansha, Tokyo, 1990).
2. F. Jiao, K. M. Shaju, P. G. Bruce, *Angew. Chem., Int. Ed.*, **44**, 6550 (2005).

3. H. S. Zhou, D. Li, M. Hibino, I. Honma, *Angew. Chem., Int. Ed.*, **44**, 797 (2005).
4. M. Okubo, E. Hosono, J. D. Kim, M. Enomoto, N. Kojima, T. Kudo, H. S. Zhou, I. Honma, *J. Am. Chem. Soc.*, **23**, 7444 (2007).
5. F. Jiao, J. Bao, A. H. Hil, P. G. Bruce, *Angew. Chem., Int. Ed.*, **47**, 9711 (2008).
6. H. Chen, C. P. Grey, *Adv. Mater.*, **20**, 2206 (2008).
7. P. Poizot, S. Laruelle, S. Grugeon, L. Dupont, J. M. Tarascon, *Nature*, **407**, 496 (2000).
8. A. R. Armstrong, G. Armstrong, J. R. Canales, P. G. Bruce, *Adv. Mater.*, **17**, 862 (2005).
9. A. S. Arico, P.G. Bruce, J. M. Trascon, W. V. Schalkwijk, *Nat. Mater.*, **4**, 266 (2005).
10. Z. S. Peng, C. R. Wang, C. Y. Jiang, *J. Power Sources*, **72**, 215 (1998).
11. J. Maier, *Physical Chemistry of Ionic Materials* (John Wiley & Sons Ltd, England, 2004).
12. J. Maier, *Nat. Mater.*, **4**, 805 (2005).
13. M. Wagemaker, J. H. B. Wouter, F. M. Mulder, *J. Am. Chem. Soc.*, **129**, 4323 (2009).
14. W. J. H. Borghols, M. Wagemaker, U. Lafont, E. M. Kelder, F. M. Mulder, *Chem. Mater.*, **20**, 2949 (2008).
15. M. Okubo, Y. Tanaka, H. S. Zhou, T. Kudo, I. Honma, *J. Phys. Chem. B*, **113**, 2840 (2009).
16. J. C. Crank, *Matermatics of Diffusion* (Oxford University Press, Oxford, 1975).
17. Y. Wang, G. Cao, *Adv. Mater.*, **20**, 1286 (2008).
18. Y. G. Guo, J. S. Hu, L. J. Wan, *Adv. Mater.*, **20**, 2878 (2008).
19. S. M. Zhu, H. S. Zhou, T. Miyoshi, M. Hibino, I. Honma, M. Ichihara, *Adv. Mater.*, **16**, 2012 (2004).
20. D. L. Li, H. S. Zhou, I. Honma, *Nat. Mater.*, **3**, 65 (2004).
21. H. S. Zhou, S. M. Zhu, M. Hibino, I. Honma, M. Ichihara, *Adv. Mater.*, **15**, 2107 (2003).
22. E. A. Ponzio, T. M. Benedetti, R. M. Torresi, *Electrochim. Acta*, **52**, 4419 (2007).
23. D. Larcher, M. R. Palacin, G. G. Amatucci, J. M. Tarascon, *J. Electrochem. Soc.*, **144**, 408 (1997).

24. T. Watanabe, H. Uono, S. W. Song, K. S. Han, M. Yoshimura, *J. Solid State Chem.*, **162**, 364 (2001).

25. S. K. Chang, H. J. Kweon, B. K. Kim, D. Y. Jung, Y. U. Kwon, *J. Power Sources*, **104**, 125 (2002).

26. A. Burukhin, O. Brylev, P. Hany, B. R. Churagulov, *Solid State Ionics*, **151**, 259 (2002).

27. H. Kobayashi, H. Shigemura, M. Tabuchi, H. Sakaebe, K. Ado, H. Kageyama, A. Hirano, R. Kanno, M. Wakita, S. Morimoto, S. Nasu, *J. Electrochem. Soc.*, **147**, 960 (2000).

28. S. T. Myung, S. Komaba, N. Kumagai, *J. Electrochem. Soc.*, **149**, 1349 (2002).

29. M. R. Palacin, D. Larcher, A. Audemer, N. Sac-Epee, G. G. Amatucci, J. M. Trascon, *J. Electrochem. Soc.*, **144**, 4226 (1997).

30. M. Tabuchi, S. Tsutsui, C. Masquelier, R. Kanno, K. Ado, I. Matsubara, S. Nasu, H. Kageyama, *J. Solid State Chem.*, **140**, 159 (1998).

31. M. Tabuchi, K. Ado, H. Kobayashi, I. Matsubara, H. Kageyama, M. Wakita, S. Tsutsui, S. Nasu, Y. Takeda, C. Masquelier, A. Hirao, R. Kanno, *J. Solid State Chem.*, **141**, 554 (1998).

32. Q. Feng, H. Kanoh, Y. Miyai, K. Ooi, *Chem. Mater.*, **7**, 1226 (1995).

33. Y. C. Zhang, H. Wang, H. Y. Xu, B. Wang, H. Yan, A. Ahniyaz, M. Yoshimura, *Solid State Ionics*, **158**, 113 (2003).

34. Y. Y. Liang, S. J. Bao, B. L. He, W. J. Zhou, H. L. Li, *J. Electrochem. Soc.*, **152**, A2030 (2005).

35. K. Kanamura, K. Dokko, T. Kaizawa, *J. Electrochem. Soc.*, **152**, A391 (2005).

36. H. M. Wu, J. P. Tu, Y. F. Yuan, X. T. Chen, J. Y. Xiang, X. B. Zhao, G. S. Cao, *J. Power Sources*, **161**, 1260 (2006).

37. C. H. Jiang, S. X. Dou, H. K. Liu, M. Ichihara, H. S. Zhou, *J. Power Sources*, **172**, 410 (2007).

38. S. Yang, P. Y. Zavalij, M. S. Whittingham, *Electrochem. Commun.*, **3**, 505 (2001).

39. S. Franger, F. L. Cras, C. Bourbon, H. Rouault, *J. Power Sources*, **119–121**, 252 (2003).

40. S. Tajimi, Y. Ikeda, K. Uematsu, K. Toda, M. Sato, *Solid State Ionics*, **175**, 287 (2004).

41. K. Dokko, K. Shiraishi, K. Kanamura, *J. Electrochem. Soc.*, **152**, A2199 (2005).

42. G. Meligrana, C. Gerbaldi, A. Tuel, S. Bodoardo, N. Penazzi, *J. Power Sources*, **160**, 516 (2006).
43. B. Ellis, W. H. Kan, W. R. M. Makahnouk, L. F. Nazar, *J. Mater. Chem.*, **17**, 3248 (2007).
44. J. Chen, M. J. Vacchio, S. Wang, N. Chernova, P. Y. Zavalij, M. S. Whittingham, *Solid State Ionics*, **178**, 1676 (2008).
45. L. Kavan, M. Kalbac, M. Zukalova, I. Exnar, V. Lorenzen, R. Nesper, M. Graetzel, *Chem. Mater.*, **16**, 477 (2004).
46. T. Chirayil, P. Y. Zavalij, M. S. Whittingham, *Chem. Mater.*, **10**, 2629 (1998).
47. W. S. Yoon, K. B. Kim, *J. Power Sources*, **81–82**, 517 (1999).
48. Y. Li, C. Wan, Y. Wu, C. Jiang, Y. Zhu, *J. Power Sources*, **85**, 294 (2000).
49. H. Chen, X. Qiu, W. Zhu, P. Hagenmuller, *Electrochem. Commun.*, **4**, 488 (2002).
50. V. Pralong, A. Delahaye-Vidal, B. Beaudoin, B. Gerand, J. M. Tarascon, *J. Mater. Chem.*, **9**, 955 (1999).
51. M. Morgensen, N. M. Sammes, G. A. Tompsett, *Solid State Ionics*, **127**, 63 (2000).
52. T. Chikyow, S. M. Bedair, L. Tye, N. A. El-Masry, *Appl. Phys. Lett.*, **65**, 1030 (1994).
53. A. Walkenhorst, M. Schmitt, H. Adrian, K. Petersen, *Appl. Phys. Lett.*, **48**, 324 (1957).
54. S. Tsunekawa, R. Sahara, Y. Kawazoe, K. Ishikawa, *Appl. Surf. Sci.*, **152**, 53 (1999).
55. S. Tsunekawa, K. Ishikawa, Z. Q. Li, Y. Kawazoe, Y. Kasuya, *Phys. Rev. Lett.*, **85**, 3440 (2000).
56. F. Zhang, S. W. Chan, J. E. Spanier, E. Apak, R. D. Robinson, I. P. Herman, *Phys. Rev. B*, **64**, 245407 (2001).
57. X. Wang, I. Loa, K. Kunc, K. Syassen, M. Amboage, *Phys. Rev. B*, **72**, 224102 (2005).
58. E. Rossen, J. N. Reimers, J. R. Dahn, *Solid State Ionics*, **62**, 53 (1993).
59. S. Tsunekawa, S. Ito, T. Mori, K. Ishikawa, Z. Q. Li, Y. Kawazoe, *Phys. Rev. B*, **62**, 3065 (2000).
60. S. Tsunekawa, R. Sivamohan, S. Ito, A. Kasusya, T. Fukuda, *Nanostruct. Mater.*, **11**, 141 (1999).
61. G. G. Amatucci, J. M. Tarascon, D. Larcher, L. C. Klein, *Solid State Ionics*, **84**, 169 (1996).

62. R. J. Gummow, M. M. Thackeray, W. I. F. David, S. Hull, *Mater. Res. Bull.*, **27**, 327 (1992).
63. M. Carewska, S. Scaccia, F. Croce, S. Arumugam, Y. Wang, S. Greenbaum, *Solid State Ionics*, **93**, 227 (1997).
64. S. T. Myung, N. Kumagai, S. Komaba, H. T. Chung, *J. Appl. Electrochem.* **30**, 1081 (2000).
65. M. Inaba, Y. Todzuka, H. Yoshida, Y. Grincourt, A. Tasaka, Y. Tomida, Z. Ogumi, *Chem. Lett.*, **10**, 889 (1995).
66. W. Huang, R. Frech, *Solid State Ionics*, **86–88**, 395 (1996).
67. W. G. Fateley, *Infrared and Raman Selection Rules for Molecular and Lattice vibrations: The Correlation Method* (Wiley-Interscience, New York, 1972).
68. H. Richter, Z. P. Wang, L. Ley, *Solid State Commun.*, **39**, 625 (1981).
69. I. Campbell, P. M. Fauchet, *Solid State Commun.*, **58**, 739 (1986).
70. C. J. Doss, R. Zallen, *Phys. Rev. B*, **48**, 15626 (1993).
71. M. Gargouri, B. Prevot, C. Schwab, *J. Appl. Phys.*, **62**, 3902 (1987).
72. K. Nakamura, M. Fujitsuka, M. Kitajima, *Phys. Rev. B*, **41**, 12260 (1990).
73. R. J. Nemanich, S. A. Solin, R. M. Martin, *Phys. Rev. B*, **23**, 6348 (1981).
74. P. G. Bruce, *Solid State Electrochemistry* (Cambridge University Press, Cambridge, 1995).
75. R. Kikuchi, *Phys. Rev.*, **81**, 988 (1951).
76. A. J. Berlinsky, W. G. Unruh, W. R. Mckinnon, R. R. Hearing, *Solid State Commun.*, **31**, 135 (1979).
77. T. Kudo, M. Hibino, *Solid State Ionics*, **84**, 65 (1996).
78. T. Kudo, M. Hibino, *Electrochim. Acta*, **43**, 781 (1998).
79. T. Katsumata, Y. Inaguma, M. Itoh, K. Kawamura, *Solid State Ionics*, **108**, 175 (1998).
80. T. Katsumata, Y. Inaguma, M. Itoh, K. Kawamura, *Chem. Mater.*, **13**, 3930 (2002).
81. J. Yamaki, M. Makidera, T. Kawamura, M. Egashira, S. Okada, *J. Power Sources*, **153**, 245 (2006).
82. J. Jamnik, J. Maier, *Phys. Chem. Chem. Phys.*, **5**, 5215 (2003).
83. M. Okubo, J. D. Kim, T. Kudo, H. S. Zhou, I. Honma, *J. Phys. Chem. C*, **113**, 15337 (2009).

Chapter 8

$LiFePO_4$: From an Insulator to a Robust Cathode Material

Miran Gaberšček,[a] Janez Jamnik,[a] Katja Weichert,[b] Dominik Samuelis,[a] and Joachim Maier[b]

[a]*National Institute of Chemistry, Hajdrihova 19, SI-1001 Ljubljana, Slovenia*
[b]*Max-Planck-Institut für Festkörperforschung, Heisenbergstraße 1, D-70569 Stuttgart, Germany*

J.Maier@fkf.mpg.de

In this chapter the relationship between nanoarchitecture and electrochemical properties of composite cathode materials based on $LiFePO_4$ will be discussed in detail. We start out from studies of the inherent materials properties based on single crystal experiments. These will provide insight into defect chemistry, mass and charge transport as well as into phase transition behavior of the $FePO_4/LiFePO_4$ two phase system. Then the possibilities of optimizing the performance (e.g., by doping) will be discussed, out of which nanostructuring appears to be most promising. Different nanoarchitectures, their significance as well as their preparation will be discussed. The detailed understanding of the transport and storage aspects allows us to deduce appropriate nanoarchitectures that comprise both electroactive material and electrochemical wiring.

Nanomaterials for Lithium-Ion Batteries: Fundamentals and Applications
Edited by Rachid Yazami

ISBN 978-981-4316-40-8 (Hardcover), 978-981-4364-23-2 (eBook)
www.panstanford.com

8.1 Introduction

$LiFePO_4$ was first described by Padhi *et al.* in 1997 as a potential cathode material for secondary lithium-ion batteries.[1] On charging $LiFePO_4$ transforms into $FePO_4$. The two-phase regime is characterized by a flat and stable potential of ~3.5 V (*vs.* lithium metal). The process is very reversible as both compounds exhibit the same crystal structure which results in an excellent cycling stability. The cell parameters differ only slightly and the cell volume decreases by ~6% for $FePO_4$ compared to $LiFePO_4$. Beside the cycling stability the $LiFePO_4/FePO_4$ redox system has attracted much interest due to its high thermal stability, non-toxicity, environmental benignity, and low cost. The operating voltage of 3.5 V is a little smaller than in commercially available transition metal oxide batteries (e.g., $LiCoO_2$, $LiNiO_2$, $LiMn_2O_4$, operating voltage ~4 V, respectively[2]) yet it offers the advantage of greater compatibility with electrolyte systems. In addition, the theoretical capacity of 170 mAh/g is considerable.

All these advantages have stimulated a lively research on $LiFePO_4$ despite its poor transport properties[3] that lead to drastic capacity losses especially at higher current rates. Substantial efforts were made to enhance the conductivity and diffusivity of $LiFePO_4$ and hence battery performance. The approaches taken include cation doping on lithium[4–11] or iron[5–7,12–14] site which attracted a lot of attention after the report of Chung *et al.* in 2002.[7] They reported an increase of the electronic conductivity by eight orders of magnitude as well as significantly better performance while doping with aliovalent cations (e.g., Ti, Zr, Nb, Mg) on the lithium site. Later it was suggested that the conductivity increase was not due to the doped bulk but due to high electronically conductive phases formed during synthesis.[15] Another concept to improve battery performance is coating of $LiFePO_4$ particles with electronically conductive media. In 1999 Ravet *et al.* could achieve almost the theoretical capacity of the $LiFePO_4$ cathode during cycling at 80°C, when the particles of the active material were coated with carbon.[16] Carbon coating is now a widely used method,[17,18] whereas coating with metals is also possible.[19,20] Hu *et al.* could improve the effect of carbon coating by additional treatment with nano-sized RuO_2.[21] The oxidic RuO_2 particles wet areas of the active material where the electronically conductive carbon network is incomplete, leading to much higher capacities at high current rates in comparison with only carbon coated $LiFePO_4$.[21] Recently, Kang *et al.* reported about ultrafast charging and discharging of $LiFePO_4$ coated with an amorphous lithium phosphate phase.[22] They suppose that Li^+ transfer to/from the particles of the active material is extremely facilitated by this surface layer, allowing extraordinary fast charge and discharge rates. Even though a coating phase such as RuO_2 can catalyze the transfer reaction, the observation in Ref. 22 that a further huge performance gain is obtained by increased volume fraction of carbon points toward the importance of the network.

All coating procedures do not only allow for creating conducting networks around the particles, but can also inhibit particle growth. As a matter of fact, particle size reduction is one of the most promising methods to improve battery performance by shortening the diffusion paths for lithium insertion and extraction.[23–25] Yet this efficient reduction of the storage kinetics requires Li transport, that is, transport of Li^+ and e^-, to the individual crystallites be fast enough. This can be guaranteed by the formation of effectively mixed conducting networks, that is, electronic wiring reaching the small particles plus a structure open enough to allow for electrolyte access.[26,27] While electron transfer from the electronic network to the particles is not expected to be problematic, it is the ion transfer from the electrolyte phase to the $LiFePO_4$ which might cause additional problems. For very high currents electrolyte polarization and rheological insufficiencies might lead to further problems.

8.2 Transport Properties of $LiFePO_4$ Single Crystals

To investigate the inherent properties of a material, single crystals are quite useful, since they allow studies of the kinetic and thermodynamic parameters along the different crystallographic directions of the material. This is of special interest for highly anisotropic materials where crystal shape and orientation can be crucial regarding the properties. In addition grain boundary effects have not to be taken into account. In case of $LiFePO_4$ only a few single crystal studies were reported.[28–32] For precise electrochemical investigations of the anisotropic transport properties large single crystals of mm^3 size are required which were only recently available by the optical floating zone[33] or flux technique.[34] In the following we discuss measurements and anisotropies of single crystals grown by the optical floating zone technique.[3,35–38] The defect model described in Section 8.3 is largely based on such single crystal studies.

8.2.1 Electrochemical Investigations

For the electrochemical measurements the $LiFePO_4$ single crystals were cut and oriented along the three main crystallographic directions [h00], [0k0], and [00l] with edge lengths between 1 and 3 mm, respectively. To determine the electronic and ionic conductivities electrochemical impedance spectroscopy (AC) and stoichiometric polarization (DC) experiments on cells with blocking electrodes have been performed. The electronic conductivity was determined by using titanium electrodes, which are highly reversible for electrons but block lithium ions. Representative impedance spectra and a plot of a DC polarization measurement for the ion-blocking cell $Ti/LiFePO_4/Ti$ are shown in Fig. 8.1a and b. Over the whole frequency range (10^7–10^{-2} Hz) the impedance spectra show only one semicircle which corresponds to the bulk ($C \sim 10^{-12}$ F, dielectric constant ~60). During DC polarization the voltage increases in a steplike manner to a stationary value after switching on a constant current, and it decreases in the same way to zero after switching off the current. Such behavior is characteristic if electronic conduction dominates in the material. In fact, the electronic conductivity calculated from the polarization experiment is the same as the total conductivity observed with impedance spectroscopy.

For the electron-blocking cell the $LiFePO_4$ sample was sandwiched between lithium iodide pellets which block the electrons but are good lithium-ion conductors. In addition lithium–aluminum alloy pellets were attached as reversible electrodes for lithium. Impedance measurements with the electron-blocking cell $LiAl/LiI/LiFePO_4/LiI/LiAl$ give the same total conductivity (calculated from the high frequency semicircle, short time polarization) as observed with the ion-blocking cell $Ti/LiFePO_4/Ti$. Some representative spectra can be seen in Fig. 8.1c. The high frequency semicircle is followed by a Warburg response at low frequencies. The Warburg part results from the diffusion limited motion of the Li ions, since the current cannot be maintained by the electrons due to the electron-blocking electrodes. To detect the ionic conductivity long-time DC polarization experiments have been performed. A constant current was set which leads to an exponential increase of the voltage (see Fig. 8.1d). While initially both electronic and ionic charge carriers move through the sample, the partial current of the electronic charge carriers decreases with time and eventually only the non-blocked Li ions carry the current. From the short time DC polarization the diffusion coefficient is available (see Section 8.2.5). A detailed description of the technique that also takes account of internal reactions is given in Ref. 39.

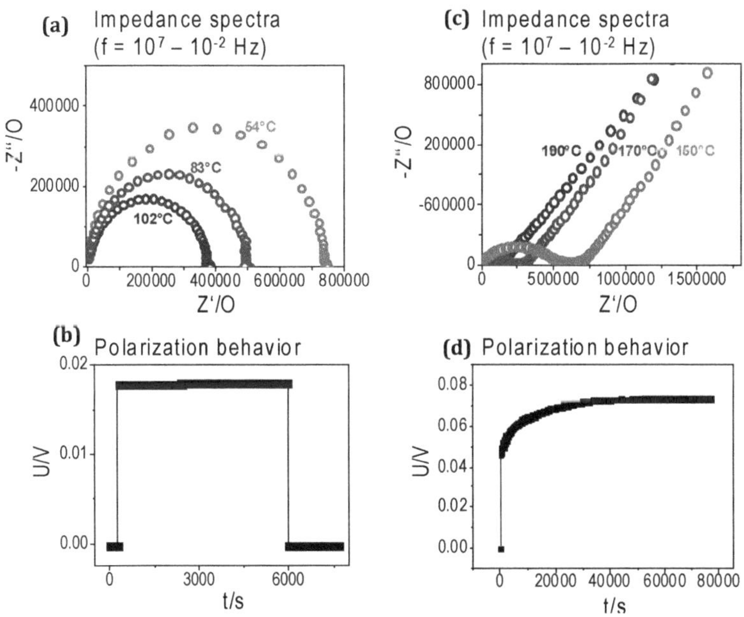

Figure 8.1 Representative spectra of impedance and polarization measurements of (a,b) the ion-blocking cell $Ti/LiFePO_4/Ti$ and (c,d) the electron-blocking cell $LiAl/LiI/LiFePO_4/LiI/LiAl$.

8.2.2 Anisotropy

The experimental results on $LiFePO_4$ single crystals demonstrate a quasi two-dimensional electronic as well as ionic transport in the b–c plane. This is in contrast to the anisotropy expected from the crystal structure (see Fig. 8.2) and predicted by theoretical modeling.[40–42] $LiFePO_4$ belongs to the olivine family and crystallizes in the typical orthorhombic cell in the space group *Pnma*. Corner-shared FeO_6 octahedra form a network in alternate b–c planes, whereas LiO_6 octahedra occupy the other b–c planes (Fig. 8.2). Along the b-axis these LiO_6 octahedra form a chain by sharing edges (highlighted blue in Fig. 8.2).

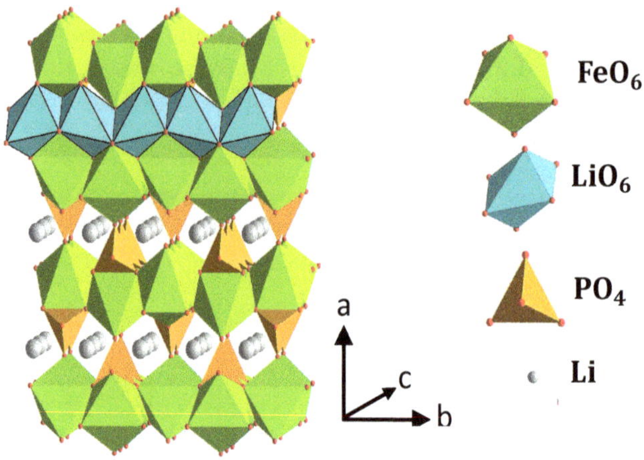

Figure 8.2 Crystal structure of $LiFePO_4$.

Computational studies favor only the b-axis for effective Li-ion conduction whereas the activation energy for conduction along c-axis is higher and even higher along a-axis. The same is true for the experiment of Nishimura *et al.* who visualized the diffusion path of lithium along the b-direction.[43] Since the distance between two Li sites is 3.0 Å along the b-axis, 4.7 Å along the c-axis, and 5.7 Å along the a-axis these results seem unsurprising.

Experimentally this anisotropy was not found for the single crystals grown by optical floating zone technique[35] (b-axis ≈ c-axis >> a-axis). This is probably due to the fact that modeling considers only short range motion, whereas in large single crystals as used for the electrochemical investigations long range motion is crucial. The degree of disorder (point defects, dislocations, domain boundaries) and the probability of the one-dimensional channels to be blocked depend on preparation as well as on size. So it cannot be excluded that very small crystals with an extremely low defect density behave differently. Malik *et al.* have analyzed the availability of continuous 1d transport pathways as a function of particle size.[44] Their computational study proposes Fe_{Li}–Li_{Fe} antisite pairs as point defects with lowest formation energy; an Fe_{Li} antisite defect would block a 1d transport channel and force the Li^+ ions to circumvent the defect, changing the diffusion mechanism into a 2d or 3d one, as already concluded by Amin *et al.*[3] Indeed, Malik *et al.* had already predicted that for Fe_{Li} antisite concentrations of 0.1, anisotropy of the diffusivities in a, b, and c directions is reduced to all three diffusion coefficients being within two orders of magnitude. Due to high preparation temperature and rapid quenching thereafter, such high antisite concentrations can easily occur in crystals prepared by optical floating zone technique.

8.2.3 Electronic Conductivity

The electronic conductivity of nominally pure $LiFePO_4$ measured by means of DC polarization in the ion-blocking cell is the same as the total conductivity measured by impedance spectroscopy.[35] This again demonstrates a predominant electronic conductivity in $LiFePO_4$ single crystals which was shown to be *p*-type.[45] The use of oriented single crystals allows conductivity measurements along the three main crystallographic directions. It was found that the electronic conductivity along the b- and c-axis is almost the same and distinctly higher than that for the a-axis (Fig. 8.3).[35] While the conductivity is anisotropic, the activation energies (E_{eon}) for each direction are almost the same, about 0.6 eV, respectively.[35]

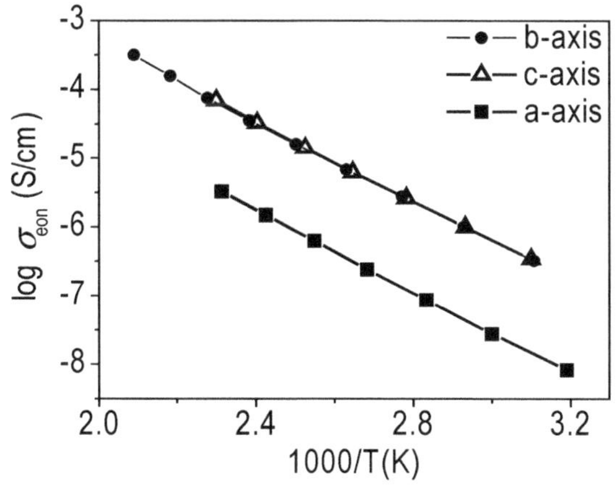

Figure 8.3 Arrhenius plot for electronic conductivity of nominally pure $LiFePO_4$.

The activation energy E_i is given by the sum of formation energy and migration energy, whereas the first term refers to the charge carrier concentration c_i and the second one to the mobility u_i of the respective charge carriers i:

$$E_i = -R\frac{\partial ln\sigma_i}{\partial 1/T} = -R\frac{\partial lnc_i}{\partial 1/T} + \left(-R\frac{\partial lnu_i}{\partial 1/T}\right) \tag{8.1}$$

The formation term was found to dominate and to be given by the trapping energy of the charged lithium vacancies to form neutral ones.[46] The migration energy of holes is small (0.05 eV,[46] 0.215 eV[47] within b–c plane, respectively) and hence not greatly varied by the anisotropic crystal structure.[48]

8.2.4 Ionic Conductivity

The ionic conductivity of $LiFePO_4$ single crystals was measured to be up to four orders of magnitude lower than the electronic conductivity.[3] The anisotropy already observed for the electronic conduction is reproduced for ionic transport. Similar values were found for b- and c-direction whereas the a-axis shows much less conductivity values (Fig. 8.4).

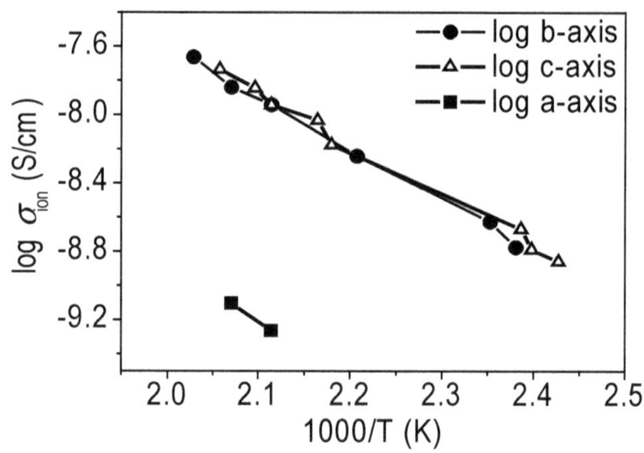

Figure 8.4 Arrhenius plot for ionic conductivity of nominally pure $LiFePO_4$.

For b- and c-axis the activation energies are 0.62 eV respectively, whereas it is 0.74 for the a-axis.[3] In contrast to the electronic charge carriers the migration term (see Eq. 8.1) dominates now.[46] It was shown by theoretical modeling that the migration energy should be highly anisotropic.[40,42] Our findings of similar conductivity values for b- and c-axis are probably due to disorder (i.e., Fe on Li sites)[3,31] which blocks the preferred Li-ion migration path. A more difficult point is the similar activation energy along a-axis (this trend is also seen in donor doped crystals).[38] How far this is due to some misorientated domains in the large crystal or due to mechanical peculiarities is open at the moment.

8.2.5 Chemical Diffusion Coefficient of Lithium

Long-time polarization measurements with the electron-blocking cell provide the steady state voltage from which the ionic conductivity can be calculated. From the transient behavior also the diffusion coefficient is available based on the following equation:[48]

$$U_{\text{ion}} = \frac{i_p L}{\sigma} + \frac{\sigma_{\text{eon}}}{\sigma}\frac{i_p L}{\sigma_{\text{ion}}}\left\{1 - \frac{8}{\pi^2}\exp\left[-\frac{t}{\tau^\delta}\right]\right\} \tag{8.2}$$

A plot of $\ln(U\text{-}U_\infty)$ t should give a straight line, with the time constant τ^δ as a slope. The calculation of τ^δ yields the chemical diffusion coefficient of lithium:

$$\tau^\delta = \frac{L^2}{\pi^2 D_{\text{Li}}^\delta}$$

Figure 8.5 shows a typical evaluation based on Eq. 8.2.

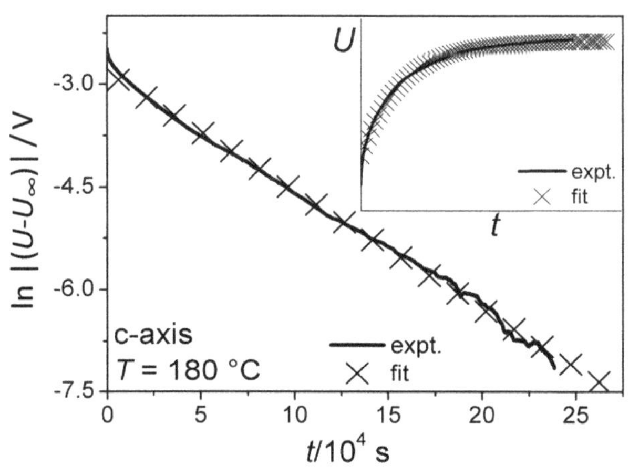

Figure 8.5 Extraction of ionic conductivity (inset) and diffusivity by fitting the obtained cell voltage. Reprinted from Ref. 38. Copyright 2009, WILEY-VCH Verlag GmbH & Co. KGaA, Weinheim.

The short-time behavior was also evaluated and showed consistent values. Since the ionic conductivity in $LiFePO_4$ single crystals was found to be effectively two-dimensional in the b–c plane,[3] a similar behavior for the diffusion coefficient was expected. The results are shown in Fig. 8.6.

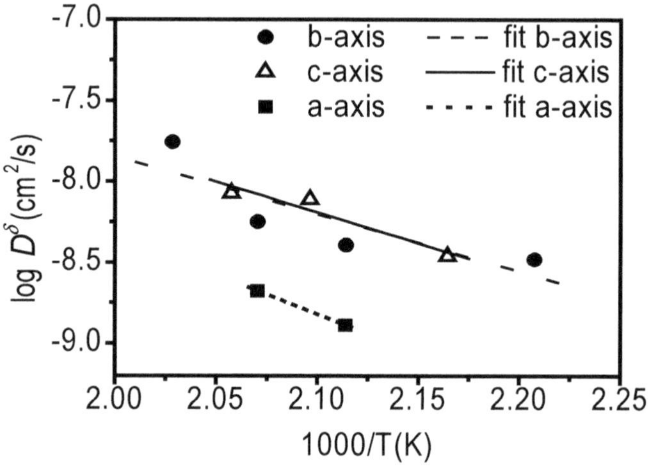

Figure 8.6 Chemical diffusion coefficient of lithium along the three main crystallographic directions of nominally pure $LiFePO_4$.

Indeed the transport parameters are almost the same along the b- and c-direction whereas a smaller value was found for the a-direction.[3] Due to the long time needed to reach the steady state long-time polarization measurements were performed at higher temperatures. The extrapolated values for the diffusion coefficient at room temperature are ~10^{-13} cm^2/s for b- and c-direction and ~10^{-15} cm^2/s along a-axis. This corresponds very well with the value of ~10^{-14} cm^2/s for polycrystalline $LiFePO_4$ reported by Prosini *et al.* who used the galvanostatic intermittent titration technique.[49] The chemical diffusion of lithium is activated by ca. 0.7 eV along b- and c-axis and 0.95 eV along a-axis.[3]

8.3 Defect Model of $LiFePO_4$

A defect model specifies the nature of the relevant point defects in a material, their internal interactions as well as the interaction with the neighboring phases. The control parameters for the variations of the ionic and electronic charge carriers are temperature, component partial pressures—fixing the stoichiometry of the pure material—as well as the doping content (content of positively or negatively charged impurity defects).[46] The methods described in Section 8.2.1 allow the measurements of electronic and ionic conductivities. Their analysis in the framework of a defect model then yields the defect chemical parameters, such as reaction enthalpies, interaction enthalpies, as well as migration energies.

8.3.1 General Remarks on the Nature of Point Defects

While it is the bulk defect chemistry that determines the efficiency of homogeneous doping, it is the defect chemistry of boundary regions that determines charge carrier variations in the vicinity of the interface. The defect model to be described below is able to solve the first problem, the elucidation of the boundary problems deserves much more further work.[48] If we ignore effects at very small sizes at which interfacial effects dominate the whole carrier chemistry within the particle, the material's properties are expected to be dominated by bulk phenomena.

The statistical mechanical treatment solves, by also taking into account electroneutrality, the concentration of any carrier as a function of the aforementioned control parameters. Not all component partial pressures can be varied *in situ*. Besides the lithium partial pressure, defining the lithium activity, effects on the iron or phosphate sublattices are considered to be frozen. The respective defects are then considered as *ex situ* parameters[50] and are to be taken account of in the effective dopant concentration. The same is assumed to be valid for defects changing the P/O ratio (a small fraction of phosphate groups (PO_4^{3-}) may be condensed to $P_2O_7^{4-}$, $P_2O_7^{4-}$ substituting for two PO_4^{3-} being equivalent to an oxygen vacancy $V_O^{\bullet\bullet}$). In line with experiments and theory the major mobile ionic defect in $LiFePO_4$ is considered to be the lithium vacancy $V_L^{'}$ which in the ideal case of pure material needs to be compensated by holes $h^{\bullet}$ (Fe^{3+} on Fe^{2+} sites) resulting in a *p*-type electron conduction. Important frozen defect concentrations are anti-site defects (iron on a lithium site and *vice versa*) in agreement with the similar ionic radiuses and with modeling.[51] In order to derive the consequences of introducing dopants, it is not decisive which element is substituted or if the dopant is interstitial, it is (for low concentrations) only the effective charge and concentration that matters in the defect chemical computation. It is straightforward to show that defect concentrations follow to a good approximation power laws in component partial pressure, doping content and mass action constant, leading to Brouwer and Arrhenius diagrams.

8.3.2 Dependence on Lithium Activity

The lithium activity was tuned by annealing polycrystalline $LiFePO_4$ in argon atmosphere at elevated temperatures at which Li is driven to leave the sample.*[,45] Connected with this is an increased concentration of lithium vacancies (V'_{Li}) and holes ($h^{\bullet}$):

$$\begin{aligned} &Li^{x}_{Li} \rightleftharpoons Li_{(g)} + V'_{Li} + h^{\bullet} \\ &K_1 = a_{Li}[V'_{Li}][h^{\bullet}] \end{aligned} \tag{8.3}$$

(with $[Li_{(g)}] = a_{Li}$). As a consequence the product $[V'_{Li}][h^{\bullet}]$ depends on the lithium activity. Reaction shown in Eq. 8.3 is only reversible for very high temperatures, very small samples, and/or long waiting times.

The increased electronic and ionic charge carrier concentrations go along with increased electronic and ionic conductivities. This could be shown experimentally for the electronic conductivity, which increases after every annealing step.[45] As a consequence of the change of lithium activity the material transfers from one defect chemical regime to another one as shown in the ideal Brouwer diagram in Fig. 8.7.[46] Such regimes are characterized by different majority defects which dominate the defect chemistry at this point. At high enough lithium deficiencies (P-regime) the intrinsic ionic and electronic carriers dominate the unavoidable impurity content that dominates at low lithium deficiencies (D-regime). In Fig. 8.7, a donor impurity was assumed. Such background impurity content can be natively realized by a slight iron excess compared to lithium accommodated by anti-site disorder on lithium sites.[31]

*Most probably lithium leaves the crystal under formation of surface Li_2O_2 via getting trapped by O_2 traces. It is also conceivable that a reaction to Li_3PO_4 and Fe_2O_3 takes place.

It is easy to show that for simple dilute cases each defect regime is characterized by linear log[defect] *vs.* $\log a_{Li}$ relations (power laws). Using the mass action law of reaction (Eq. 8.3), $\partial\log[V'_{Li}]/\partial\log a_{Li} = -1/2$ results for the P-regime where $[V'_{Li}] = [h^\bullet]$. In the regime of donor doping (D-regime) the concentration of $[V'_{Li}]$ is kept constant by the impurity $[D^\bullet]$ and therefore $\partial\log[V'_{Li}]/\partial\log a_{Li} = 0$ and $\partial\log[h^\bullet]/\partial\log a_{Li} = -1$. The neutral lithium vacancy concentration (V^{x}_{Li}) increases with −1 slope (sum of slopes for $h^\bullet$ and V'_{Li} according to reaction shown in Eq. 8.4).

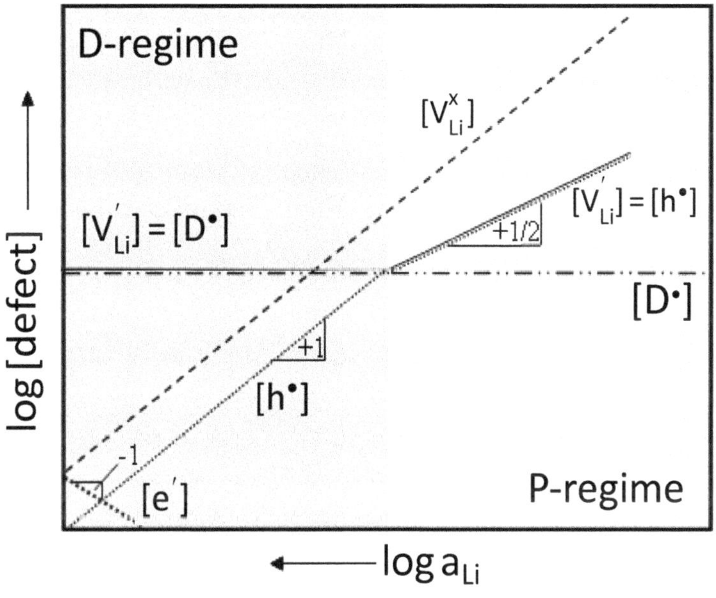

Figure 8.7 Brouwer diagram showing the dependence of the defect concentration on lithium activity.

8.3.3 Temperature Dependence

While the lithium activity can be varied at higher temperatures as described in the previous paragraph, the lithium content is frozen-in (in particular for larger crystals) at lower temperatures where conductivity measurements were performed (40–200°C). In the ideal Brouwer diagram in Fig. 8.8 the P- and D-regimes are reflected by different temperature dependencies that can be directly explained on the basis of the previous considerations.

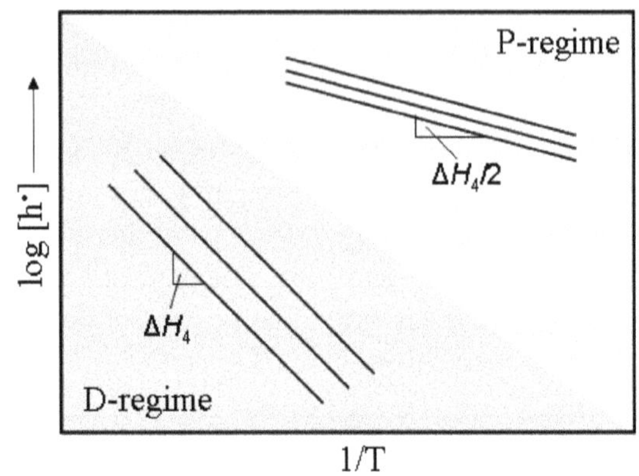

Figure 8.8 Temperature dependence of the hole concentration in terms of the ideal Brouwer diagram.

Under conditions in which reaction (3) is not reversible, the temperature dependence of the defect concentrations is rather determined by internal interactions, which is in particular the association between lithium vacancies and holes to neutral lithium vacancies (V_{Li}^{x}):[46]

$$V_{Li}' + h^{\bullet} \rightleftharpoons V_{Li}^{x}$$
$$K_4 = \frac{[V_{Li}^{x}]}{[V_{Li}'][h^{\bullet}]} \quad (8.4)$$

From the mass action law the activation energies can be easily calculated for two conditions: (i) an almost perfect lithium stoichiometry $Li_{1-\delta}FePO_4$ with δ close to zero and (ii) a high lithium deficiency with $\delta >> 0$. The first case represents necessarily the regime in which impurities dominate (D-regime) with a constant concentration of ionized lithium vacancies (V_{Li}'). The second case reflects the P-regime where the concentrations of ionized lithium vacancies and holes are equal. Simple expressions result if $[V_{Li}^{x}]$ is large and constant what we will assume here. For the more general situation see Ref. 46. For the D-regime the activation energy for electronic and ionic conduction (E_{eon} and E_{ion}, respectively) yields

$$E_{eon} = -R\frac{\partial \ln \sigma_{eon}}{\partial \ln(1/T)} = -\Delta H_4 + h_h \quad (8.5a)$$

$$E_{\text{ion}} = -R\frac{\partial \ln\sigma_{\text{ion}}}{\partial \ln(1/T)} = h_V \tag{8.5b}$$

with ΔH_4 being the reaction enthalpy of the trapping reaction (Eq. 8.4) and h_h as well as h_V being the migration energy of holes and ionized lithium vacancies, respectively. For the P-regime it holds that

$$E_{\text{eon}} = -R\frac{\ln\sigma_{\text{eon}}}{\ln(1/T)} = -\frac{\Delta H_4}{2} + h_h \tag{8.6a}$$

$$E_{\text{ion}} = -R\frac{\ln\sigma_{\text{ion}}}{\ln(1/T)} = -\frac{\Delta H_4}{2} + h_V \tag{8.6b}$$

With the reasonable assumption that the value for the migration energy of the holes will be small, the defect model predicts a drop of the activation energy from D- to P-regime by a factor of two. This excellently reproduces the experimental findings for the electronic conductivity measured on polycrystalline $LiFePO_4$, where an activation energy of 0.65 eV was determined for non-annealed samples (almost stoichiometric lithium content, D-regime) and of 0.30 eV for high temperature annealed samples (high lithium deficiency, P-regime),[45] as sketched in Fig. 8.8. The more precise analysis then takes account of the slight discrepancy and allows us to extract the hole migration energy (see section 8.3.5). In Fig. 8.8 it can also be seen that the parallel lines in the P-regime are closer to each other than in the D-regime. This follows directly from the mass action constant.[46]

8.3.4 Dependence on Dopant Content

Extrinsic dopants are crucial for tuning the properties of a material. In case of $LiFePO_4$ the effects of donor and acceptor doping are demonstrated in Fig. 8.9 by the ideal Brouwer diagrams. Donor doping results in the formation of positively charged defects $D^{\bullet}$ (e.g., Al^{3+} on Fe^{2+} sites = $Al_{Fe}^{\bullet}$) which leads to an increased concentration of negatively charged defects—here lithium vacancies V_{Li}'. Since the concentrations of lithium vacancies and holes are always coupled (see trapping reaction in Eq. 8.4), the hole concentration will decrease at the same time as shown in Fig. 8.9a.

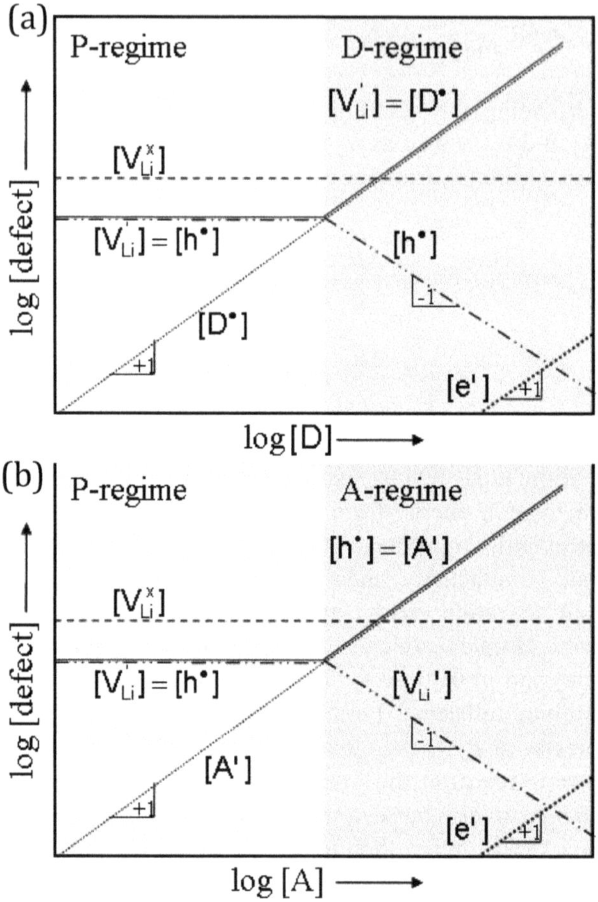

Figure 8.9 Brouwer diagrams showing the dependence of the defect concentration on (a) donor dopant content and (b) acceptor dopant content.

As to transport this means enhanced ionic conductivity while the electronic conductivity is depressed. In terms of ambipolar lithium conductivity, where a decisive quantity for the coupled motion of Li^+ and e^- is required for filling the electrode particles, this means a rather modest effect. This also holds for acceptor doping with the roles of V'_{Li} and $h^\bullet$ being swapped. Here doping produces negatively charged defects (A') which will raise the hole concentration and decrease the concentration of V'_{Li} (Fig. 8.9b, A-regime). With increasing acceptor dopant concentration the electronic conductivity is enhanced along with a depressed ionic conductivity.

The predictions by the Brouwer approximation could be proven experimentally with aluminum and silicon doped $LiFePO_4$.[36–38] In case of aluminum doped $LiFePO_4$ a further effect was observed besides the changes of the conductivities due to the doping. Here the electronic conductivity flattens at low temperatures, while the decrease of the ionic conductivity becomes steeper at low temperatures (the starting temperatures being the same).[36,37] The quantitative analysis succeeds by taking account of defect association:

$$Al^{\bullet}_{Fe} + V'_{Li} \rightarrow Al_{Fe}V_{Li} \tag{8.7}$$

Calculations of the activation energies of ionic and electronic conduction for both cases, (i) full association and (ii) no association lead to the following:[46]

Electronic conductivity:

$$\text{case 1: } E_{eon} = \frac{\Delta H_7}{2} - \Delta H_4 + h_h$$

$$\text{case 2: } E_{eon} = -\Delta H_4 + h_h \tag{8.5a}$$

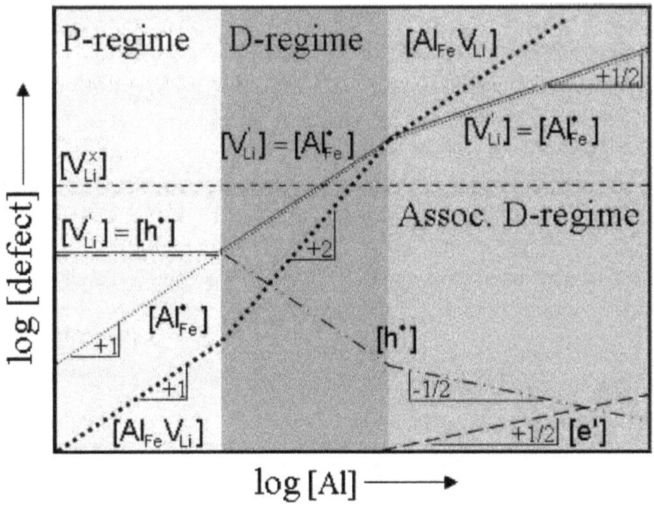

Figure 8.10 Dependence of defect concentration on donor dopant content ([Al]) including association.

Ionic conductivity:

$$\text{case 1: } E_{\text{ion}} = -\frac{\Delta H_7}{2} + h_v$$

$$\text{case 2: } E_{\text{ion}} = h_V \tag{8.5b}$$

with ΔH_7 being the reaction enthalpy for the association between lithium vacancies and aluminum defects according to Eq. 8.7, ΔH_4 being the trapping energy between lithium vacancies and holes, h_V and h_h being the migration enthalpies for lithium vacancies and holes, respectively. Qualitatively the effect of associa-tion in the D-regime is depicted in the Brouwer diagram in Fig. 8.10.

8.3.5 Extraction of Formation Enthalpies, Interaction Enthalpies, and Migration Energies

The defect chemical analysis does not only allow the construction of Brouwer diagrams and the prediction of how the transport properties develop, it also allows the determination of thermodynamic parameters (Table 8.1). Conductivity measurements on polycrystalline $LiFePO_4$ revealed activation energies for electronic conduction of 0.65 eV in the D-regime and 0.3 eV in the P-regime.[45] From Eqs. 8.5a and 8.6a the reaction enthalpy for the association of ionized lithium vacancies and holes (trapping energy) and the migration energy of holes can be estimated to be $\Delta H_4 = -0.6$ eV and $h_h = 0.05$ eV, respectively. Whereas ΔH_4 is in good agreement with values revealed by theoretical modeling (−0.4 eV[42] and > −0.5 eV,[47] the magnitude of migration enthalpy for the holes is some what lower than predicted by modeling studies (0.215 eV),[47] but is rather typical for ceramic insulators.[52]

Table 8.1 Extracted parameters from the defect chemical analysis

	b, c-axis
Migration energy of holes h_h	0.05 eV[a]
Migration energy of Li vacancies h_V	0.6 eV[b]
Trapping energy ΔH_4	−0.6 eV
Association energy ΔH_7 (1% Al doped $LiFePO_4$)	~ −1.0 eV

For a-axis: [a]0.2 eV; [b]0.74 eV.

From single crystal experiments on pure $LiFePO_4$ the migration energy for lithium vacancies h_v amounts to 0.6 eV along the b-axis[46] which is in excellent agreement with the value reported by Islam *et al.* but somewhat higher compared to the results of Morgan *et al.* (0.55 eV[42] and 0.3 eV,[40] respectively, both from theoretical modeling). The migration energy for lithium vacancies is the same along c-axis but higher along a-axis (0.7 eV).[46] For the silicon doped $LiFePO_4$ single crystals the migration energy amounts to ~0.7 eV for b, c-axis and 0.9 eV along a-axis.[38]

In the case of the aluminum doped $LiFePO_4$ sample the association energy between lithium vacancies and the aluminum defects $Al_{Fe}^{\cdot}$ can be determined to $\Delta H_7 \sim -1$ eV.[36,37] The difference between the activation energies in the fully and non-associated regime is for both, ionic and electronic conductivity, half of the association energy ΔH_7, which is in very good agreement with the experimental results.[36,37]

8.3.6 Comprehensive defect model of the $LiFePO_4$ / $FePO_4$ system

In a two-phase insertion mechanism, in order to correlate electrochemical measurements with defect models, the defect chemistry of both end members of the two-phase system needs to be established. Zhu *et al.* have investigated electronic transport in heterosite $FePO_4$.[53] Based on those transport measurements, a comprehensive defect model covering the $LiFePO_4/FePO_4$ system was derived (Fig. 8.11).

The striking inversion symmetry of the defect concentrations in $LiFePO_4$ and $FePO_4$ may be surprising at a first glance, but can be explained in a simple Gedanken experiment. Assuming for a moment that there is a continuous solid solution series spanning from fully delithiated $FePO_4$ to fully lithiated $LiFePO_4$ (vertical axis in the defect diagram), excess Li in $FePO_4$ can be considered as interstitials, while in almost fully occupied $LiFePO_4$ the defects are vacancies even though we refer to the same crystallographic sites. Excess electrons brought in by Li excess in the $FePO_4$ populate the almost empty conduction band. In slightly Li-deficient $LiFePO_4$ this band is almost full and the electronic defects will be referred to as holes in a valence band, even though we refer to the same band (Fig. 8.12).

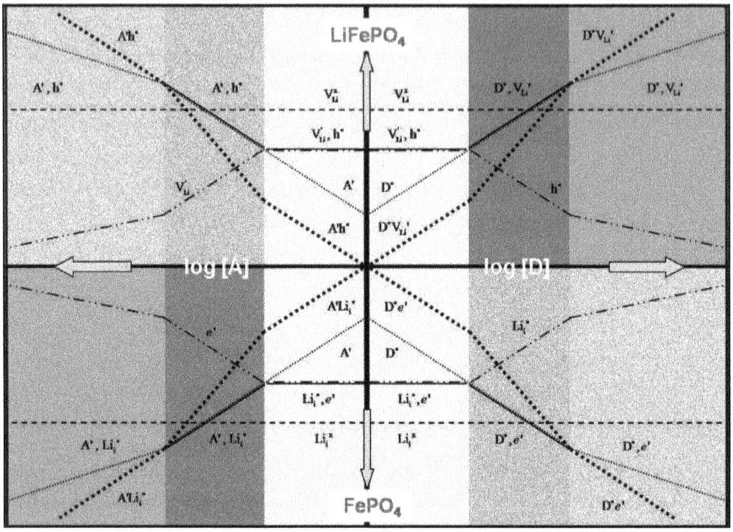

Figure 8.11 Dependence of defect concentrations in $LiFePO_4$ and $FePO_4$ on acceptor (A) and donor (D) content. Please note the symmetry that is highlighted by the colors. Reprinted with permission from Ref. 54. Copyright 2013, W. De Gruyter, Berlin.

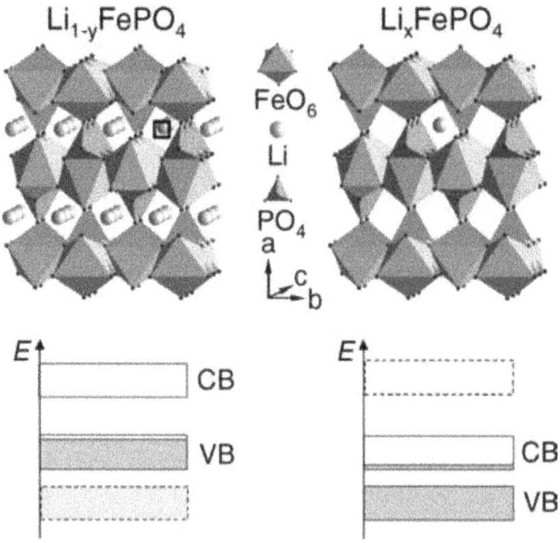

Figure 8.12 Band model of the electronic defects in slightly lithium-deficient $Li_{1-y}FePO_4$ and lithium-containing $FePO_4$. Upon changing the defect chemical reference point from $FePO_4$ to $LiFePO_4$, former conduction band now becomes the valence band. Reprinted with permission from Ref. 53, copyright 2011, John Wiley and Sons.

8.4 The $LiFePO_4/FePO_4$ Phase Transformation

As much as the field of defect chemistry has been terra incognita, as little is settled in terms of phase transition kinetics.[55–60] This latter point is particularly aggravated by the fact that the phase transition mechanism appears to be a sensitive function of size. While in large crystals a chemical delithiation leads to the formation of a porous $FePO_4$ layer growing with a parabolic rate law (the effective rate constant of which comprises solid and liquid contributions) at the surface of $LiFePO_4$ and in contact with the electrolyte. In the case of small crystals however, $FePO_4$ has been observed to grow within $LiFePO_4$, presumably due to strain effects.[57,61,62] This astonishing fact was corroborated recently by comparing small crystallites with large single crystals.[61] In the latter case the kinetics is determined by ambipolar lithium motion. Owing to the large volume variation, both solid state transport as well as lithium-ion conduction via the electrolyte through cracks can play a decisive role. In the model discussed by Delmas *et al.* lithium is not carried off by transport through the $FePO_4$ phase rather by lateral transport in $LiFePO_4$ along the fast b-axis.[59] The complexity of the phenomenon is underlined by the stripe pattern of the two co-existing phases formed in the experiments of Chen *et al*,[58] and even more strikingly by the recent finding of long-range ordering (staging).[63] Depending on size and other parameters, interfaces within particles have been observed as well as interface free single phase transitions and digital distributions (particles with zero and full Li-content). A key issue for this puzzling situation is the low excess free energies of the intermediate compositions.[64]

Here we will only stress four points:

(i) Before the phase transformation starts, the lithium content in the $LiFePO_4$ phase has to reach the saturation limit. The compositions of the solid solutions $Li_{1-\beta}FePO_4$ and $Li_\alpha FePO_4$ at room temperature vary ($0.04 \le \beta \le 0.17$, $0.02 \le \alpha \le 0.12$) and depend on particle size.[65–69] This process is determined by the chemical diffusion coefficient of lithium (D^{δ}_{Li}).

(ii) The propagation of the phase boundary is either surface or bulk controlled. We ignore here the possibility revealed by advanced TEM studies that an actual $Li_{0.5}FePO_4$ phase forms an interphase between $LiFePO_4$ and $FePO_4$. In the case that bulk transport is limiting, the decisive quantity is the ambipolar conductivity σ^{δ} of lithium defined as:

$$\sigma^{\delta} = \frac{\sigma_{\text{ion}}\sigma_{\text{eon}}}{\sigma_{\text{ion}} + \sigma_{\text{eon}}} \tag{8.8}$$

with the ionic conductivity σ_{ion} and the electronic conductivity σ_{eon}. Because of the significance of D^{δ}_{Li} and σ^{δ} we will shortly consider them from a defect-chemical point of view.

As already set out above electronic and ionic conductivities of $LiFePO_4$ can be influenced by doping with aliovalent cations. But due to the opposite effect of doping on positive ($h^{\bullet}$) and negative (V'_{Li}) charge carriers neither donor nor acceptor doping is able to greatly increase σ^{δ} as shown in Fig. 8.13.

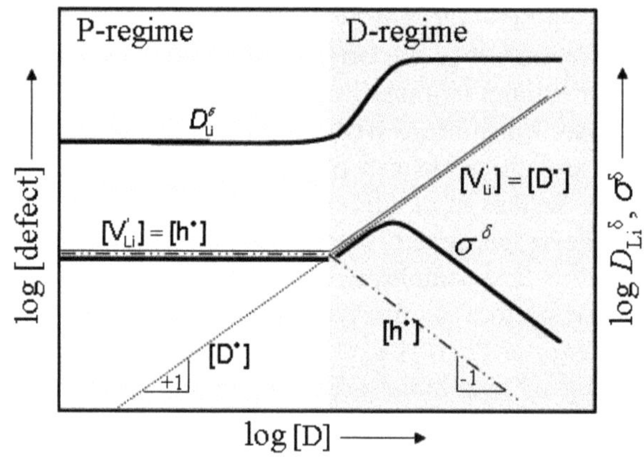

Figure 8.13 Brouwer diagram showing the dependence of σ^{δ} and D^{δ}_{Li} on donor doping content (for simplicity, dynamic trapping ignored).

The same holds true for D^{δ}_{Li}. In Fig. 8.13 where for the sake of simplicity dynamic trapping effects (see Refs. 46 and 70) are ignored switches to the constant diffusivity of the holes, when $[D^{\bullet}]$ is increased, which becomes limiting due to the depression of the hole concentration. The detailed consideration of D^{δ}_{Li} is more complicated than indicated as the interaction between V'_{Li} and $h^{\bullet}$ leads to severe differential trapping effects (see Ref. 48).

(iii) At extremely small particle sizes or 1d morphologies, such as electrospun fibers, the miscibility gap might disappear completely, and the insertion mechanism changes from two-phase to single-phase.[71,72] This may be caused by a varied Li potential (cf. capillary pressure[73]), but also by the marked energetic cost of interface formation. Also, high current densities in discharge can make the spinodal disappear, leading to apparent single-phase behavior[72,74] or amorphization.[75]

(iv) It has to be noted that a multiparticle situation may be very different from a single-phase situation. It has been shown by Dreyer *et al.* that the non-monotonic chemical potential variation on Li-content necessarily leads to a favored digital ($x = 0, 1$) distribution in a multi-particle ensemble.[76]

Irrespective of all uncertainties it seems clear that a particle size reduction down to some tens of a nanometer leads to a quick lithium infiltration provided the lithium (Li^+ and e^-) has been transported rapidly to the particle surface (and provided the transfer reaction is quick enough). Hence it is the network addressing these small particles and for high currents the Li^+ transfer resistance which become decisive for the performance of the electrode. The second half of the contribution is devoted to exactly this theme.

8.5 Size Effects

There are two important consequences of reducing the particle size in our context:[23,77–82] (i) faster charging and discharging and (ii) lower electrode polarization. Independently of the charging mechanism (insertion or bi-phasic), the shortest time τ needed to fully charge or discharge the active particles scales with the particle size squared ($\tau \propto L^2$). This quantity is much easier to define than to measure.

Measurements of electrode performance are usually carried out in the galvanostatic mode, therefore, the charging time also depends on the magnitude of the current. If the current value is relatively high, then the particles will not be fully charged at all (within a given potential window).

Figure 8.14a shows how the experimentally obtained capacity values vary with particle size at a current value of ca. 170 mAg^{-1} (C-rate = 1 C). In the case of larger particles, the diffusion front obviously did not reach the center of the particles–effectively only the particle shells (of a thickness l) were charged. Reducing the particle size down to l (the previous "shell" now representing the whole particle), however, results in increase of capacity which eventually approaches the theoretical value.

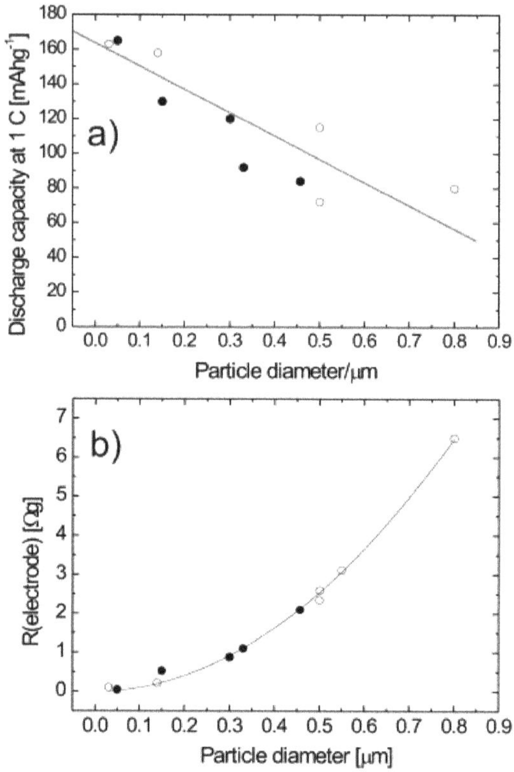

Figure 8.14 (a) Reversible capacity and (b) electrode resistance as functions of $LiFePO_4$ particle size. Full circles–coated $LiFePO_4$, open circles–uncoated $LiFePO_4$. The solid line in graph (b) is a fit with a function $R_{electrode}$ = *Const.* × R^n. The fit gives for n a value of 1.99 ± 0.05. Data taken from Refs. 14, 23, 24, and 83–89.

A more quantitative analysis is possible if we consider the size effect on the overpotential, that is, on the particle resistance. Because in a typical powder-based electrode all individual particles are supposed to be contacted by both Li^+- and e^--conducting pathways, we can evaluate the particle resistance simply from the resistance of the whole electrode. Experimentally, the whole electrode resistance can be estimated by dividing the overvoltage (deviation of the actual potential from the equilibrium potential) by the current flowing through the electrode.[90,91] Fig. 8.14 shows a collection of data gathered from the literature[14,23,24,83–89] including our own results. It is very interesting that despite the different preparation routes and different carbon decoration, all $LiFePO_4$ samples obey the same type of dependence of the electrode resistance on particle size. Moreover, the analysis shows that this dependence corresponds almost exactly to a square law.† The analysis given below shows that both the results of Fig. 8.14 as well as all other available literature data can be explained using relatively simple theoretical concepts.

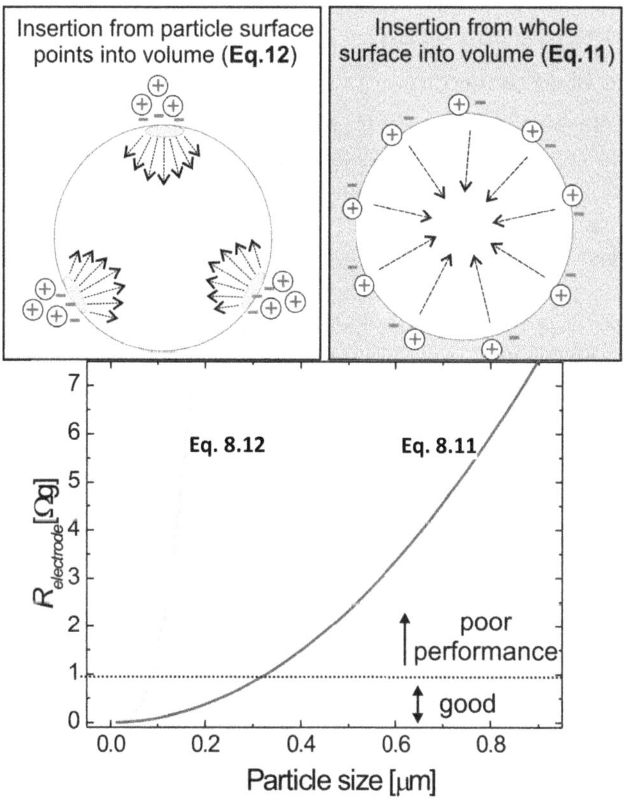

Figure 8.15 Schematic presentation of two extreme types of electrochemical contacting. Left curve: point-like electrochemical contacting corresponding to Eq. 8.10. Right curve: ideal electrochemical contacting corresponding to Eq. 8.9.

†Later on, this square dependence is also treated theoretically. At this point, we only wish to stress that such dependence is only valid for relatively low rates (≤1 C) which were used in all cases (at higher rates Eqs. 11 and 12 are not applicable anymore, see Section 8.6).

First, we explain theoretically the origin of the square law observed experimentally in Fig. 8.14b. We show that such a law naturally occurs when solving the equations for transport of mass and charge inside a particle. The detailed analysis depends on the boundary conditions, that is, on the way and geometry the particles are contacted. For the present purposes we will only consider two limiting cases: ideal contact (particles are uniformly "coated" by an ionic and electronic conductor) and point-like electrochemical contacting (Fig. 8.15, on left).

Case A: Ideal contacting

The transport equation for chemical diffusion inside a solid particle can be analytically solved for the case of homogeneous insertion. In $LiFePO_4$, homogeneous insertion occurs at both ends of typical charge/discharge curves,[67] in cases involving very small particles[60] or at sufficiently high temperatures.[92] In other cases, the intra-transport also involves a phase transition.[55,65,67] While in the first case the transport mechanism is diffusion (at each locus within the particle transport and storage take place simultaneo usly), in the second case the transport mechanism is permeation (storage occurs at the phase boundary that accordingly moves through the lattice). In both cases the "transport resistance" varies with the state of charge/discharge. We consider here the diffusion controlled mechanism and determine the resistance which refers to the fully charged/discharged system, that is, the diffusion front spreads over the whole particle. The same result is obtained also in the case of permeation if one refers to the same state of charge (phase boundary is assumed to be close to the particle centre, thus the permeation length becomes approximately equal to the particle radius) and assumes that the phase boundary mobility is very high.

$$R_m = \frac{1}{60}\frac{\rho}{\sigma^\delta}d^2 \tag{8.9}$$

where R_m is the electrode resistance, ρ is the active particle bulk density, d is the particle diameter and the total (ambipolar) conductivity, σ^δ, is defined

$$\sigma^\delta = \frac{\sigma_{\text{ion}}\sigma_{\text{eon}}}{\sigma_{\text{ion}} + \sigma_{\text{eon}}} \tag{8.8}$$

Here σ_{ion} and σ_{eon} are the ionic and electronic conductivity of the particle, respectively.

Case B: Point-like contacting

Another extreme case is met when the ionic and electronic conductor contact the active particle only at one or more points on the surface whereby the typical contact diameter is smaller than the typical distance between the contacts (Fig. 8.15, on left). In this case, the mass specific electrode resistance adopts a different, cubic dependence on the particle size:[23,77,93]

$$R_{m,\text{pt.contact}} = \frac{\pi}{12}\frac{\rho}{\sigma^{\delta} n d_{p.c.}} d^3 \tag{8.10}$$

where n is the number of contacts at a single particle and $d_{p.c.}$ the contact (wetting) diameter.

Very importantly, Eq. 8.10 also well describes another case of interest: the asymmetrical case where only the carrier with lower bulk $LiFePO_4$ conductivity point-contacts the $LiFePO_4$ surface (while the wetting topology by the carrier with the higher conductivity is optional).

Note the key difference between the ideal and point-like contacting: while Eq. 8.9 predicts a square law with respect to particle dimension d, the law becomes cubic in Eq. 8.10.

Other cases of possible interests

For the sake of completeness, we consider a third possible case that often appears to be of interest when discussing insertion systems. In this case we assume that the rate limiting step is not solid state diffusion (as in both cases above) but the surface step it self. Let us assume that the insertion/extraction proceeds homogeneously over the whole surface area of each active particle whereby the resistance due to insertion per unit surface area is R_{ins} (a convenient practical unit could be $\Omega\,cm^2$). Then the electrode resistance as a function of particle size is obtained in a straight forward way:

$$R_{m,\text{insertion}} = \frac{R_{ins}\rho}{6} d$$

As expected intuitively, the dependence between the electrode resistance and particle size is now linear—simply because the specific surface area increases linearly with decreasing particle size. Of course, the real measurement (Fig. 8.14b) is far from linearity, indicating that in the case of $LiFePO_4$ the insertion reaction is relatively fast so that $R_{m,\text{insertion}} << R_m$.

Practical consequences of Eqs. 8.9–8.10

The theoretical cases A and B shown above, in fact, illustrate quite well the conceptual dilemma that has served as inspiration for many experimental works during the past years. This dilemma is about the actual role of carbon coating around $LiFePO_4$ particles. Such carbon coating was initially proposed by Ravet *et al.*,[94] first clearly demonstrated by Dominko *et al.*[95] and then further developed in more than hundred works (typical examples can be found in Refs. 26, 84, 96–120). The dilemma about the coatings is the following: in vast majority of cases, the samples containing carbon coatings have shown a better electrochemical performance (higher capacity at given C-rate) than the samples without such coatings. If so, does this mean that carbon coating helps to improve the electrochemical contacting of $LiFePO_4$ particles? Or in terms of the above modeling: Does carbon coating perhaps shift the $LiFePO_4$ behavior from Eq. 8.12 to Eq. 8.11?

In the light of the above theoretical models, the answer to this dilemma is the following. First of all, carbon suppresses particle growth during heat treatment; so if we prepare coated and non-coated samples the particle sizes of the two samples will, very likely, be very different. In addition, the carbon coating prevents particle agglomeration.

Only such non-agglomerated particles can experience full electrochemical contacting (such as shown on the right of Fig. 8.15). On the contrary, the absence of native carbon may result in significant particle agglomeration during heating. Particles within agglomerates may have problems with being wetted by either the electrolyte (ions) and/or the electron source (carbon black additives cannot enter into agglomerated region, etc.). This limited access of either carrier to certain particles will effectively increase the particle size–the effective size will be the size of agglomerate rather than the size of individual primary particles constituting the agglomerate.

A deeper analysis of Eqs. 8.9 and 8.10) and comparison with the graph in Fig. 8.14b leads to a further, even more surprising conclusion: If the electrochemical contacting is independent of the presence or absence of an electronic conductor (carbon) around the whole particle surface, this must mean that the electronic conductivity of $LiFePO_4$ is significantly higher than the ionic conductivity. This appeared completely unexpected to the lithium battery community as it was general conviction that the problem of $LiFePO_4$ is the very low electronic conductivity (10^{-9} S cm^{-1}).[7,15,91] The present results indicate that the ionic conductivity is even lower, thus determining the overall kinetics. Although surprising, this finding is in nice agreement with the ionic conductivity measurements on single crystals of $LiFePO_4$ (see Section 8.2 and Ref. 35). Due to experimental limitations, such measurements were only possible at high temperatures (near or above 100°C), so direct comparison with the powder data obtained at room temperature is difficult. Nevertheless, the single crystals measurements have revealed undoubtedly that at temperatures between 140 and 152°C[35] the ionic conductivity is at least one to two orders of magnitude lower that the electronic conductivity (if the sample is not acceptor doped). Extrapolating to room temperature the ionic conductivity of $LiFePO_4$ would be in the range 10^{-10}–10^{-11} S cm^{-1}. This is the same order of magnitude as can be deduced from a fit of curve in Fig. 8.16 using Eq. 8.10.[78]

8.6 Network Effects

In Section 8.5 we focused on cases in which the overall kinetics were controlled exclusively by solid state transport inside active particles (intra-particle transport). We assumed—without actually proving it—that this would be the case at least at relatively low C-rates (say below 1 C). Our assumption was based on the phenomena known from the well-explored conventional electrochemical systems involving the transport of active species toward and their reaction at flat metallic electrodes. It is widely known that in such conventional systems, the overall kinetics at low overvoltages (small currents) are determined by the electrochemical reaction at the metal surface. Only at high enough overpotentials (high currents) the transport from or to the electrode will become rate determining. Unlike in conventional electrochemistry, complete kinetic analysis in terms of series and parallel steps has hardly been reported in battery research (the rare examples of such studies are Refs. 55, 65, 121, 122). In our opinion it is the lack of systematic experiments and the pronounced morphological complexity that is responsible for this unsatisfactory situation. In the following we present several approaches that can prove useful in further attempts to experimentally discern between the reaction and transport kinetics in battery insertion electrodes.‡

To understand the diversity of the phenomena that might occur at intermediate or high charge-discharge rates, it is useful to take a closer look at the distribution topology of the various phases that constitute a typical state-of-the-art $LiFePO_4$ cathode. Such a cathode is usually composed of 80 wt.% of active particles ($LiFePO_4$) with a mean particle size of 50–200 nm,[14,83–89,123] 10 wt.% of carbon black and 10 wt.% of a polymeric additive. This composite is pressed on an aluminum foil so that the final thickness of the composite is between 10 and 50 μm. The projected surface area (usually referred to as the geometric surface area) is about 1 cm^2. This means that the number of active particles in a typical $LiFePO_4$ cathode is between 10^{11} and 10^{12}. The ultimate goal in the course of electrode preparation is to connect all these particles in such a way that they will be simultaneously supplied by electrons and ions from the respective sources (sometimes we term this simultaneous connection of all particles to ionic and electronic source as "parallel electrochemical wiring", see Section 8.5). Of course, ideal parallel wiring of all 10^{11}–10^{12} particles is impossible to achieve in practice: there will always be a portion of particles only partly supplied either by ions and/or electrons. Some particles will probably be even disconnected from one or both charge sources or will disconnect upon cycling. At high charge/discharge rates the problem will only become worse. In fact, most improvements in the rate capability of $LiFePO_4$ cathodes that we have witnessed during the past years have explicitly or implicitly addressed the quality of electrochemical wiring. Here it is worth stressing that further reduction of particle size will inevitably lead to further complications in electrochemical wiring—the smaller the particles the higher their number and the more difficult it becomes to control the precise cathode nanoarchitectures (i.e., mutual spatial arrangement of the electrode nanophases).

To better understand the problems of charge and mass transport in a composite $LiFePO_4$ cathode, let us use a schematics showing how electrons and ions are actually supplied to a given $LiFePO_4$ active particle (Fig. 8.16). Upon insertion, the electrons will move along the easiest electronic paths starting from the aluminum

‡In our terminology between steps II and I, respectively (see further text and Fig. 8.16).

collector to the given $LiFePO_4$ active particle. The easiest paths will probably be along interconnected carbon black particles because their electronic conductivity (ca. 1 $S\,cm^{-1}$)[124,125] is a billion times higher than the conductivity of $LiFePO_4$ (ca. $10^{-9}\,S\,cm^{-1}$).[7,15,91] At the same time, the ions will move from the bulk electrolyte along the electrolyte in the pores between the $LiFePO_4$ particles to finally reach the given $LiFePO_4$ particle. Let us designate this part of transport (from collector or bulk electrolyte to the vicinity of selected particle) as transport step I (or "charge delivery", see arrows in Fig. 8.16, on left). The next major step (step II) refers to the active particles themselves (Fig. 8.16, on right) and has to some extent been treated in Section 8.5.

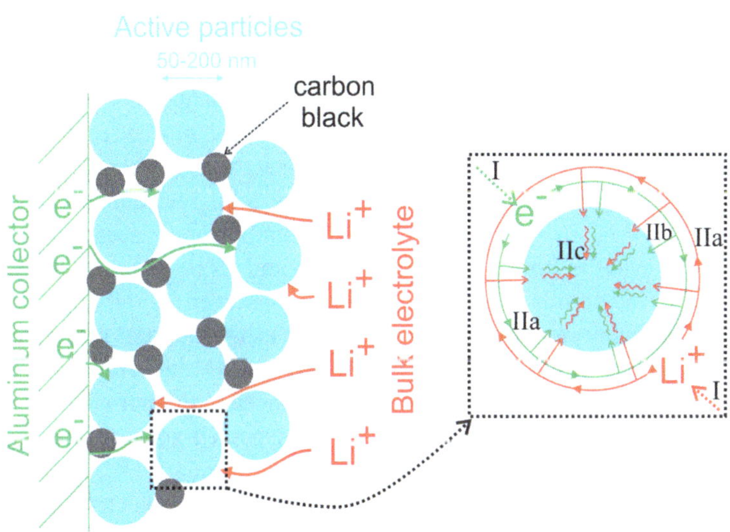

Figure 8.16 Schematics showing two major transport steps in a composite insertion cathode (e.g., $LiFePO_4$). Step II can be divided into several further substeps (IIa, IIb, and IIc) as discussed in certain sections of Section 8.6.

Let us first show how one can conveniently observe a transition from the case where step II is determining the transport to the case where step I becomes important as well. Such a transition can be observed by an experiment in which we prepare a set of electrodes from the same composite material but of different thicknesses. All electrodes are then galvanostatically charged/discharged in a wide range of different C-rates.[90] For each measurement the electrode resistance is determined as explained in the previous section. The electrode resistance is then plotted as a function of electrode thickness[§] for different C-rates. Examples of such sets of curves are shown in Fig. 8.17. We can see that the shape of the Electrode resistance-*vs.*-thickness (mass) curve greatly depends on the C-rate. Analysis shows that at low rates the dependence is exactly an inverse proportionality ($R_{electrode} \propto 1/L$, where L is the electrode thickness). At high enough rates, however, $R_{electrode}$ becomes virtually independent of the electrode thickness (mass). Let us try to understand this result by referring to the typical electrode structuring as described in the beginning of this chapter.

Using the division into two steps, we may represent the overall transport in terms of an equivalent circuit such as shown in Fig. 8.17, bottom. At low C-rates, step II is assumed to be rate limiting, as discussed above and in Section 8.5. This means that resistors $R_{I,i}$ in the equivalent circuit can be neglected. Now, the total resistance will depend on the number of the remaining resistors $R_{II,i}$. This number, in turn, is exactly proportional to the electrode mass (or thickness). Thus, the total resistance will be $R_{electrode} = R_{II,i}/N \propto 1/L$, where N is the number of resistors. This is exactly what is observed at rates C/20 and C/10 and, roughly, at C/5. However, this is not the only effect observed in Fig. 8.17. The other, equally remarkable phenomenon is that the total resistance is dropping significantly as we increase the C-rate. This drop of resistance is relatively much more pronounced in thinner electrodes (low masses) than in thicker. This is unexpected and not yet fully understood.

A resume would be the following: When the electrodes are thin enough and/or the rates small enough, the rate-determining step in the overall charge and mass transport will be step II, depending on network. On the contrary, when the electrodes are rather thick and/or the C-rates high, the overall transport will eventually become determined by step I (delivery).[||] The latter case resembles more the behavior of a supercapacitor rather than of a battery. Based on our experience we would propose that thin electrodes are such whose thickness is in the order of

[§]Alternatively, instead of thickness, one can plot electrode mass which is easier to measure.

[||]It would certainly be interesting to examine the recent reports on very high-rate $LiFePO_4$ (see Ref. 22 also in the light of the present findings.

10–20 μm. Low current rates in the case of $LiFePO_4$ are probably those below C/10. At the other extreme, electrodes are already considered thick when above 50–60 μm. High C-rates would be somewhere above 10 C.

A similar conclusion could also be drawn if the rate-determining step was the surface incorporation rather than bulk diffusion. As indicated in the equivalent circuit in Fig. 8.15, in the present context both substeps are described by a single resistor ($R_{II,i}$).

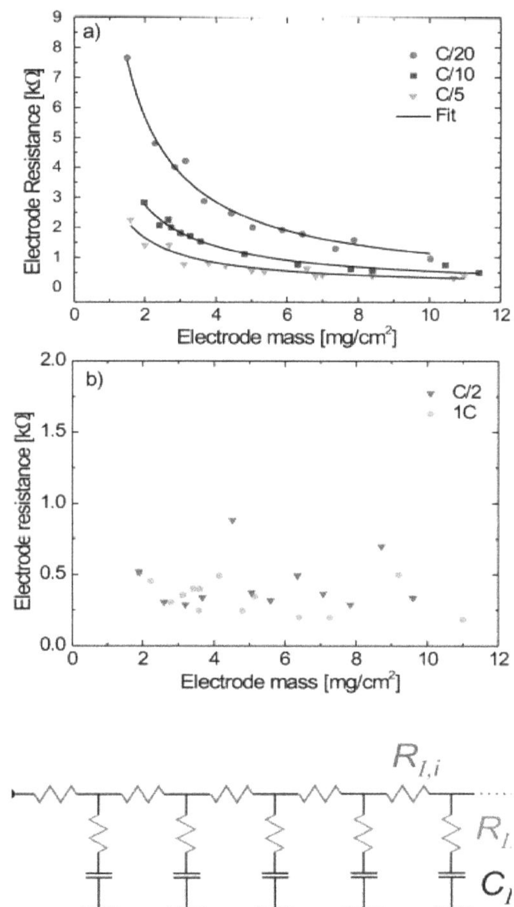

Figure 8.17 Electrode resistance as a function of electrode mass (thickness). Bottom: equivalent circuit taking into account resistance due to the both major steps (I and II). Data taken from Ref. 90.

8.6.1 Investigation of Substeps Determining the Rate of Step I

Due to the importance of step I at high rates and in thick electrodes, it is worth trying to find out which particular process within step I is the most critical. Let us first remember which processes are actually contained in step I. First, there is diffusion of ions from the bulk electrolyte along the electrolyte that is captured in the pores of $LiFePO_4$ active matter.[#] Under high current densities the concentration gradient may saturate and thus limit the C-rate. Finally, the transfer step of Li^+ from the electrolyte into the solid matter occurs. The transfer involves partial or complete desolvation, transfer through the solid electrolyte interface (SEI), etc. Second, step I also contains movement of electrons from the current collector to the surface of active particle. This movement can be quite complex: the electron has to jump from the metal into the nearby conductive carbon particle. Then it has to jump from one to another conductive carbon particle until reaching the active $LiFePO_4$ surface where it is incorporated. Alternatively, when carbon coating is also present (beside the carbon black particles), the electron may also jump from carbon black particle onto carbon coating or *vice versa*. Let us try to pinpoint which of the various proposed processes is actually the slowest.

The most accurate method that can discriminate between various kinetic processes is impedance spectroscopy.[126–128] A typical impedance spectrum of a composite $LiFePO_4$ electrode is shown in Fig. 8.18.[129,130] The spectrum consists of a high-frequency intercept (HI) followed by a clearly resolved high-frequency arc (HA). At even lower frequencies we sometimes observe a poorly resolved medium-frequency hump (MH) which is followed by the typical inclined spike (IS) at lowest frequencies. HI can easily be identified as the sum of bulk electrolyte and metal connector resistances. On the other side, IS is easily associated with the diffusion inside the $LiFePO_4$ active matter. The other two features, MH and HA are more difficult to identify.[130]

Especially the meaning of HA is very important as it represents a substantial contribution to the overall impedance. Traditionally, this arc has been ascribed to the incorporation resistance due to

[#]The pores could be those between the active nanoparticles or, alternatively, within inherently porous microsized particles.

step IIb (see Fig. 8.16) combined with the capacitance due to the double layer formed at the interface between $LiFePO_4$ particles and the surrounding electrolyte. However, already a very approximate calculation reveals a huge discrepancy between such an explanation and the measured value of arc capacity. Given the large surface area of active matter in a typical electrode ($\geq$200 cm^2) and noting that the typical capacity due to a double layer at the solid-liquid interface is about 10^{-5} Fcm^{-2}, we get an expected arc capacitance of $\geq$0.002 F.** This, however, is at least 2 orders of magnitude higher than typically observed. This discrepancy is too large to be ascribed to experimental uncertainty. If so, there is a need to find alternative explanation for HA. One suggestion could be that this arc is due to a process within step I. As described above, step I contains various processes where electrons jump from one phase to another. Such jumps are rate-determining, and thus reflected in HA, then the magnitude of HA should depend on the external mechanical pressure (the pressure should improve inter-phase contacting and thus facilitate the movement of electrons). Following this basic idea, we have performed several sets of experiments on different cell designs (coffee bag type) in which the external mechanical pressure was systematically varied.[130] In all cases, the magnitude of HA consistently followed the variation of pressure. A closer analysis of these experiments allowed us to identify the precise location of the rate-determining process: viz. the contact between aluminum current collector and the electrode composite. This assumption was then further checked by improving this particular contact using silver conductive paste or other conductive composites—after all such special treatments the magnitude of HA shrank virtually to zero.[130] In this way, we confirmed the origin of HA beyond any doubt. Most importantly, the value of the capacita-nce is also consistent with the properties of the aluminum-electrode interface (about 10^{-5} F/cm^2).

Finally, we also checked the correlation between the magnitude of HA and the galvanostatic performance. It was found consistently that the smaller the HA the better the electrochemical performance.[130]

Based on the extensive impedance measurements under various conditions we have also attempted a construction of the

**The same result is obtained if the interfacial capacity is ascribed to any other interfacial phase with a thickness of ca. 1 nm (e.g., electrolyte-soaked carbon nanofilm or eventual passive layer).

corresponding general equivalent circuit that takes account of all possible complications due to transport step I.[130] Analyzing this circuit, we could also identify the possible meaning of the poorly expressed MH—it is either due to interparticle jumps of electrons or due to the incorporation reaction (transport step IIb, see Fig. 8.16). In any case, the contribution of MH to the total transport impedance is usually very small and can be neglected in the case of usual $LiFePO_4$. This gives some further credit to our modeling of processes II where we decided implicitly to neglect process IIb.

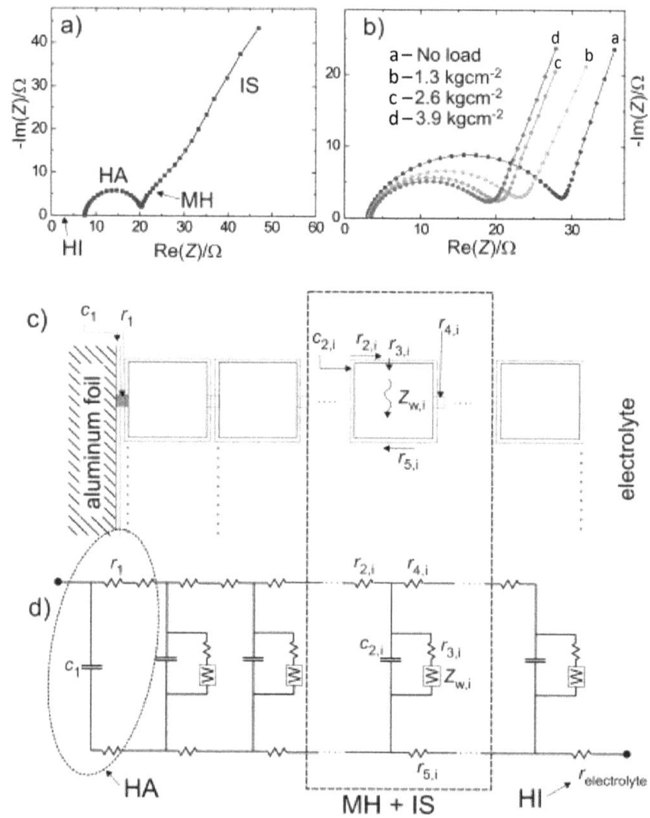

Figure 8.18 (a) Typical complex plane impedance plot of a $LiFePO_4$ electrode. (b) Dependence of impedance arc on external mechanical loading (the experiments were performed on a symmetrical two-electrode cell as described in Ref. 130. (c) Schematic presentation of contacting in a polyparticle electrode system. (d) Equivalent circuit roughly describing the schematics shown in graph (c).

The above results reveal that inter-phase contacting may have a considerable contribution to step I and thus to the overall transport in a $LiFePO_4$ cathode, especially at higher rates or in thicker electrodes. It is thus of high importance to find ways how to improve this contacting. Introduction of new, conductive binders instead of the present inert insulating polymers is certainly an idea to be considered. Similarly, carbon black additives could be improved so that they would adhere more tightly onto active particles.

In a perspective, even if we succeed to get rid of the impedance due to interphase contacting, there will always remain the contribution due to electrolyte—both bulk electrolyte (HI) and the electrolyte captured in the pores of electrode material (another process hidden in MH). Minimization of these contributions has actually already been subject of intensive investigations, however, not in batteries but in the field of supercapacitors. At least when trying to create high-power (high C-rate) batteries, it would be advisable to make better use of this knowledge.

8.6.2 Porous $LiFePO_4$

At least some problems discussed above (contacts, wiring) can be avoided using another concept of electrode structuring, *viz.* porous electrode structures.

In Section 8.5 we showed that one of the serious problems that can severely affect the kinetics of $LiFePO_4$ is particle agglomeration. Due to limited electrochemical wetting of particles that are captured in the interior of agglomerates, agglomeration could be viewed as nothing else but an effective increase of mean particle size. The crucial role of mean particle size was clearly demonstrated both experimentally (Fig. 8.14) and theoretically (Eqs. 8.9–8.10).

If agglomeration is recognized as a problem, then it is reasonable to seek for further ways to avoid it. An obvious and conceptually elegant approach is to use porous rather than particulate materials architectures. For example, a porous $LiFePO_4$ particle with sufficient open porosity could accommodate a significant amount of electrolyte which would then ionically wet the huge internal surface area. The fact that porous structures are expected to be inherently (meta-) stable and should efficiently sustain the mechanical pressures during insertion/extraction, only adds to their potential attractiveness. The problem, of course, could be electronic contacting of the huge internal surface area. Given the very low electronic conductivity of $LiFePO_4$ (10^{-9} S cm^{-1}) one assumes that we need to introduce at least some amount of electronic conductor into the interior of porous structure to provide suitable electronic paths from the outside to the interior surface where the insertion reaction takes place.

We have solved the problem of internal electronic wiring of porous $LiFePO_4$ by developing a special sol–gel procedure where by, apart from using reactants for formation of $LiFePO_4$, the crucial additive was a citrate precursor.[27,131–133] Upon heat treatment under appropriate conditions, this precursor is decomposed into solid carbon plus gases (CO_2, CO, etc.).[27] If the gases evolve just prior to solidification of the active compound ($LiFePO_4$), they can create interconnected porosity throughout the soft pre-solid matter. Simultaneously, the solid carbon is deposited on the walls of solidifying matter. Thus, after solidification we end up with interconnected porous system whereby the walls of pores are decorated with carbon. After adding electrolyte we have thus a good electrochemical wetting of all internal surfaces of the porous particles: the electrons are accommodated on the carbon phase while ions are provided by the electrolyte.

After introducing the above concept of porous $LiFePO_4$[27,131] we were interested whether this system could be further optimized. In particular, we were interested in two fundamental aspects: (i) whether the internal electronic contacting due to deposited carbon film was really successful or could be further improved and (ii) whether the porosity was sufficient to allow good ionic wetting of the interior of porous $LiFePO_4$. Both issues are discussed separately in continuation of this chapter.

8.6.2.1 Optimization of electronic contacting

Although our citrate-based synthesis[27,131–133] provides significant amounts of solid carbon material that is deposited on all surfaces, including those in the interior of porous $LiFePO_4$ particles, it is possible that this carbon deposits are not distributed optimally. Any discontinuities in carbon paths, however, might lead to non-optimal electronic wetting of the huge internal surface area. For this reason we tried to find methods to further improve the electronic wetting of porous $LiFePO_4$. A good candidate was found in RuO_2 with metallic-like conductivity. Pore metallization by RuO_2 has proven successful for optimizing the anode performance of anatase[21] and has been successfully exploited previously for decoration of porous systems intended for catalysis.[134] Using a similar approach we filled our porous, carbon decorated $LiFePO_4$ with a certain amount of RuO_2.[21] As shown in Fig. 8.19, at high rates where an efficient wiring becomes indispensable the performance of RuO_2 treated $LiFePO_4$-C was greatly improved. We explain this improvement by assuming that after treatment with RuO_2, owing to the better wetting, the discontinuities in carbon deposits were repaired and, perhaps in addition, the Li^+ transfer was facilitated. It is also remarkable that the benefit of RuO_2 shows at high rates where fast charge carrier availability is more crucial.

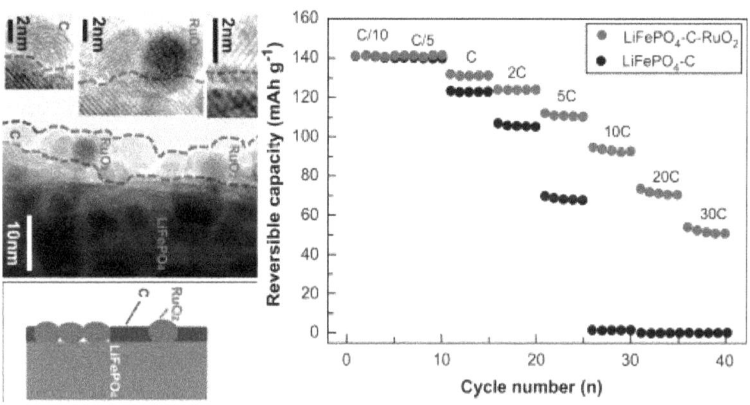

Figure 8.19 Restoration of citrate derived carbon coating on porous $LiFePO_4$ (prepared according to Ref. 27) by addition of RuO_2. The latter fills the cracks in the native carbon and thus improves the rate capability. Adapted from Ref. 21.

8.6.2.2 Optimization of porosity

In the course of synthesis development we found that the heating regime (heating rate, the final temperature, etc.) had extreme influence on the final porosity of $LiFePO_4$ particles (the size of the latter was typically several micrometers). Of course, the heating rate affects the rate of gas evolution but also the rate of solidification of $LiFePO_4$. Both processes combined will determine the final sample porosity. The porosity, in turn, will have important impact on the electrochemistry. The dependence of electrochemical performance on the porosity of $LiFePO_4$ micrometer-sized particles can be demonstrated on three samples with different distribution of pore sizes and different total volume of pores (Fig. 8.20). Obviously, the impact of total pore volume (and pore size distribution) on the $LiFePO_4$ capacity at constant C-rate is tremendous.

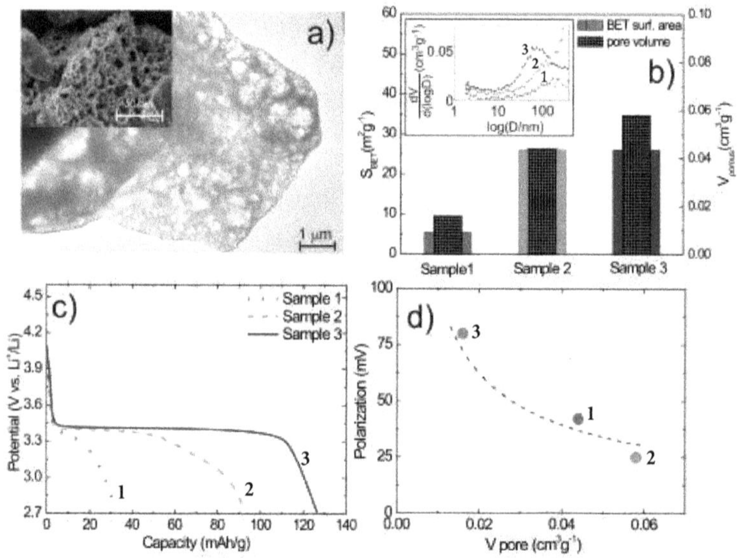

Figure 8.20 (a) SEM photographs showing a typical porous structure of $LiFePO_4$ as discussed in detail in Ref. 27. (b) Pore volume and Brunauer–Emmett–Teller (BET) surface area for three samples with different porosities; inset: the corresponding distribution of pore sizes. (c) 2nd discharge curves for all three $LiFePO_4$ samples at C/2 (85 mAh g^{-1}). (d) Voltage difference (proportional to electrode resistance) between charge and discharge as a function of pore volume.

Figure 8.20d also reveals a correlation between the pore volume (or pore size distribution) and the electrode polarization, the latter being proportional to the electrode resistance because the C-rate is always the same. A closer inspection of microstructural features of all three samples offers the following explanation of these results: sample 2 works better than sample 1 either because it has a larger BET surface area (more surface is electrochemically wetted) and/or because it has greater pore volume (allowing better delivery of ions, see next paragraph). Sample 3 works better than sample 2 primarily because it has greater pore volume (better delivery of ions, see above) while the BET surface area is comparable. This explanation already indicates that good electrochemical wetting is only one condition that leads to improved electrochemistry. There are certainly cases where step I (delivery of charges) is equally if not more important. In fact, such cases have been tackled in some detail above (see Fig. 8.17 and the corresponding explanation).

Given that electrode porosity is obviously very important for the kinetics, one might ask how the porosity could be optimized in the case of particulate materials. This problem becomes especially relevant when the particle size becomes very small, say in the range below 20 nm. Namely, in such cases carbon black and other additives are too big to serve as spacers between individual active nanoparticles. Unfortunately, no experimental studies covering this topic are available for $LiFePO_4$. Erjavec *et al.* have tackled this problem, however, on the example of anatase (TiO_2) with typical particle size between 5 to 10 nm.[135–137] They have found that the porosity of material consisting of such small nanoparticles can be improved by surface functionalization of particles using silica or other similar ceramic materials (alumina, niobia). Here it is necessary to stress that this functionalization means covalent attachment of extremely small amounts of silica rather than formation of ceramic films. Such surface functionalization is sufficient to prevent active particle (TiO_2) agglomeration which, consequently, increases the meso- and microporosity of the final electrode material.[136] It can be advised to try a similar approach in $LiFePO_4$ particulate materials with typical particle sizes < ca. 50 nm.

Recently, the Ceder group reported considerably improved performance of $LiFePO_4$ at high rates.[22] The authors explained this by a glassy coating that apparently improves Li^+ delivery to the $LiFePO_4$. Due to the high Li^+ mobility in the glassy layer, delivery to the (010) facet of the particles is enhanced, reducing the overall anisotropy in Li^+ transport. An alternate explanation of the improved transport properties is possible within the framework of heterogeneous doping[138]: Adams proposes Li^+ redistribution at the $LiFePO_4$|glassy $Li_4P_2O_7$ interface, with 10–60 fold increase of Li+ diffusivity in vicinity of the interface.[139,140] Another alternative interpretation comes from the enormous increase in rate performance when the carbon content is increased from 15% to 60%. This suggests that, obviously, a very good wiring was achieved and that this point seems more important than the catalytic/doping effect proposed.

Another recent development are reports on flow batteries based on $LiFePO_4$ electrode material.[141,142] Here, the active material is not attached to metallic current collectors, as in a regular lithium ion battery, but is dispersed in a colloidal suspension, that can be replenished from a storage tank, allowing one to decouple energy density and power density of the cell. Transport networks between active material particles, current collector, and electrolyte form *in situ* as the dispersion also contains carbon particles. Challenges are the complex rheologic behavior of such suspensions, and percolation issues in the transport networks.

A final practical recipe that can be extracted from the above results could be condensed into two short statements. The first is that at charge/discharge rates near or below 1 C given standard conditions (carbon admixing, binder . . .) the electrode resistance (and the performance) of $LiFePO_4$ is basically determined only by the particle size. This is because the effect of wiring can be neglected at such conditions and the transport is merely determined by the transport into and/or inside the solid $LiFePO_4$ particles. The second conclusion would be that the main role of carbon coatings is to prevent particle growth and also to prevent particle agglomeration during heat treatment; this way, an intimate electrochemical contact of each individual particle is made possible. In the absence of carbon, however, individual $LiFePO_4$ nanoparticles will not only grow during heat treatment but will also form agglomerates during and after heating; the individual particles that are captured within such agglomerates are not wired well neither to the ionic nor to the electronic reservoir.

References

1. Padhi, A. K., Nanjundaswamy, K. S. & Goodenough, J. B. Phospho-olivines as positive-electrode materials for rechargeble lithium batteries. *J. Electrochem. Soc.*, **144**, 1188–1194 (1997).
2. Ohzuku, T. & Brodd, R. J. An overview of positive-electrode materials for advanced lithium-ion batteries. *J. Power Sources*, **174**, 449 (2007).

3. Amin, R., Maier, J., Balaya, P., Chen, D. P. & Lin, C. T. Ionic and electronic transport in single crystalline $LiFePO_4$ grown by optical floating zone technique. *Solid State Ionics*, **179**, 1683 (2008).

4. Yao, J., Konstantinov, K., Wang, G. & Liu, H. Electrochemical and magnetic characterization of $LiFePO_4$ and $Li_{0.95}Mg_{0.05}FePO_4$ cathode materials. *J. Solid State Electrochem.*, **11**, 177 (2007).

5. Meethong, N., Kao, Y.-H., Speakman, S. A. & Chiang, Y.-M. Aliovalent substitutions in olivine lithium iron phosphate and impact on structure and properties. *Adv. Funct. Mater.*, **19**, 1060–1070 (2009).

6. Wagemaker, M., Ellis, B. L., Lützenkirchen-Hecht, D., Mulder, F. M. & Nazar, L. F. Proof of supervalent doping in olivine $LiFePO_4$. *Chem. Mater.*, **20**, 6313–6315 (2008).

7. Chung, S.-Y., Bloking, J. T. & Chiang, Y.-M. Electronically conductive phospho-olivines as lithium storage electrodes. *Nature Mater.*, **1**, 123–128 (2002).

8. Wang, G. X. *et al.* Physical and electrochemical properties of doped lithium iron phosphate electrodes. *Electrochim. Acta* **50**, 443 (2004).

9. Wang, G., Cheng, Y., Yan, M. & Jiang, Z. $Li_{0.99}Ti_{0.01}FePO_4$/C composite as cathode material for lithium ion battery. *J. Solid State Electrochem.*, **11**, 457–462 (2007).

10. Shi, S. *et al.* Enhancement of electronic conductivity of $LiFePO_4$ by Cr doping and its identification by first-principles calculations. *Phys. Rev. B*, **68**, 195108 (2003).

11. Ni, J. F., Zhou, H. H., Chen, J. T. & Zhang, X. X. $LiFePO_4$ doped with ions prepared by co-precipitation method. *Mater. Lett.*, **59**, 2361 (2005).

12. Wang, G. X. *et al.* Synthesis and characterization of $LiFePO_4$ and $LiTi_{0.01}Fe_{0.99}PO_4$ cathode materials. *J. Electrochem. Soc.*, **153**, A25 (2006).

13. Wang, D., Li, H., Shi, S., Huang, X. & Chen, L. Improving the rate performance of $LiFePO_4$ by Fe-site doping. *Electrochim. Acta*, **50**, 2955–2958 (2005).

14. Liu, H. *et al.* Doping effects of zinc on $LiFePO_4$ cathode material for lithium ion batteries. *Electrochemistry Communications*, **8**, 1553 (2006).

15. Herle, P. S., Ellis, B., Coombs, N. & Nazar, L. F. Nano- *Electrochem. Commun.* Netword electronic conduction in iron and nickel olivine phosphates. *Nature Mater.*, **3**, 147–152 (2004).

16. Ravet, N., Goodenough, J. B., Besner, S., Gauthier, M. & Armand, M. Additives improving phospho-olivines conductivity. Abs 127, The

Electrochemical Society and the Electrochemical Society of Japan meeting Abstracts, Vol. 99-2, Honolulu, HI, Oct 17–22 (1999).

17. Cheruvally, G. in *Materials Science Foundations Series*, Vol. 38, Trans Tech Publications, Switzerland (2008).
18. Murugan, A. V., Muraliganth, T., Ferreira, P. J. & Manthiram, A. Dimensionally modulated, single-crystalline $LiMPO_4$ (M = Mn, Fe, Co, and Ni) with nano-thumblike shapes for high-power energy storage. *Inorg. Chem.* **48**, 946–952, doi:10.1021/ic8015723 (2009).
19. Park, K. S. *et al.* Surface modification by silver coating for improving electrochemical properties of $LiFePO_4$. *Solid State Commun.*, **129**, 311 (2004).
20. Croce, F. *et al.* A novel concept for the synthesis of an improved $LiFePO_4$ lithium battery cathode. *Electrochem. Solid-State Lett.*, **5**, A47–A50 (2002).
21. Hu, Y.-S. *et al.* Improved electrode performance of porous $LiFePO_4$ using RuO_2 as an oxidic nanoscale interconnect. *Adv. Mater.*, **19**, 1963–1966 (2007).
22. Kang, B. & Ceder, G. Battery materials for ultrafast charging and discharging. *Nature*, **458**, 190–193 (2009).
23. Delacourt, C., Poizot, P., Levasseur, S. & Masquelier, C. Size effects on carbon-free $LiFePO_4$ powders—The key to superior energy density. *Electrochem. Solid-State Lett.*, **9**, A352–A355 (2006).
24. Kim, D.-H. & Kim, J. Synthesis of $LiFePO_4$ nanoparticles in polyol medium and their electrochemical properties. *Electrochem. Solid-State, Lett.* **9**, A439–A442 (2006).
25. Prosini, P. P. *et al.* A new synthestic route for preparing $LiFePO_4$ with enhanced electrochemical performance. *J. Electrochem. Soc.*, **149**, A886–A890 (2002).
26. Gaberscek, M., Dominko, R., Bele, M., Remskar, M. & Jamnik, J. Mass and charge transport in hierarchically organized storage materials. Example: Porous active materials with nanocoated walls of pores. *Solid State Ionics*, **177**, 3015 (2006).
27. Dominko, R. *et al.* Wired porous cathode materials: A novel concept for synthesis of $LiFePO_4$. *Chem. Mater.*, **19**, 2960–2969, doi:10.1021/cm062843g (2007).
28. Streltsov, V. A., Belokoneva, E. L., Tsirelson, V. G. & Hansen, N. K. Multipole analysis of the electron density in triphylite, $LiFePO_4$, using x-ray diffraction data. *Acta Cryst. B*, **49**, 147–153 (1993).

29. Liang, G. *et al.* Anisotropy in magnetic properties and electronic structure of single-crystal $LiFePO_4$. *Phys. Rev., B* **77**, 064414 (2008).

30. Chen, D. P. *et al.* Magnetic anisotropy in doped and undoped $LiFePO_4$ single crystals. *J. Appl. Phys.*, **101**, 09N512 (2007).

31. Chen, J. *et al.* The hydrothermal synthesis and characterization of olivines and related compounds for electrochemical applications. *Solid State Ionics*, **178**, 1676 (2008).

32. Li, J., Garlea, V. O., Zarestky, J. L. & Vaknin, D. Spin-waves in antiferromagnetic single-crystal $LiFePO_4$. *Phys. Rev. B*, **73**, 024410 (2006).

33. Chen, D. P., Maljuk, A. & Lin, C. T. Floating zone growth of lithium iron(II) phosphate single crystals. *J. Cryst. Growth*, **284**, 86–90 (2005).

34. Li, J., Yao, W., Martin, S. & Vaknin, D. Lithium ion conductivity in single crystal $LiFePO_4$. *Solid State Ionics*, **179**, 2016–2019 (2008).

35. Amin, R., Balaya, P. & Maier, J. Anisotropy of electronic and ionic transport in $LiFePO_4$ single crystals. *Electrochem. Solid-State Lett.*, **10**, A13–A16 (2007).

36. Amin, R., Lin, C. & Maier, J. Aluminium-doped $LiFePO_4$ single crystals Part I. Growth, characterization and total conductivity. *Phys. Chem. Chem. Phys.*, **10**, 3519–3523 (2008).

37. Amin, R., Lin, C. & Maier, J. Aluminium-doped $LiFePO_4$ single crystals Part II. Ionic conductivity, diffusivity and defect model. *Phys. Chem. Chem. Phys.*, **10**, 3524–3529 (2008).

38. Amin, R. *et al.* Silicon-doped $LiFePO_4$ single crystals: Growth, conductivity behavior, and diffusivity. *Adv. Funct. Mater.*, **19**, 1697–1704 (2009).

39. Maier, J. Evaluation of electrochmical methods in solid state research and their generalization for defects with variable charges. *Z. Phys. Chem.*, **140**, 191–215 (1984).

40. Morgan, D., Van der Ven, A. & Ceder, G. Li Conductivity in Li_xMPO_4 (M = Mn, Fe, Co, Ni) olivine materials. *Electrochem. Solid-State Lett.*, **7**, A30–A32 (2004).

41. Ouyang, C., Shi, S., Wang, Z., Huang, X. & Chen, L. First-principle study of Li ion diffusion in $LiFePO_4$. *Phys. Rev. B*, **69**, 104303 (2004).

42. Fisher, C. A. J., Prieto, V. M. H. & Islam, M. S. Lithium battery materials $LiMPO_4$ (M = Mn, Fe, Co and Ni): Insights into defect association, transport mechanism and doping behavior. *Chem. Mater.*, **20**, 5907–5915 (2008).

43. Nishimura, S.-I. *et al.* Experimental visualization of lithium diffusion in Li_xFePO_4. *Nature Mater.*, **7**, 707–711 (2008).

44. Malik, R., Burch, D., Bazant, M. & Ceder, G. Particle size dependence of the ionic diffusivity. *Nano Lett.*, **10**, 4123–4127, doi:10.1021/nl1023595 (2010).

45. Amin, R. & Maier, J. Effect of annealing on transport properties of $LiFePO_4$: Towards a defect chemical model. *Solid State Ionics*, **178**, 1831 (2008).

46. Maier, J. & Amin, R. Defect chemistry of $LiFePO_4$. *J. Electrochem. Soc.*, **155**, A339 (2008).

47. Maxisch, T., Zhou, F. & Ceder, G. Ab initio study of the migration of small polarons in olivine Li_xFePO_4 and their association with lithium ions and vacancies. *Phys. Rev. B*, **73**, 104301 (2006).

48. Maier, J. Physical chemistry of ionic materials: Ions and electrons in solids (2004).

49. Prosini, P. P., Lisi, M., Zane, D. & Pasquali, M. Determination of the chemical diffusion coefficient of lithium in $LiFePO_4$. *Solid State Ionics*, **148**, 45–51 (2002).

50. Maier, J. Complex oxides: High temperature defect chemistry vs. low temperature defect chemistry. *Phys. Chem. Chem. Phys.*, **5**, 2164–2173 (2003).

51. Islam, M. S., Driscoll, D. J., Fisher, C. A. J. & Slater, P. R. Atomic scale investigation of defects, dopants, and lithium transport in the $LiFePO_4$ olivine-type battery material. *Chem. Mater.*, **17**, 5085–5092 (2005).

52. Müller, J., Kreuer, K. D., Maier, J., Matsuo, S. & Ishigame, M. A conductivity and thermal gravimetric analysis of a Y-doped $SrZrO_3$ single crystal. *Solid State Ionics*, **97**, 421 (1997).

53. Zhu, C., Weichert, K. & Maier, J. Electronic conductivity and defect chemistry of heterosite $FePO_4$. *Adv. Funct. Mater.*, **21**, 1917–1921, doi:10.1002/adfm.201002059 (2011).

54. Samuelis, D. & Maier, J. in *Chemical Energy Storage* (ed. Robert Schlögl) Ch. 3.5, 225–248 (De Gruyter, 2013).

55. Srinivasan, V. & Newman, J. Existence of path-dependence in the $LiFePO_4$ electrode. *Electrochem. Solid-State Lett.*, **9**, A110–A114 (2006).

56. Andersson, A. S. & Thomas, J. O. The source of first-cycle capacity loss in $LiFePO_4$. *J. Power Sources*, **97–98**, 498–502 (2001).

57. Laffont, L. *et al.* Study of the $LiFePO_4/FePO_4$ two-phase system by high-resolution electron energy loss spectroscopy. *Chem. Mater.*, **18**, 5520–5529 (2006).

58. Chen, G., Song, X. & Richardson, T. J. Electron microscopy study of the $LiFePO_4$ to $FePO_4$ phase transition. *Electrochem. Solid-State Lett.*, **9**, A295 (2006).

59. Delmas, C., Maccario, M., Croguennec, L., Le Cras, F. & Weill, F. Lithium deintercalation in $LiFePO_4$ nanoparticles via a domino-cascade model. *Nat. Mater.*, **7**, 665 (2008).

60. Gibot, P. *et al.* Room-temperature single-phase Li insertion/extraction in nanoscale Li_xFePO_4. *Nat. Mater.*, **7**, 741 (2008).

61. Sigle, W., Amin, R., Weichert, K., van Aken, P. A. & Maier, J. Delithiation study of $LiFePO_4$ crystals using electron energy-loss spectroscopy. *Electrochem. Solid-State Lett.*, **12**, A151 (2009).

62. Weichert, K. *et al.* Phase boundary propagation in large $LiFePO_4$ single crystals on delithiation. *J. Am. Chem. Soc.*, **134**, 2988–2992, doi:10.1021/ja207124a (2012).

63. Gu, L. *et al.* Direct observation of lithium staging in partially delithiated $LiFePO_4$ at atomic resolution. *J. Am. Chem. Soc.*, **133**, 4661 (2011).

64. Malik, R., Zhou, F. & Ceder, G. Kinetics of non-equilibrium lithium incorporation in $LiFePO_4$. *Nat. Mater.*, **10**, 587–590, doi:10.1038/nmat3065 (2011).

65. Srinivasan, V. & Newman, J. Discharge model for the lithium iron-phosphate electrode. *J. Electrochem. Soc.*, **151**, A1517–A1529 (2004).

66. Yamada, A., Koizumi, H., Sonoyama, N. & Kanno, R. Phase change in Li_xFePO_4. *Electrochem. Solid-State Lett.*, **8**, A409–A413 (2005).

67. Yamada, A. *et al.* Room-temperature miscibility gap in Li_xFePO_4. *Nat. Mat.*, **5**, 357–360 (2006).

68. Meethong, N., Huang, H.-Y. S., Carter, Y. C. & Chiang, Y.-M. Size-dependent lithium miscibility gap in nanoscale $Li_{1-x}FePO_4$. *Electrochem. Solid-State Lett.*, **10**, A134–A138 (2007).

69. Wagemaker, M. *et al.* Dynamic solubility limits in nanosized olivine $LiFePO_4$. *J. Am. Chem. Soc.*, **133**, 10222–10228, doi:10.1021/ja2026213 (2011).

70. Maier, J. Mass transport in the presence of internal defect reactions—Concept of conservative ensembles: I, Trapping effect of dopants on chemical diffusion. *J. Am. Ceram. Soc.*, **76**, 1223–1227 (1993).

71. Burch, D. & Bazant, M. Z. Size-dependent spinodal and miscibility gaps for intercalation in nanoparticles. *Nano Lett.*, **9**, 3795–3800 (2009).

72. Burch, D., Singh, G., Ceder, G. & Bazant, M. Z. Phase-transformation wave dynamics in $LiFePO_4$. *Def. Diff. Forum*, **B139**, 95–100 (2008).

73. Maier, J. Thermodynamics of Electrochemical Lithium Storage. *Angewandte Chemie International Edition*, (in print) doi:10.1002/anie.201205569 (2013).

74. Bai, P., Cogswell, D. A. & Bazant, M. Z. Suppression of phase separation in $LiFePO_4$ nanoparticles during battery discharge. *Nano Letters*, **11**, 4890–4896, doi:10.1021/nl202764f (2011).

75. Tang, M., Carter, W. C., Belak, J. F. & Chiang, Y.-M. Modeling the competing phase transition pathways in nanoscale olivine electrodes. *Electrochim. Acta*, **56**, 969–976, doi:10.1016/j.electacta.2010.09.027 (2010).

76. Dreyer, W. *et al.* The thermo-dynamic origin of hysteresis in insertion batteries. *Nature Mater.*, **9**, 448–453 (2010).

77. Gaberscek, M. & Jamnik, J. Impact of electrochemical wiring topology on the kinetics of insertion electrodes. *Solid State Ionics*, **177**, 2647 (2006).

78. Gaberscek, M., Dominko, R. & Jamnik, J. Is small particle size more important than carbon coating? An example study on $LiFePO_4$ cathodes. *Electrochem. Commun.*, **9**, 2778 (2007).

79. Drezen, T. *et al.* Effect of particle size on $LiMnPO_4$ cathodes. *J. Power Sources*, **174**, 949 (2007).

80. Bramnik, N. N. & Ehrenberg, H. Precursor-based synthesis and electrochemical performance of $LiMnPO_4$. *J. Alloys Compd.*, **464**, 259 (2008).

81. Murugan, A. V., Muraliganth, T. & Manthiram, A. Comparison of microwave assisted solvothermal and hydrothermal syntheses of $LiFePO_4$/C nanocomposite cathodes for lithium ion batteries. *J. Phys. Chem.*, **112**, 14665 (2008).

82. Benoit, C. & Franger, S. Chemistry and electrochemistry of lithium iron phosphate. *J. Solid State Electrochem.*, **12**, 987 (2008).

83. Sides, C. R., Croce, F., Young, V. Y., Martin, C. R. & Scrosati, B. A High-rate, nanocomposite $LiFePO_4$/carbon cathode. *Electrochem. Solid-State Lett.*, **8**, A484–A487 (2005).

84. Shin, H. C., Cho, W. I. & Jang, H. Electrochemical properties of carbon-coated $LiFePO_4$ cathode using graphite, carbon black, and acetylene black. *Electrochim. Acta*, **52**, 1472 (2006).

85. Yu, D. Y. W. *et al.* Study of $LiFePO_4$ by cyclic voltammetry. *J. Electrochem. Soc.*, **154**, A253–A257 (2007).
86. Wang, D. *et al.* New solid-state synthesis routine and mechanism for $LiFePO_4$ using LiF as lithium precursor. *J. Solid State Chem.*, **177**, 4582 (2004).
87. Dominko, R. *et al.* The role of carbon black distribution in cathodes for Li ion batteries. *J. Power Sources*, **119–121**, 770 (2003).
88. Kim, J.-K. *et al.* Effect of mechanical activation process parameters on the properties of $LiFePO_4$ cathode material. *J. Power Sources*, **166**, 211–218 (2007).
89. Fey, G. T.-K., Chen, Y. G. & Kao, H.-M. Electrochemical properties of $LiFePO_4$ prepared via ball-milling. *J. Power Sources*, **189**, 169 (2009).
90. Gaberscek, M., Kuzma, M. & Jamnik, J. Electrochemical kinetics of porous, carbon-decorated $LiFePO_4$ cathodes: Separation of wiring effects from solid state diffusion. *Phys. Chem. Chem. Phys.*, **9**, 1815–1820 (2007).
91. Delacourt, C. *et al.* Toward understanding of electrical limitations (electronic, ionic) in $LiMPO_4$ (M = Fe, Mn) electrode materials. *J. Electrochem. Soc.*, **152**, A913–A921 (2005).
92. Delacourt, C., Poizot, P., Tarascon, J.-M. & Masquelier, C. The existence of a temperature-driven solid solution in Li_xFePO_4 for $0 < x < 1$. *Nat. Mat.*, **4**, 254–260 (2005).
93. Jamnik, J. & Maier, J. Generalised equivalent circuits for mass and charge transport: chemical capacitance and its implications. *Phys. Chem. Chem. Phys.*, **3**, 1668–1678 (2001).
94. Ravet, N., Besner, S., Simoneau, M., Valee, A. & Armand, M. *Canadian Patent*, No. 2.270.771 (1999).
95. Dominko, R. *et al.* Impact of the carbon coating thickness on the electrochemical performance of $LiFePO_4$/C composites. *J. Electrochem. Soc.*, **152**, A607–A610 (2005).
96. Cao, Y. L., Yu, L. H., Li, T., Ai, X. P. & Yang, H. X. Synthesis and electrochemical characterization of carbon-coated nanocrystalline $LiFePO_4$ prepared by polyacrylates-pyrolysis route. *J. Power Sources*, **172**, 913 (2007).
97. Gabrisch, H., Wilcox, J. D. & Doeff, M. M. Carbon surface layers on a high-rate $LiFePO_4$. *Electrochem. Solid-State Lett.*, **9**, A360–A363 (2006).
98. Xu, Y., Lu, Y., Yan, L., Yang, Z. & Yang, R. Synthesis and effect of forming Fe_2P phase on the physics and electrochemical properties of $LiFePO_4$/C materials. *J. Power Sources*, **160**, 570 (2006).

99. Yun, N. J., Ha, H.-W., Jeong, K. H., Park, H.-Y. & Kim, K. Synthesis and electrochemical properties of olivine-type $LiFePO_4/C$ composite cathode material prepared from a poly(vinyl alcohol)-containing precursor. *J. Power Sources*, **160**, 1361 (2006).

100. Zaghib, K. *et al.* Optimized electrochemical performance of $LiFePO_4$ at 60°C with purity controlled by SQUID magnetometry. *J. Power Sources*, **163**, 560 (2006).

101. Huang, H., Yin, S. C. & Nazar, L. F. Approaching theoretical capacity of $LiFePO_4$ at room temperature at high rates. *Electrochem. Solid-State Lett.*, **4**, A170–A172 (2001).

102. Chen, Z. H. & Dahn, J. R. Reducing carbon in $LiFePO_4/C$ composite electrodes to maximize specific energy, volumetric energy, and tap density. *J. Electrochem. Soc.*, **149**, A1184–A1189 (2002).

103. Wang, G. X. *et al.* Electrochemical properties of carbon coated $LiFePO_4$ cathode materials. *J. Power Sources*, **146**, 521 (2005).

104. Spong, A. D., Vitins, G. & Owen, J. R. A Solution–precursor synthesis of carbon-coated $LiFePO_4$ for Li-ion cells. *J. Electrochem. Soc.*, **152**, A2376–A2382 (2005).

105. Mi, C. H., Zhao, X. B., Cao, G. S. & Tu, J. P. *In situ* synthesis and properties of carbon-coated $LiFePO_4$ as Li-Ion Battery Cathodes. *J. Electrochem. Soc.*, **152**, A483-A487 (2005).

106. Liu, H. *et al.* Effects of carbon coatings on nanocomposite electrodes for lithium-ion batteries. *Electrochem. Solid-State Lett.*, **9**, A529–A533 (2006).

107. Yang, M.-R., Teng, T.-H. & Wu, S.-H. $LiFePO_4$/carbon cathode materials prepared by ultrasonic spray pyrolysis. *J. Power Sources*, **159**, 307 (2006).

108. Shin, H. C., Cho, W. I. & Jang, H. Electrochemical properties of the carbon-coated $LiFePO_4$ as a cathode material for lithium-ion secondary batteries. *J. Power Sources*, **159**, 1383 (2006).

109. Yang, J. & Xu, J. J. Synthesis and characterization of carbon-coated lithium transition metal phosphates $LiMPO_4$ (M = Fe, Mn, Co, Ni) prepared via a nonaqueous sol-gel route. *J. Electrochem. Soc.*, **153**, A716–A723 (2006).

110. Gao, F., Tang, Z. & Xue, J. Preparation and characterization of nano-particle $LiFePO_4$ and $LiFePO_4/C$ by spray-drying and post-annealing method. *Electrochim. Acta*, **53**, 1939 (2007).

111. Dominko, R., Gaberscek, M., Bele, M., Mihailovic, D. & Jamnik, J. Carbon nanocoatings on active materials for Li-ion batteries. *J. Eur. Ceram. Soc.*, **27**, 909 (2007).

112. Gu, H.-B., Jun, D.-K., Park, G.-C., Jin, B. & Jin, E. M. Nanosized $LiFePO_4$ cathode materials for lithium ion batteries. *J. Nanosci. Nanotechnol.*, **7**, 3980 (2007).

113. Hui, L., Tang, Z. L., Luo, S. H. & Zhang, Z. T. Sol-gel synthesis of $LiFePO_4$ cathode. *Prog. Chem.*, **19**, 1460–1466 (2007).

114. Xu, Z., Xu, L., Lai, Q. & Ji, X. Microemulsion synthesis of $LiFePO_4$/C and its electrochemical properties as cathode materials for lithium-ion cells. *Mater. Chem. Phys.*, **105**, 80 (2007).

115. Kim, K., Jeong, J. H., Kim, I.-J. & Kim, H.-S. Carbon coatings with olive oil, soybean oil and butter on nano- $LiFePO_4$. *J. Power Sources*, **167**, 524 (2007).

116. Xu, Z., Xu, L., Lai, Q. & Ji, X. A PEG assisted sol-gel synthesis of $LiFePO_4$ as cathodic material for lithium ion cells. *Mater. Res. Bull.*, **42**, 883 (2007).

117. Kim, J.-K., Cheruvally, G., Ahn, J.-H. & Ahn, H.-J. Electrochemical properties of $LiFePO_4$/C composite cathode material: Carbon coating by the precursor method and direct addition. *J. Phys. Chem. Solids*, **69**, 1257 (2008).

118. Shanmukaraj, D., Wang, G. X., Murugan, R. & Liu, H. K. Electrochemical studies on $LiFe_{1-x}Co_xPO_4$/carbon composite cathode materials synthesized by citrate gel technique for lithium-ion batteries. *Mater. Sci. Eng. B*, **149**, 93 (2008).

119. Oh, S. W. *et al.* The effect of morphological properties on the electrochemical behavior of high tap density C—$LiFePO_4$ prepared via coprecipitation. *J. Electrochem. Soc.*, **155**, A414–A420 (2008).

120. Zaghib, K., Mauger, A., Gendron, F. & Julien, C. M. Surface effects on the physical and electrochemical properties of thin $LiFePO_4$ particles. *Chem. Mater.*, **20**, 462 (2007).

121. Roberts, M. R., Spong, A. D., Vitins, G. & Owen, J. R. High throughput screening of the effect of carbon coating in $LiFePO_4$ electrodes. *J. Electrochem. Soc.*, **154**, A921–A928 (2007).

122. Roberts, M. R., Vitins, G. & Owen, J. R. High-throughput studies of $Li_{1-x}Mg_{x}/2FePO_4$ and $LiFe_{1-y}Mg_yPO_4$ and the effect of carbon coating. *J. Power Sources*, **179**, 754 (2008).

123. Kim, D.-H. & Kim, J. Synthesis of $LiFePO_4$ nanoparticles in polyol medium and their electrochemical properties. *Electrochem. Solid-State Lett.*, **9**, A439 (2006).

124. Moskon, J., Dominko, R., Cerc-Korosec, R., Gaberscek, M. & Jamnik, J. Morphology and electrical properties of conductive carbon coatings for cathode materials. *J. Power Sources*, **174**, 683 (2007).

125. Moskon, J., Dominko, R., Gaberscek, M., Cerc-Korosec, R. & Jamnik, J. Citrate-derived carbon nanocoatings for poorly conducting cathode. *J. Electrochem. Soc.*, **153**, A1805–A1811 (2006).

126. Kawada, T. *et al.* Determination of oxygen vacancy concentration in a thin film of $La_{0.6}Sr_{0.4}CoO_3$-δ by an electrochemical method. *J. Electrochem. Soc.*, **149**, E252–E259 (2002).

127. Lai, W. & Haile, S. M. Impedance spectroscopy as a tool for chemical and electrochemical analysis of mixed conductors: A case study of ceria. *J. Am. Ceram. Soc.*, **88**, 2979–2997 (2005).

128. Jamnik, J., Guo, X. & Maier, J. Field-induced relaxation of bulk composition due to internal boundaries. *Appl. Phys. Lett.*, **82**, 2820–2822 (2003).

129. Gaberscek, M., Dominko, R. & Jamnik, J. The meaning of impedance measurements of $LiFePO_4$ cathodes: A linearity study. *J. Power Sources*, **174**, 944 (2007).

130. Gaberscek, M., Moskon, J., Erjavec, B., Dominko, R. & Jamnik, J. The importance of interphase contacts in Li ion electrodes: The meaning of the high-frequency impedance arc. *Electrochem. Solid-State Lett.*, **11**, A170–A174 (2008).

131. Gaberscek, M. *et al.* Porous, carbon-decorated $LiFePO_4$ prepared by sol–gel method based on citric acid. *Solid State Ionics*, **176**, 1801 (2005).

132. Dominko, R. *et al.* Impact of $LiFePO_4$ composites porosity on their electrochemical performance. *J. Electrochem. Soc.*, **152**, A858–A863 (2005).

133. Dominko, R. *et al.* Porous olivine composites synthesized by sol-gel technique. *J. Power Sources*, **153**, 274 (2006).

134. Ryan, J. V. *et al.* Electronic connection to the interior of a mesoporous insulator with nanowires of crystalline RuO_2. *Nature*, **406**, 169 (2000).

135. Erjavec, B. *et al.* RuO_2-wired high-rate nanoparticulate TiO_2 (anatase): Suppression of particle growth using silica. *Electrochem. Commun.*, **10**, 926 (2008).

136. Jamnik, J. *et al.* Stabilizers of particle size and morphology: A road towards high-rate performance insertion materials. *Adv. Mater.* **21**, 2715–2719 (2009).

137. Erjavec, B. *et al.* Tailoring nanostructured TiO_2 for high power Li-ion batteries. *J. Power Sources*, **189**, 869 (2009).

138. Maier, J. Ionic conduction in space charge regions. *Prog. Solid State Chem.*, **23**, 171 (1995).

139. Adams, S. Lithium ion pathways in $LiFePO_4$ and related olivines. *J. Solid State Electrochem.*, 1787–1792, doi:10.1007/s10008-010-1012-1 (2010).

140. Adams, S. Ultrafast lithium migration in surface modified $LiFePO_4$ by heterogeneous doping. *App.l Energy*, **90**, 323–328, doi:http://dx.doi.org/10.1016/j.apenergy.2011.04.053 (2012).

141. Duduta, M. *et al.* Semi-solid lithium rechargeable flow battery. *Adv. Energy Mater.*, **1**, 511–516, doi:10.1002/aenm.201100152 (2011).

142. Huang, Q., Li, H., Gratzel, M. & Wang, Q. Reversible chemical delithiation/lithiation of $LiFePO_4$: Towards a redox flow lithium-ion battery. *Phys. Chem. Chem. Phys.*, **15**, 1793–1797, doi:10.1039/c2cp44466f (2013).

Chapter 9

Redox Reaction in Size-Controlled Li_xFePO_4

Atsuo Yamada
The University of Tokyo, Building 5-607, 7-3-1 Hongo, Bunkyo-ku, Tokyo 113-8656, Japan
yamada@chemsys.t.u-tokyo.ac.jp

Crystalline size has tremendous effect on the thermodynamics and kinetics in intercalation compounds. This includes diffusion/transport length, effective surface area for exchange current, surface energy, and interphase energy. In this chapter, we focus on the compound Li_xFePO_4, in which phase changes by simple two-phase separation but with controllable miscible character and by raising temperature and/or reducing crystalline size provides a simple model system to rationalize thermodynamics and electrochemistry in electrode reaction. Systematic experimental results on this issue are reviewed and summarized in this chapter along with the negative aspects of nanoparticle, including surface poisoning by air contact and its diagnosis.

Nanomaterials for Lithium-Ion Batteries: Fundamentals and Applications
Edited by Rachid Yazami

ISBN 978-981-4316-40-8 (Hardcover), 978-981-4364-23-2 (eBook)
www.panstanford.com

9.1 Background

The crystal chemistry of phospho-olivine Li_xFePO_4 ($0 < x < 1$) has received much attention ever since it was introduced as an alternative cathode material for new-generation lithium-ion batteries. The material provides several advantages over the conventional cathodes; lower cost, much safer performance, lower toxicity, and extremely flat charge-discharge profile at reasonably high potential of *ca.* 3.4 V *vs.* Li/Li^+.[1,2] Nevertheless, at an early stage of the development, its poor intrinsic electronic conductivity, on the order of 10^9 $S{\cdot}m^{-1}$, and small tap density had been addressed as major problems to be solved before it can be deployed on a commercial scale.[3–5] Now, at least the first problem has been overcome by using small particles of <100 nm and the efficient formation of composites with conductive phase, typically carbon. Surprisingly high specification for the power density of more than 5 kW/kg is reported on the commercial cells with $LiFePO_4$ cathode.

One of the essential reasons for the very low conductivity measured for the stoichiometric $LiFePO_4$ seems to be in the pre-factor rather than in the activation energy. In addition, a key feature of the Li_xFePO_4 cathode is its extremely flat charge/discharge profile, typical of two-phase reaction systems. Conventional belief based on the Gibbs phase rule was that the reaction proceeds in a topotactic manner, varying the $FePO_4/LiFePO_4$ ratio throughout the range of x in Li_xFePO_4.[1] However, carrier (mobile electron or hole) densities in the monovalent stoichiometric end members should be negligible and this evidently accounts for the very low electronic conductivity of $<10^{-9}$ $S\ cm^{-1}$ in $LiFePO_4$, despite its relatively small binding energy of 390–500 meV for polaron localization.

It is true that, within the polaronic conduction mechanism and the conventional $FePO_4/LiFePO_4$ two phase model, the number of charge carriers in monovalent systems such as $LiFePO_4$ or $FePO_4$ is determined extrinsically by impurities. Defect chemistry studies of related $Li_{1-3x}Fe_xNiPO_4$ phases suggest that cation vacancies can be accommodated in the olivine structure.[6] However, without any external doping in $LiFePO_4$, a rapid electrochemical response as well as a good rate capability has been achieved in our lab. with an optimized electrode composite of pure $LiFePO_4$. These motivated a closer look at the phase change of Li_xFePO_4 at room temperature to determine whether the creation of charge carrier is spontaneous in non-doped Li_xFePO_4 during the electrochemical charge/discharge process (i.e., does lithium or lithium defect function as a carrier-doping impurity).

9.2 Binary Phase Diagram

As in many binary systems, thermodynamics suggests there are two solid solution regions outside the miscibility gap at finite temperatures. Generally, these two solid solution regions grow as a function of temperature finalized by the continuous solid solution over the whole composition between the two end members, as there is an increasing contribution of entropy to the total free energy at elevated temperatures. This is typically seen in spinodal decomposition in binary liquid systems with an inverse-parabolic phase boundary in the temperature-composition diagram. For intercalation solid electrode materials, this behavior was intensively studied in the 1980s using a lattice gas model based on mean field theory, where the attractive interaction between lithium ions is essential to induce phase separation. Onset temperature of the miscibility gap varies from *ca.* 270 K for metallic Sheverel $Li_xMo_6Se_8$ ($0 < x < 1$)[7,8] to *ca.* 1200 K for insulating NASICON-type $Na_{1+x}Ti_2(PO_4)_3$ ($0 < x < 2$).[9] Analogous to these compounds, the intermediate phase in the Li_xFePO_4 cathode at room temperature may be dominated by the mixture of the $Li_\alpha FePO_4$ and $Li_{1-\beta}FePO_4$ phases with the narrow monophase regions, $0 < x < \alpha$ and $1-\beta < x < 1$, close to the stoichiometric end members of $LiFePO_4$ and $FePO_4$ at each end of the region.

This was first proved by the refinement of X-ray diffraction (XRD) profiles measured for Li_xFePO_4 ($0 \leq x \leq 1$) phases at room temperature.[10] In the bi-phase reaction of Li_xFePO_4 at room temperature, the orthorhombic lattice constants in each phase slightly deviate from those of the stoichiometric end components of $LiFePO_4$ and $FePO_4$. The deviation can be explained by assuming a model in which most of the intermediate phase in the Li_xFePO_4 cathode is dominated by the mixture of $Li_\alpha FePO_4$ and $Li_{1-\beta}FePO_4$ phases (Fig. 9.1).

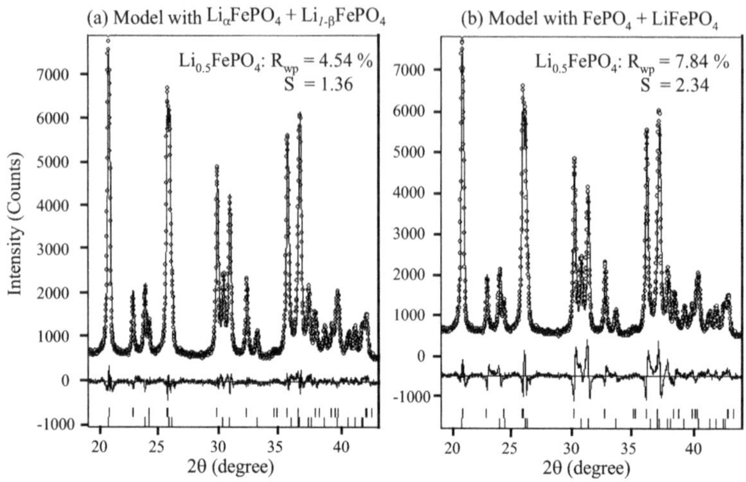

Figure 9.1 Difference in the goodness of fit in the Reitveld refinement of $Li_{0.5}FePO_4$ under the model with (a) a mixture of the stoichiometric $FePO_4$ and $LiFePO_4$, and (b) a mixture of the nonstoichiometric $Li_\alpha FePO_4$ and $Li_{1-\beta}FePO_4$ (α = *ca.* 0.030, β = *ca.* 0.036) phases. Reproduced with permission from Ref. 10. Copyright 2005, The Electrochemical Society.

9.3 Size Effect *vs.* Surface Effect

Meethong *et al.* have suggested that the miscibility gap shrinks systematically with decreasing particle size and increasing temperature.[11] Such a size-dependent miscibility gap is a general phenomena observed in other two-phase intercalation reactions such as Li_xTiO_2 or MgH_x systems.[12,13] Thermodynamically, the dependence can be explained by the increasing contribution of the elastic energy induced along coherent two-phase interface in the smaller particles by a factor of $r \sim A(x)/V(x)$, where r is the particle radius, $A(x)$ is the interface area, and $V(x)$ is the particle volume. The coexistence of two crystallographic phases within one particle leads to a domain wall–related energy penalty, which is determined by the strain mainly due to a difference in lattice parameters. Such a strain induced energy penalty can destabilize the two-phase coexistence in smaller particles. As a result, the energy gain due to phase separation, represented by the free energy of mixing ΔG_{mix}, will decrease for smaller particle sizes. This will gradually close the miscibility gap for decreasing particle size, which explains the enhanced solubility of Li in smaller particles. More recently, Gibot *et al.* reported that the inherent crystalline disorder coupled with nanocrystallization below 40 nm leads to the peculiar situation where solid solution dominates most of the lithium compositional region in Li_xFePO_4 even at room temperature.[14] Such a thermodynamic concept has been verified in detail by Wagemaker *et al.*,[15] including the effect of particle shape. Meethong *et al.*[16] related these intriguing findings to the kinetics of phase-boundary movement, where the reduction in lattice strain caused by the reduction in lattice mismatches in smaller particles was claimed to be suitable to form coherent interface between $Li_\alpha FePO_4$ and $Li_{1-\beta}FePO_4$ so as to enhance the cooperative phase-boundary movement upon electrochemical reaction with minimum formation of dislocation or clacks. This logic was regarded to give comprehensive understanding of the enhanced electrochemical activity of the nanosized $LiFePO_4$ particle as compared to the other slower two-phase electrode system with larger mismatch.

However, attention should be paid that the surface portion of the particle abruptly increases when the size is smaller than 100 nm. A simple estimation based on the sphere particle gives the surface (one unit cell) portions to be *ca.* 4%, 7%, and 17% for the particles with 100 nm, 50 nm, and 20 nm diameters, respectively. In such nanoscale regime, there is an increasing possibility to observe the simultaneous effects both from bulk crystal and from particle surface, which may lead to some confusion. For example, surface redox potential was calculated to be dependent on the crystallographic plane and was shown a wide dispersion over the range of 2.95–3.84 V.[17] Both of the surface impurities and the redox on the specific crystallographic facet plane would be important as the perturbation may cause parasitic capacity below/above the bulk two-phase equilibrium potential in the voltage curve. The capacity outside the two-phase region for Li_xFePO_4 have been used as one of the indicator of the bulk solubility limit of lithium and lithium defect in $Li_\alpha FePO_4$ and $Li_{1-\beta}FePO_4$, and then may need careful reinvestigation.

More importantly, the side reaction upon ambient exposure is known to be pronounced for smaller particle leading to spontaneous lithium extraction and surface oxidation.[18,19] Before going to any further discussions, such a side reaction should be removed, which is explained in the next section.

9.4 Air Poisoning

XRD has been carried out to investigate the effect of different atmosphere exposure conditions. Table 9.1 summarizes the lattice dimensions. In all cases, the diffractograms display a single phase with an ordered olivine structure indexed to the orthorhombic Pnmb space group. The Rietveld refinement permits to approximate the olivine crystallites size to 75 nm, which is verified on the SEM image of the powder presented.

Systematic decrease of the cell volume upon atmospheric exposure was clearly observed together with the corresponding changes in lattice parameters. It begins to be visible even after a short time (1 h) of ambient air exposure at room temperature and becomes more evident after longer time (10 days) or higher temperature (120°C) exposure. Water and oxygen should be included in this side reaction, as the effect is significantly suppressed by using dry O_2 20%/N_2 80% and becomes negligible by using dry CO_2. These volume changes can be linked to the solid solution compositional domain in $Li_{1-x}FePO_4$ ($x < \beta$),[10,11,19,20] in which olivine can accommodate the corresponding lithium defect without first-order phase transformation.

Figure 9.2 demonstrates the linear relative dependence of the *a* and *b* lattice constants following the Vegard's law, in the hypothesis of a solid solution between $LiFePO_4$ prepared with no contact with air and $FePO_4$. These results show that some lithium is certainly extracted when $LiFePO_4$ is put in contact with atmosphere. The Vegard's law gives 2.3% of missing lithium after ambient air exposure for 24 hours at 120°C. It has to be noted that the lattice constants obtained by avoiding air contact are very close to the ones reported for pure, much larger particles,[21] which present less possibilities of reactivity with air due to smaller solid solution domain and smaller exposed surface. It tends to prove that the sample has been well protected during our experiment and that possible air pollution during storage in glove box has been minimized.

For a subsequent study of size-dependent phase diagram, we made effort to eliminate such fluctuant factor in the pristine samples, because it would degrade the quality of any crystallographic and/or electrochemical data and hence mislead the subsequent quantitative discussions. The controlled pristine $LiFePO_4$ samples used in this study are free from such poisoning by air exposure and the particle

Table 9.1 Parameters and cell volumes of $LiFePO_4$ calculated from XRD refinement after various atmosphere exposure conditions

Storage condition		a(Å)	b(Å)	c(Å)	V(Å^3)	x
Dry air (standard)	Room temperature	10.3308(2)	6.00813(11)	4.69444(10)	291.378(6)	0
Ambient air	Room temperature, 1 h	10.3275(2)	6.00656(13)	4.69271(12)	291.1040(12)	0.007
	Room temperature, 1 day	10.3254(2)	6.00573(11)	4.69329(11)	291.037(10)	0.011
	Room temperature, 21 day	10.3227(2)	6.00428(15)	4.69346(14)	290.903(14)	0.017
	120°C 1 day	10.3188(2)	6.00303(15)	4.69451(14)	290.79(4)	0.023
Dry air (N_2/O_2)	Room temperature, 1 day	10.3294(4)	6.0072(2)	4.6932(2)	291.22(2)	0.004
	120°C, 1 day	10.3254(3)	6.0054(2)	4.69373(19)	291.048(19)	0.012
Dry CO_2	Room temperature, 1 h	10.3302(4)	6.0078(2)	4.6936(2)	291.29(2)	0.001
	120°C, 1 day	10.3290(3)	6.00716(18)	4.696361(17)	291.231(16)	0.004

Source: Reproduced with permission from Ref. 10. Copyright 2005, The Electrochemical Society.

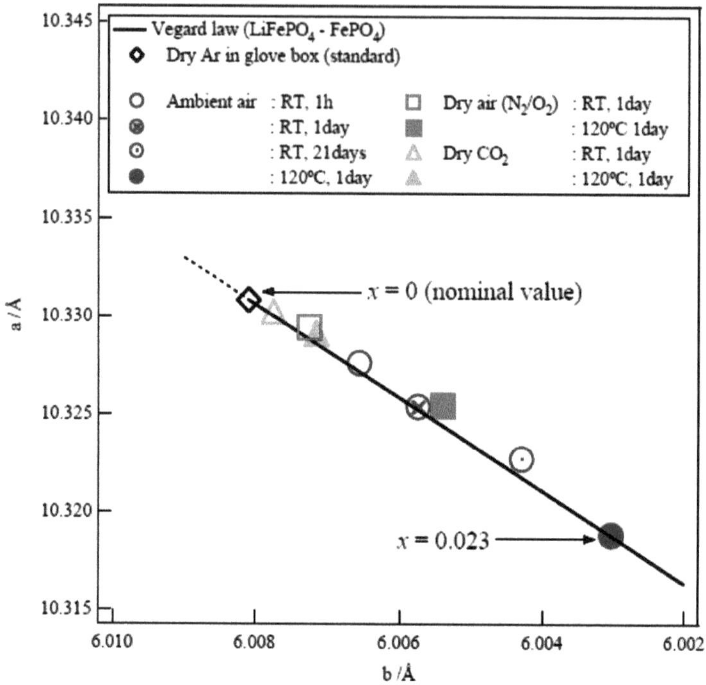

Figure 9.2 Lattice constants changes for $Li_{1-x}FePO_4$ after the different air-contact experiments compared to Vegard's law (arrow). The values for x are calculated using Vegard's law and $FePO_4$ cell parameters (a = 9.8142 Å and b = 5.7893 Å). Reproduced with permission from Ref. 18. Copyright 2008, The Electrochemical Society.

size is the only varied parameter. The samples used in this study are free from any disorder as the lattice dimensions are almost identical to the single crystal values [a = 10.332(4) Å, b = 6.010(5) Å, c = 4.692(2) Å, V = 291.4(3) Å^3][21] and showed negligible dependence on the particle's size, while air-exposed samples suffered from significant volume contraction particularly on smaller particles, as in Table 9.2. With the same trend in variations of the lattice dimensions, Mössbauer spectra for the controlled pristine samples were analyzed by single Fe^{2+} doublet inherent to $LiFePO_4$, while parasitic Fe^{3+} sub-signal was always observed for the air-exposed samples increasing its intensity for smaller particles, as shown in Fig. 9.3.

Table 9.2 Lattice dimensions of pristine and air-exposed $LiFePO_4$ with various particle sizes

No Air Exposure				
Particle size (nm)	a(Å)	b(Å)	c(Å)	V(Å^3)
200	10.33.46(13)	6.00846(7)	4.69238(7)	291.257(7)
80	10.3303(2)	6.00764(12)	4.69313(11)	291.259(11)
40	10.3301(5)	6.0052(3)	4.6947(3)	291.23(2)
Air Exposure; Room Temperature; 1 day				
Particle size (nm)	a(Å)	b(Å)	c(Å)	c(Å^3)
200	10.32839(15)	6.00767(8)	4.69228(8)	291.157(8)
80	10.3253(2)	6.00590(12)	4.69335(11)	291.047(11)
40	6.0015(3)	6.00590(3)	4.6955(3)	290.88(2)

Source: Reproduced with permission from Ref. 18. Copyright 2008, The Electrochemical Society.

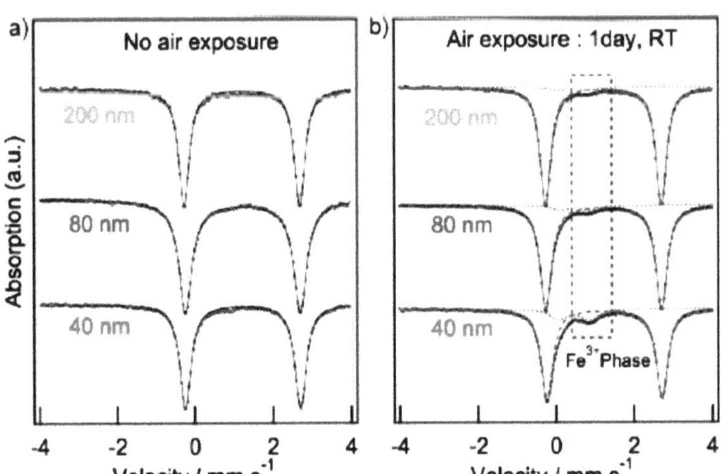

Figure 9.3 Mössbauer spectra measured for (a) pristine (b) and air-exposed $LiFePO_4$ with various particle sizes of 40 nm, 80 nm, and 200 nm.

9.5 Careful Verification of Phase Diagram

9.5.1 Electrochemical Approach

The first step for the analysis is the accurate determination of the open-circuit voltage (OCV) profile *vs.* lithium composition x for three samples with different particle sizes of 40 nm, 80 nm, and 200 nm. Initially, the cell was charged galvanostatically at C/20 rate at 25°C followed by maintenance at 4.5 V for 24 hours forming $FePO_4$. Capacity close to the theoretical capacity of 170 mAh/g was confirmed for all samples. Then, it was discharged under a rate of C/20 for 12 minutes, which corresponds to *ca.* 1of the theoretical capacity. The OCV at 25°C was measured after 24 hours. Again, the cell was discharged under a rate of C/20 for 12 minutes. These procedures were repeated several times until the OCV at 25°C became constant at *ca.* 3.42 V in the bi-phasic region. A similar experiment was performed starting from $LiFePO_4$ toward the anodic direction and confirmed the identical OCV values at same lithium compositions. Size-dependent OCV curves for Li_xFePO_4 are given in Fig. 9.4, where the capacity observed below/above the two-phase equilibrium potential becomes larger as particle size decreases.

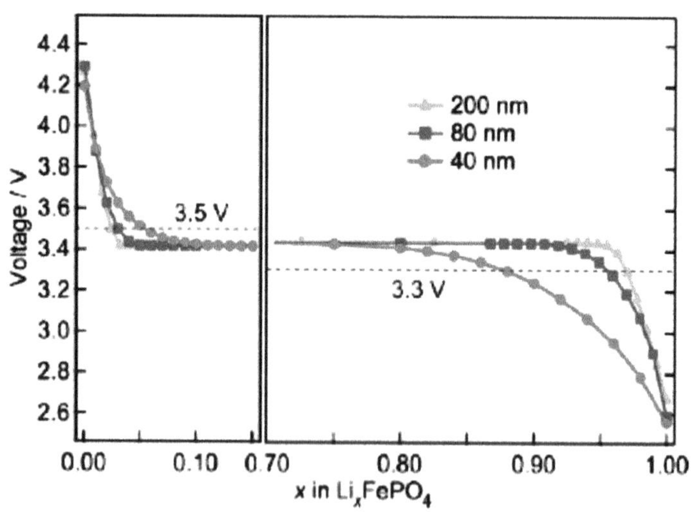

Figure 9.4 OCV curves measured for Li_xFePO_4 with various mean particles sizes of 200 nm, 80 nm, and 40 nm.

9.5.2 Structural Approach

Two target lithium compositions of x in Li_xFePO_4 ($x = 0.6$, $x = 0.93$) were chosen and were adjusted by the chemical way, allowing more homogeneous Li^+ incorporation over electrochemical reactions. Based on the room temperature phase diagram we reported previously,[10,19,20] the composition near an end member, $Li_{0.93}FePO_4$, was selected as an experimental criterion phase, since it would be around the phase transition zone between the two-phase region and the single-phase region in the $FePO_4$–$LiFePO_4$ binary phase diagram, where the two-phase (larger particle) to single-phase (smaller particle) transformation is expected as a function of the particle size. The composition $x = 0.6$, $Li_{0.60}FePO_4$, was selected as a reference phase, because it should be composed of a mixture of two intermediate phases, analysis of which should lead to the definitive positions of the solid solubility limits, α and β, in $Li_\alpha FePO_4$ and $Li_{1-\beta}FePO_4$. In order to set the lithium composition in Li_xFePO_4 to $x = 0.6$ or $x = 0.93$, all of the lithium in pristine $LiFePO_4$ was first removed to form isostructural $FePO_4$ using a strong oxidizing agent NO_2BF_4 in acetonitrile. Then, $FePO_4$ was reacted with the specific amount of mild reducing agent LiI in acetonitrile to form Li_xFePO_4 with $x = 0.6$ or $x = 0.93$. The chemical treatment was performed for each pristine $LiFePO_4$ sample with three different particle sizes of 40 nm, 80 nm, and 200 nm for the subsequent high-resolution synchrotron XRD.

In order to overview the change in the room-temperature phase diagram with decreasing particle size, the lattice parameters refined for $Li_{0.6}FePO_4$ (two-phase mixture of $Li_\alpha FePO_4$ and $Li_{1-\beta}FePO_4$) were analyzed. For all particle sizes, $Li_{0.6}FePO_4$ is evidently in two-phase state as typically seen in the two diffraction peaks clearly observed at around 13.5° and 14.5°, corresponding to the (200) reflection of $Li_{1-\beta}FePO_4$ and the (200) reflection of $Li_\alpha FePO_4$, respectively. All of the lattice parameters refined for the two separated phases in $Li_{0.6}FePO_4$ deviates from those of stoichiometric $FePO_4$ and $LiFePO_4$, and can be explained by assuming partial lithium occupancy α to $FePO_4$ and partial lithium deficiency β from $LiFePO_4$ as well as Vegard's law along $FePO_4$–$LiFePO_4$ solid solution.

Schematic derivation of the size-dependent $FePO_4$–$LiFePO_4$ binary phase diagram using the XRD data for $Li_{0.6}FePO_4$ is illustrated in Fig. 9.5, which highlights the fact that the solid solubility limits α, β, in $Li_\alpha FePO_4$ and $Li_{1-\beta}FePO_4$ systematically increases along with the reduction in particle size. Most probably, these systematic variations originate from the elastic thermodynamic effects on the miscibility gap as discussed in the various other binary systems. The solid–solid phase transitions are thermodynamically first order and involve a change in enthalpy (ΔH), which is highly related to the particle size as depicted by $\Delta H = \Delta H_{\text{intrinsic}} + \Delta H_D$, where ΔH_D represents the particle size contribution of opposite sign.[15] If the particle size decreases, the Gibbs free energy (ΔG) of mixing would be decreased by the increasing contribution of the penalty in elastic energy at the two-phase boundary, resulting in extension of the single phase regions near the end members on the composition phase diagram.

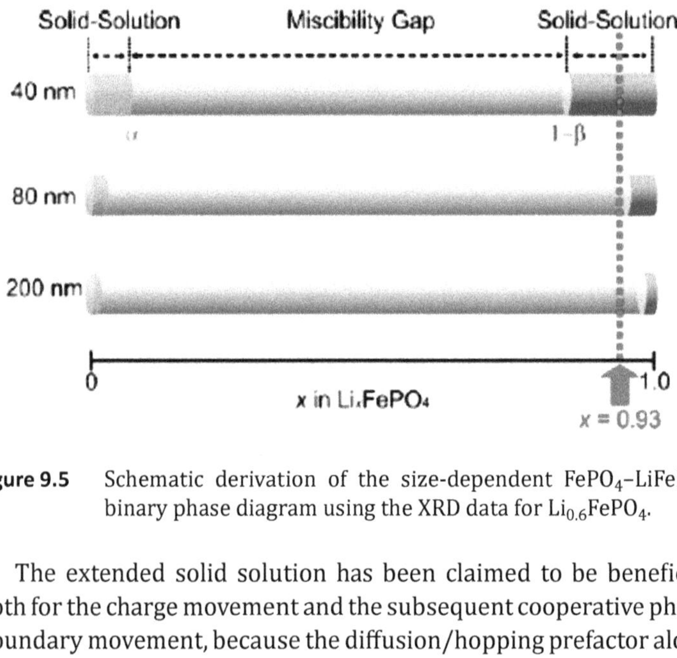

Figure 9.5 Schematic derivation of the size-dependent $FePO_4$–$LiFePO_4$ binary phase diagram using the XRD data for $Li_{0.6}FePO_4$.

The extended solid solution has been claimed to be beneficial both for the charge movement and the subsequent cooperative phase boundary movement, because the diffusion/hopping prefactor along the one-dimensional diffusion path is increased while the lattice mismatch along the two-phase boundary is reduced. However, experimental evidences given so far for the existence of the small solid solution compositional domain are not direct ones and lead to some controversy whether they really originate from the bulk effect or from the artifact by the enhanced surface effect in nanoregime.[17] Then, isolation of the solid solution phase would give more definitive and straightforward experimental evidence for the size-dependent miscibility gap model.

Towards this end, we now focus on the sample with x = 0.93, which is around the phase transition compositional zone between the two-phase region and the single-phase region in the $FePO_4$–$LiFePO_4$ binary phase diagram (see Fig. 9.5). At the composition of x = 0.93, the two-phase (larger particle) to single-phase (smaller particle) transformation is expected as a function of the particle size as easily speculated from the phase behavior along the vertical compositional line at x = 0.93 in Fig. 9.5. The change in the lattice parameter a are summarized in Fig. 9.6 together with the low-angle peaks of $Li_{0.93}FePO_4$ indexed as $(200)_T$ (lithium-rich triphylite phase) and $(200)_H$ (lithium-poor heterosite phase) at different particle sizes. The intensity of $(200)_H$ is gradually decreased when the particle size is reduced, and finally disappeared for the 40 nm particle. More importantly, the deviation of the lattice constant a at x = 0.93 from those of the phase-separated $Li_{0.6}FePO_4$ was observed only for 40 nm particle. Furthermore, it is noteworthy that the lattice constant a at x = 0.93 for 40 nm particle positions is strictly on the line of Vegard's law assuming the complete solid solution of $FePO_4$ and $LiFePO_4$. The same behavior was identified also for the b- and c- axes. Clearly, the immiscible–miscible transition is induced by the particle size reduction at the identical lithium composition x = 0.93, and the solid solution in Li_xFePO_4 is isolated at x = 0.93 in the 40 nm particle, details of which are described below.

As for the 200 nm particle at x = 0.93, two independent peaks were clearly observed as indexed $(200)_T$ and $(200)_H$, respectively, which indicates that the $Li_{0.93}FePO_4$ phase consists of two intermediate phases, $Li_\alpha FePO_4$ and $Li_{1-\beta}FePO_4$. This is a reasonable result because the lithium composition for the immiscible-miscible transition was estimated to be $1-\beta_{200}$ = 0.98, which is much larger than the present lithium composition, x = 0.93. The lattice parameters are almost identical with those of $Li_{0.6}FePO_4$, which also reveal that the $Li_{0.93}FePO_4$ with a particle size of 200 nm is inside the miscibility gap at room temperature.

When the particle size is reduced to 80 nm, the phase at x = 0.93 is still in the two-phase separation, but the intensity of the $(200)_H$ peak has significantly decreased in the synchrotron XRD profile, which can be explained by the reduction of the $Li_\alpha FePO_4$ phase fraction in the two-phase mixture of $Li_\alpha FePO_4$ and $Li_{1-\beta}FePO_4$, following the size-dependent miscibility gap model.[11] The estimated lithium composition for the immiscible–miscible transition, $1-\beta_{80}$, has now

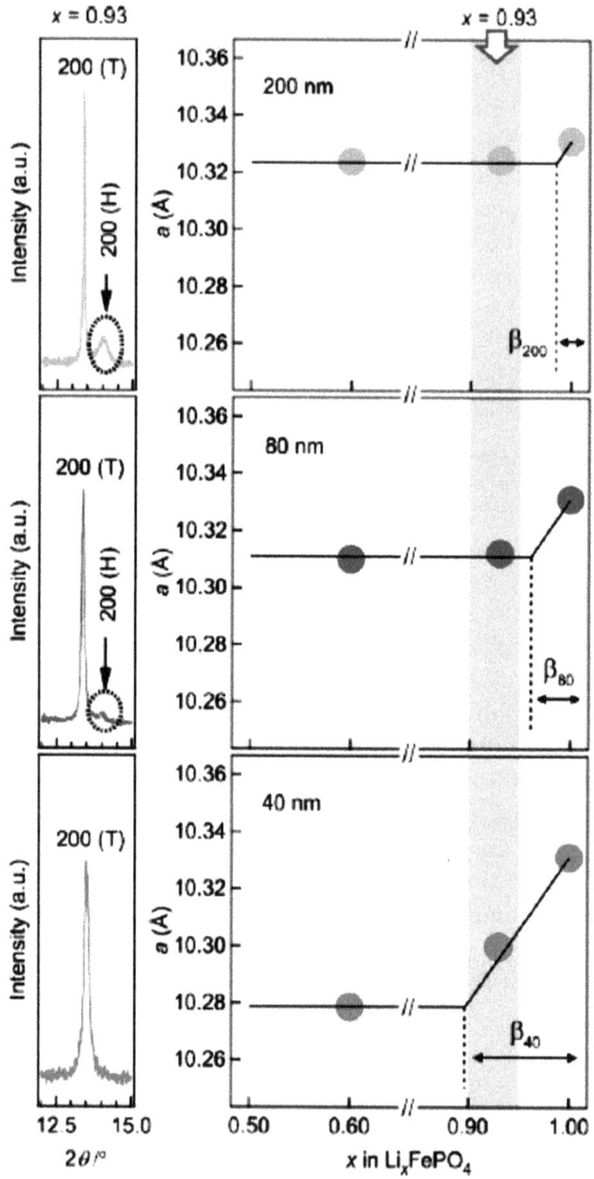

Figure 9.6 Left: Magnified view of the 200 peak measured for $Li_{0.93}FePO_4$ with different particle sizes of 200 nm, 80 nm, and 40 nm at room temperature. Right: Derivation of changes in *a* axis lattice parameter as a function of lithium composition *x* for different particle sizes.

decreased to 0.96 and is closer to the present lithium composition, $x = 0.93$. The $Li_{0.93}FePO_4$ phase with an average particle size of 80 nm is close to the transition zone from immiscible to miscible state and contains smaller amount of $Li_{\alpha}FePO_4$ at room temperature. The lattice constants are still identical with those of bi-phased $Li_{0.6}FePO_4$; the $Li_{0.93}FePO_4$ with a particle size of 80 nm is still inside the miscibility gap at room temperature but in closer access to the immiscible–miscible transition zone.

At the 40 nm particle size with $x = 0.93$, the $(200)_H$ peak has disappeared, and the discrepancies in the lattice parameters from the two-phase values are prominent. Instead, the lattice constants are exactly on the line of Vegard's law assuming the complete solid solution between $FePO_4$ and $LiFePO_4$. This behavior proves that the $Li_{0.93}FePO_4$ phase with 40 nm size is in the single phase solid solution range and allows to conclude that a solid solution of Li_xFePO_4 could be isolated as a single phase with the solid solubility of $\beta_{40} > 0.1$.

For a direct comparison between the stoichiometric $LiFePO_4$ and the isolated solid solution of $Li_{0.93}FePO_4$ with an identical particle size of 40 nm, synchrotron XRD profiles of the two samples are shown in Fig. 9.7 with some magnified presentations of the typical

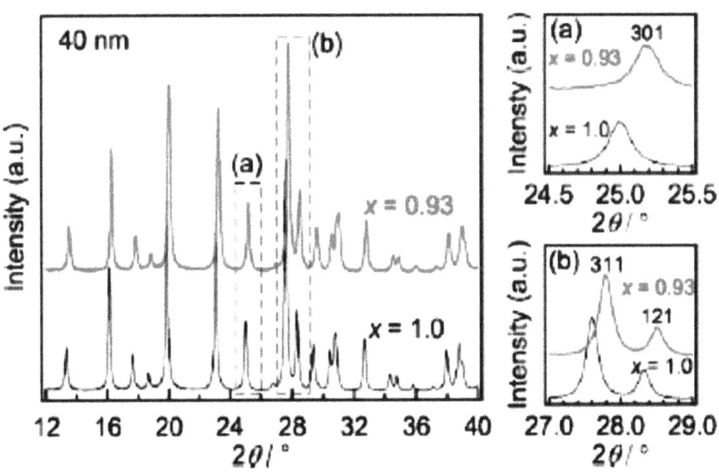

Figure 9.7 Comparison of the synchrotron XRD profiles measured for stoichiometric $LiFePO_4$ and the isolated solid solution of $Li_{0.93}FePO_4$ with an identical particle size of 40 nm with magnified presentations of 301, 311, and 121 as the typical peaks.

peaks. Peak shifts are visually evident and large enough to confirm the wide lithium compositional range of at least 7% for solid solution. The shift observations without appearing of any second phase further support the room-temperature isolation of solid solution in Li_xFePO_4.

9.5.3 Consistency between Electrochemical and Structural Data

For a further confirmation of the inverse relationship between the solid solubility and particle size, the composition x has been electrochemically controlled by galvanostatically charging the cells in a potential range of 2.0 V to 4.5 V and then potentiostatically discharging them at a constant potential below the equilibrium potential at 3.30 V, which corresponds to $Li_{0.98}FePO_4$ (200 nm), $Li_{0.96}FePO_4$ (80 nm), and $Li_{0.88}FePO_4$ (40 nm) in OCV curves in Fig. 9.4. Similar experiment was performed with higher constant potential above equilibrium potential at 3.50 V, which corresponds to $Li_{0.02}FePO_4$ (200 nm), $Li_{0.03}FePO_4$ (80 nm), and $Li_{0.06}FePO_4$ (40 nm) in OCV curves in Fig. 9.4. The cell was dismantled in an Ar glove box, and then the cathode electrode was washed in dimethyl carbonate (DMC) for subsequent XRD measurements.

Size-dependent peak shift is clear both for samples in equilibrium at 3.30 V and at 3.50 V as shown in Fig. 9.8. The lattice parameters were refined for all samples by Rietveld analyses and are summarized in Table 9.3. The extent of lithium nonstoichiometry x in Li_xFePO_4 were calculated for each samples assuming the Vegard's law between $FePO_4$ and $LiFePO_4$, showing excellent agreement with the electrochemical estimation of x, i.e., x (Vegard's law) = x (OCV), for all of the six cases with three different particle sizes (40 nm, 80 nm, and 200 nm) and two different OCV (3.30 V and 3.50 V). This provides firm experimental evidence that the capacity observed below/above the two-phase equilibrium potential is not caused by some impurities or the potential distribution dependent on the crystalline surface but largely due to the bulk lithium nonstoichiometry at least for particle sizes larger than 40 nm.

Further inspections on the solid solution compositional domain were performed for the electrochemically adjusted Li_xFePO_4 with a particle size of 40 nm, which allow largest miscibility in the present

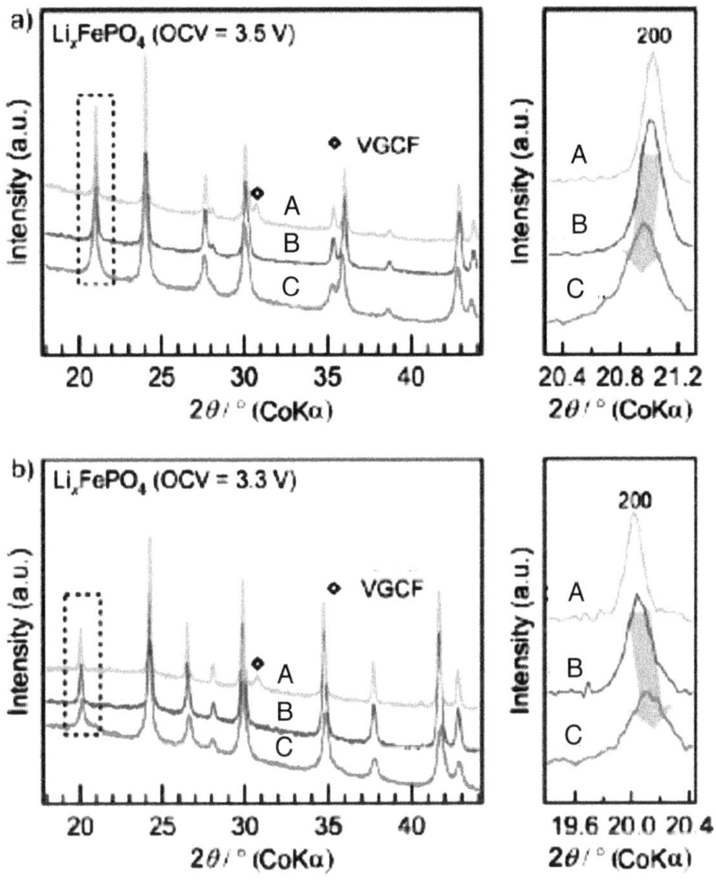

Figure 9.8 XRD profiles for Li_xFePO_4 electrode with various mean particle sizes of (A)200 nm, (B)80 nm, and (C)40 nm, equilibrated at 3.3 V (a) and 3.5 V (b), with magnified presentation of the 200 diffraction peaks.

study. The potential was set to 3.8 V, 3.6 V, 3.5 V, 3.3 V, 3.2 V, 3.1 V, 3.0 V, and 2.8 V *vs.* Li^+/Li. They correspond to the lithium composition of 0.02, 0.04, 0.06, 0.88, 0.91, 0.93, 0.95, and 0.95, respectively, with an eye to the OCV data shown in Fig. 9.9 and are inside the solid solution domains. All the samples were subjected to the XRD experiments with the subsequent Rietveld analysis and the plots of lattice parameters *vs.* lithium compositions are given in Fig. 9.9.

Table 9.3 Lattice dimensions of Li_xFePO_4 electrode with various mean particle sizes of 200 nm, 80 nm, and 40 nm, equilibrated at (a) 3.3 V and (b) 3.5 V. Lithium nonstoichiometry from $x = 1$ and $x = 0$ estimated by two different method, Vegard's law and OCV curves, show excellent agreement. The extent of the lithium nonstoichiometry is enhanced as mean particle size decreases

(a) Li-poor (heterosite phase, OCV = 3.5 V)

Particle size (nm)	*a* (Å)	*b* (Å)	*c* (Å)	*V* (Å^3)	*x*(Vegard's low)	*x* (OCV)
200	9.8159(9)	5.7891(4)	4.7814(4)	271.70(4)	£0.010	0.022
80	9.8292(19)	5.7994(5)	4.7792(11)	272.43(5)	0.030	0.029
40	9.8486(13)	5.8138(6)	4.7837(7)	273.90(6)	0.074	0.056

(b) Li-rich (triphylite phase, OCV = 3.3 V)

Particle size (nm)	*a* (Å)	*b* (Å)	*c* (Å)	*V* (Å^3)	*x*(Vegard's low)	*x* (OCV)
200	10.3224(4)	6.0039(2)	4.6909(2)	290.67(2)	0.975	0.98
80	10.3184(7)	6.0016(4)	4.6904(4)	290.46(3)	0.965	0.96
40	10.2790(13)	6.9825(6)	4.7054(7)	289.35(6)	0.891	0.88

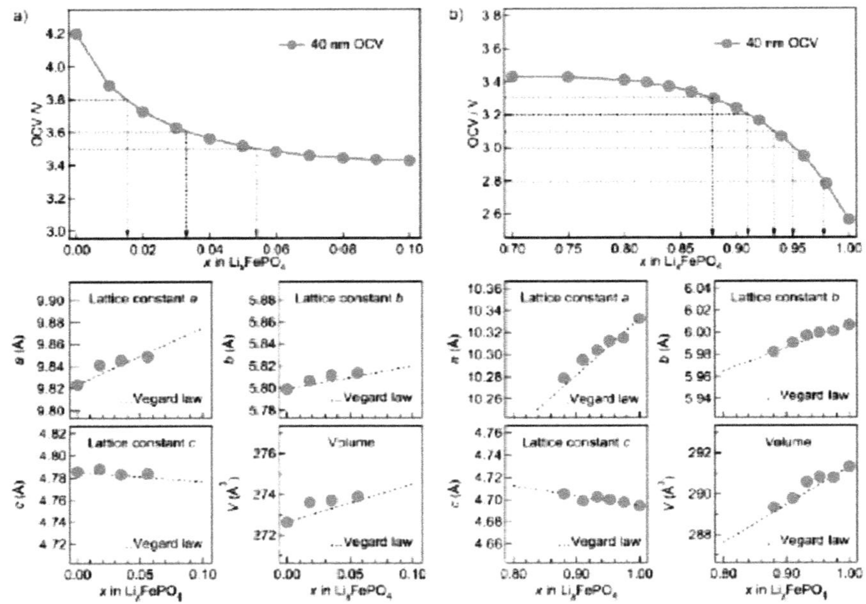

Figure 9.9 OCV curve and change in lattice dimensions (a) above and (b) below the two-phase equilibrium potential. Lithium composition was adjusted potentiostatically (equilibrium condition: <0.1 μA) and estimated by the OCV curves, as indicated by dashed lines. The linear lines along the plot of lattice constants show the Vegard's law between $FePO_4$ and $LiFePO_4$.

All the data are on the line of Vegard's law of $FePO_4$–$LiFePO_4$ solid solution without any systematic deviation and further support the two main conclusions of this paper with no ambiguity: (i) validity of the size-dependent miscibility gap model in Li_xFePO_4 and (ii) bulk miscibility dominates the electrochemical behavior of Li_xFePO_4 at least for particle sizes larger than 40 nm.

9.5.4 Thermodynamic Approach

We now address the thermodynamic experimental evidences for the solid solution outside of the miscibility gap.[20] Microcalorimetry can directly measure entropy changes *vs.* x at fixed temperature T during the electrochemical reaction. Observed heat flow P is related to entropy change ΔS by $P = -IT\Delta S/nF$, where I is the input current and F is the Faraday constant. The classical entropy of mixing $S_{cf}(x)$ can be calculated for a fully disordered lattice of lithium atoms and vacant sites, and heat flow varies linearly with $\Delta S_{cf}(x) = k_B \ln[1/x]$, with lithium atom vacancy disorder when $0 < x < 1$. Although this expression is singular at $x = 0$ and $x = 1$, an important overall trend demonstrates that the heat flow changes symmetrically from negative at initial composition $x = 0$ to positive at final composition $x = 1$. Typical anomalies can be clearly seen at the very beginning of charging $LiFePO_4$ as shown in Fig. 9.10. However, the endothermic region is narrow compared to the subsequent anomalous exothermic region. Similar anomalies were also detected at the very beginning of $FePO_4$ discharge, but again with a much narrower, negligible endothermic region. Such strong asymmetries for endo/exothermic behavior may be due to the dynamic character of calorimetric measurements. These two thermally anomalous compositional regions close to $LiFePO_4$ and $FePO_4$ in Li_xFePO_4 show good agreement with the values of α and $1-\beta$ in $Li_\alpha FePO_4$ and $Li_{1-\beta}FePO_4$ obtained by structural data also with the region in which the deviation of OCV from the constant value of $V_{oc} = 3.42$ V is evident (see Fig. 9.4). Furthermore, these initial anomalies in entropy upon charge/discharge were not observed when measurement was initiated at any point inside the miscibility gap. Therefore, it is likely they represent the change in configurational entropy in solid solution outside the miscibility gap in Li_xFePO_4.

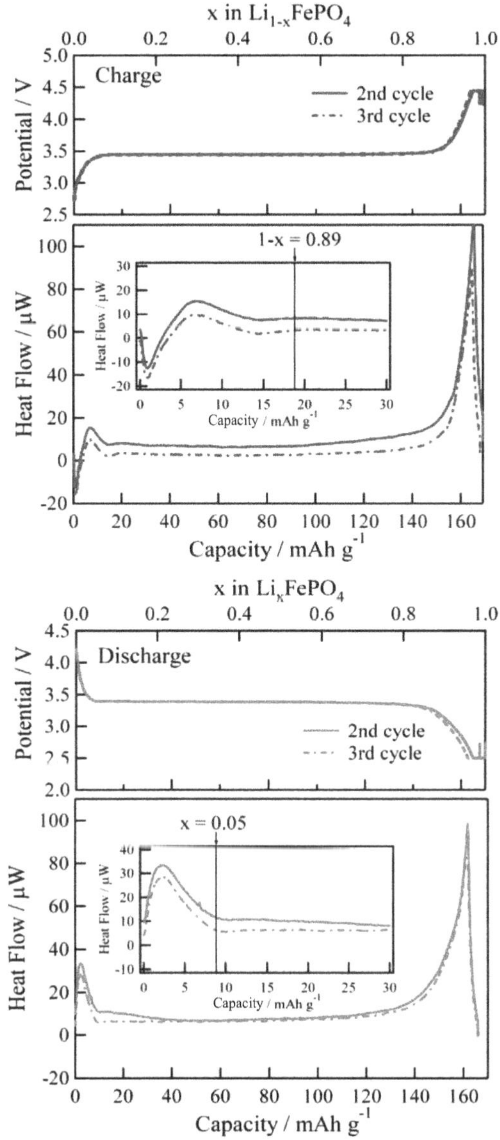

Figure 9.10 Calorimetric measurements for a 2032 type coin cell for charging and discharging processes. A clear anomaly exists at the beginning of both of charge and discharge. Vertical lines are the mono-phase/bi-phase boundary estimated from the structural data. Reprinted with permission from Ref. 20. Copyright 2006, Nature Publishing Group.

Except for these two mono-phase regions, heat flow exhibited almost constant exothermic behavior with no significant anomalies, explained by a simple two-phase reaction scheme with a continuous phase-boundary movement without any structural or electronic phase transition. Abrupt increase of the exothermic heat flow at the end of charge/discharge processes in Li_xFePO_4 is due to the increasing polarization, and it hinders the thermal anomaly for one of the two narrow mono-phase regions. It is noteworthy that the amount of heat flow during the charge-discharge reaction (<5 μW) is comparable or even smaller than those measured during similar experiments on conventional electrodes such as Li_xCoO_2 and $Li_xMn_2O_4$.

9.6 Summary

These systematic findings on nano-size effects for the phase diagram of important electrode material of Li_xFePO_4 redefine guidelines for understanding the polaronic transport, cooperative phase boundary movement, and its thermodynamics in bi-phasic intercalation processes. Polaronic insulators of the transition metal polyanion family show a wide variety of framework structure and accommodate various transition metals and guest atoms. Recent advances in carbon coating technology have been applied to many of these compounds to shed light on their potential application as intercalation battery cathodes. There is, therefore, still plenty of scope for further investigations on this issue.

References

1. A. K. Padhi, K. S. Nanjundaswamy, J. B. Goodenough, *J. Electrochem. Soc.* **144**, 1188 (1997).
2. A. Yamada, S. C. Chung, K. Hinokuma, *J. Electrochem. Soc.* **148**, A224 (2001).
3. H. Huang, S.-C. Yin, L. F. Nazar, *Electrochem. Solid-State Lett.* **4**, A170 (2001).
4. S. Y. Chung, J. T. Bloking, Y. M. Chiang, *Nat. Mater.*, **1**, 123 (2002).
5. Z. Chen, J. R. Dahn, *J. Electrochem. Soc.*, **149**, A1184 (2002).
6. A. Goñi, L. Lezama, M. I. Arriortua, G. E. Barberis, T. Rojo, *J. Mater. Chem.* **10**, 423 (2000).

7. J. R. Dahn, W. R. McKinnon, *Phys. Rev.* **B32,** 3003 (1985).
8. J. R. Dahn, W. R. McKinnon, J. J. Murray, R. R. Haering, R. S. McMillan, A. H. Rivers-Bowerman, *Phys. Rev.* **B32,** 3316 (1985).
9. C. Delmas, A. Nadiri, J. L. Soubeyroux, *Solid State Ionics* **28–30,** 419 (1988).
10. A. Yamada, H. Koizumi, N. Sonoyama, R. Kanno, *Electrochem. Solid-State Lett.*, **8**, A409 (2005).
11. N. Meethong, H.-Y. S. Huang, W. C. Carter, Y.-M. Chiang, *Electrochem. Solid-State Lett.* **10**, A134 (2007).
12. M. Wagemaker, W. J. H. Borghols, F. M. Mulder, *J. Am. Chem. Soc.*, **129**, 4323 (2007).
13. H. G. Schimmel, J. Huot, L. C. Chapon, F. D. Tichelaar, F. M. Mulder, *J. Am. Chem. Soc.*, **127**, 14348 (2005).
14. P. Gibot, M. Casa-Cabanas, L. Laffont, S. Levasseur, P. Carlach, S. Hamelet, J. M. Tarascon, C. Masquelier, *Nature Mater.* **7**, 741 (2008).
15. M. Wagemaker, *Adv. Mater.*, **21**, (2009). DOI: 10.1002/adma.200990096.
16. N. Meethong, H.-Y. S. Huang, S. A. Speakman, W. C. Carter, Y.-M. Chiang, *Adv. Funct. Mater.* **17**, 1115 (2007).
17. L. Wang, F. Zhou, Y. S. Meng, G. Ceder, *Phys. Rev. B*, **76**, 165435 (2007).
18. J. F. Martin, A. Yamada, G. Kobayashi, S. Nishimura, R. Kanno. D. Guyomard, N. Dupre, *Electrochem. Solid-State Lett.*, **11**, A12 (2008).
19. G. Kobayashi, S. Nishimura, M.-S. Park, R. Kanno, M. Yashlma, T. Ida, A. Yamada, *Adv. Funct. Mater.*, **18**, 395 (2009).
20. A. Yamada, H. Koizumi, S. Nishimura, N. Sonoyama, R. Kanno, M. Yonemura, T. Nakamura, Y. Kobayashi, *Nat. Mater.* **5**, 357 (2006).
21. V. A. Streltsov, E. L. Belokoneva, V. G. Tsirelson, N. K. Hansen, *Acta Crystallogr.*, **B49**, 147 (1993).

Chapter 10

Reduced Graphene Oxide–Based Hybrid Materials for High-Rate Lithium-Ion Batteries

Seong Min Bak, Hyun Kyung Kim, Sang Hoon Park, and Kwang Bum Kim

Department of Material Science and Engineering, Yonsei University, 134 Shinchon-dong, Seodaemun-gu, Seoul, 120-749 Republic of Korea

kbkim@yonsei.ac.kr

As a new class of nanocarbon, reduced graphene oxide (RGO) is expected to bring about a breakthrough in many applications including electrochemical energy storage devices due to its large specific surface area, high electrical conductivity, and its chemical and mechanical properties. By the benefit of their outstanding electrical and chemical properties, metal oxide/RGO hybrids have been intensively investigated for the lithium-ion battery applications. The excellent properties of metal oxide/RGO hybrids are generated from synergistic combination of RGO with metal oxide on the nanometer scale. In the hybrids, RGO not only serves as a highly conductive support material, but also provides a large 2-D surface for the well-dispersed deposition of metal oxides at the nanometer scale. Furthermore, attached metal oxides on RGO may efficiently prevent the restacking of individual RGO

Nanomaterials for Lithium-Ion Batteries: Fundamentals and Applications
Edited by Rachid Yazami

ISBN 978-981-4316-40-8 (Hardcover), 978-981-4364-23-2 (eBook)
www.panstanford.com

in hybrids. In this chapter, we report on the synthesis and electrochemical characterization of RGO and metal oxide/RGO hybrids for high-rate lithium-ion battery applications.

10.1 Introduction

Lithium ion (Li-ion) batteries are widespread application as secondary batteries in numerous applications in consumer electronics, and considered to be one of the best candidates for various forms of electric vehicle (EV).[1] However, the power density of Li-ion batteries is still low to support the industrial needs such as EV. Researches on the electrode materials have been widely investigated to improve their high-rate capability, for achieving high-power density of Li-ion batteries.[2–4]

Nanosized metal oxides are expected to enhance the rate capability due to the highly effective interfacial area between the oxide surface and the electrolyte, and the shorter travel length of lithium ions and electrons during charge/discharge.[5] Unfortunately, many of the nanosized metal oxides reported in the literature for Li-ion batteries suffer from low electrical conductivity and a tendency of particle agglomeration, resulting in a marginal improvement of rate capability.[6,7] Therefore, how to prepare well-dispersed nanosized metal oxides with high electrical conductivity is the key in achieving high-rate capability.[6,8]

One approach is to employ a hybrid composed of nanosized metal oxides and nanocarbon with high electrical conductivity. Dispersing the metal oxide nanoparticles on the highly conductive nanocarbon template hinders the agglomeration of the oxide nanoparticles and improves the electrical conductivity.[9] Reduced graphene oxide (RGO) or graphene nanosheets have rapidly emerged as potential electrode materials for electrochemical energy storage due to their large surface area high electrical conductivity, and good mechanical properties, which are comparable with, or better than, those of carbon nanotubes.[10] Therefore, RGO can be an ideal template to construct a hybrids with good dispersion of metal oxide nanoparticles and improved electrical conductivity. In the past few years, considerable efforts have been devoted to apply various metal oxide/RGO hybrids to Li-ion battery application. (e.g., Co_3O_4,[11] Fe_3O_4,[12] SnO_2,[13] $Li_4Ti_5O_{12}$,[14] and $LiMn_2O_4$[15] with RGO).

In this chapter, we first introduce the preparation and characterization of RGO and metal oxide/RGO hybrid (i.e., $Li_4Ti_5O_{12}$/RGO and $LiMn_2O_4$) for high-rate Li-ion battery application. Then we present the capacitive effect of these hybrid materials which led to faster charge–discharge rates, especially for the $LiMn_2O_4$/RGO hybrid by assuming that the current response at a fixed potential as being combination of the surface and bulk reaction.

10.2 Reduced Graphene Oxide

Graphene is 2-D carbon structure composed of sp^2 bonded carbon network with single atomic thickness. Recently, it has been actively investigated as a potential electrode material for electrochemical energy conversion and storage applications due to the large surface area, high electrical conductivity, and good mechanical properties.

In general, graphene can be divided into exfoliated graphene (prepared through top-down approach) and epitaxially grown graphene (prepared through bottom-up approach). Since the first report of the synthesis of graphene by the micromechanical cleavage of graphite, extensive efforts have been made to fabricate single or few layer graphene by epitaxial growth on SiC surfaces, solvothermal synthesis method, liquid phase exfoliation of graphite and chemical reduction or thermal exfoliation of graphite oxide (GO).[16–22] Among these methods, preparation of graphene from GO reduction stands out (i.e., RGO), since it is promising for the mass production of graphene based materials.[20–22]

In particular, for the electrochemical energy storage applications, thermal or chemical reduction process is widely used to prepare graphene flakes (i.e., RGO).

As shown in Scheme 10.1, in the thermal or chemical approach, graphite is first oxidized to GO through the oxidation treatment. Therefore, in the GO, planes of carbon atoms are decorated with oxygen-containing functional groups. They expand the interlayer distance of graphite for easy exfoliation and make the carbon layer very hydrophilic. However, conjugated structure of graphite is broken in GO by the functional groups and pi electrons are localized. Therefore, GO loses the electrical conductivity of the graphite. In order to restore the electrical conductivity, GO is converted to its reduced form, RGO, by chemical, thermal reduction or a combination of the two reduction treatment. Removal of residual oxygen introduced during the synthesis of GO is a key factor in improving electrical conductivity of RGO.[23] GO can be reduced through different methods with tailored properties by controlling the reduction conditions. In the chemical reduction or thermal reduction process, sonication or thermal treatment cannot have complete exfoliation of graphite to the level of individual graphene sheets. In addition, it has a natural tendency to stack into multilayers, thereby losing some specific surface area. Furthermore, these reduction treatments cannot remove all the functional groups and lattice defect of GO. As a result, virtually all studies in the literature, which used a variety of reduction process, report the presence of residual oxygen in RGO. This represents a fundamental limitation that must be overcome if RGO is to achieve properties that resemble those of mechanically exfoliated graphene.[24]

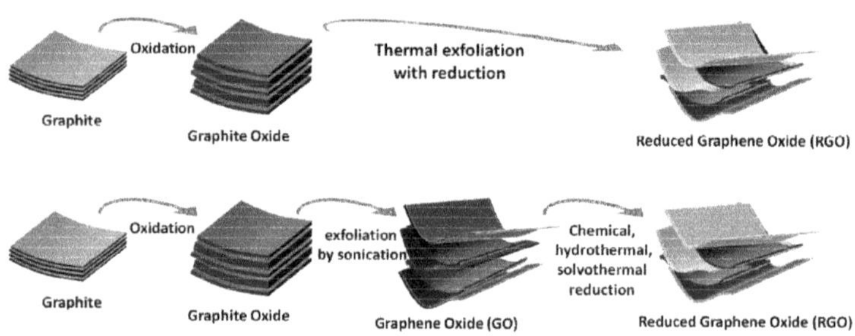

Scheme 10.1 General synthesis route of RGO.

Although chemical or thermal reduction methods have their own limitations, recent intensive research efforts have continuously made it closer to better properties of ideal graphene. Reportedly, the electrical conductivity of RGO could reach as high as 5.7×10^4 S m^{-1},[25] which is about two orders of magnitude lower than that of graphene ($5\text{–}6.4 \times 10^6$ S m^{-1}).[26] In addition, due to their processing advantages and unique properties such as huge surface area, tunable electrical and chemical properties, RGO is still very attractive for many applications. In particular, graphene-based hybrid materials with polymer, metal or metal oxides have shown unique mechanical, electrical, and electrochemical properties.[27–29] We focus on the hybrid materials made of metal oxides that could have important applications for the electrochemical energy storage. The low conductivity and poor stability of metal oxides usually required adding conductive additives to enhance electrical conductivity in electrodes of energy storage devices such as Li-ion battery and supercapacitors. In this chapter, we present the metal oxide/RGO hybrids can be applied to electrode materials for high-power Li-ion batteries and demonstrate that their electrochemical properties can be much improved.

10.3 Design of the Metal Oxide/Reduced Graphene Oxide Hybrid

The superior electrochemical properties of metal oxide/RGO hybrids are generated from synergistic combination of RGO with metal oxide on the nanometer scale. To achieve the high-rate performance of metal oxide in hybrid material, the metal oxide/RGO hybrid was designed and synthesized to meet the following requirements:

(1) The nanosized metal oxide particles should be evenly dispersed on the RGO template, which prevents agglomeration of individual metal oxide nanoparticles and consequently maintains their inherent highly active surface area.

(2) The nanosized metal oxide particles on the RGO sheets should have high crystallinity and well-ordered local atomic structure, whereas nanosized metal oxide particles reported in the literature often show poor crystallinity.

(3) The RGO sheets in the hybrid should have a high electrical conductivity in order to serve as a conductive template for the metal oxide nanoparticles, which decrease the resistance of the hybrid and electrode, thereby enabling the excellent high-rate capability.

10.4 $Li_4Ti_5O_{12}$/Reduced Graphene Oxide Hybrid Material[14]

One of key safety issues of Li-ion battery is dendritic lithium growth on the graphite anodes surface at high-charging currents at potentials approaching 0 V *versus* Li at the end of Li insertion.[30] A solution to this could be the use of electrochemical redox couples with higher equilibrium potentials, which make Li formation thermodynamically less favorable. Spinel $Li_4Ti_5O_{12}$ has attracted attention for this purpose as a potential anode material because of its good reversibility of Li-ion intercalation/deintercalation and its near-zero structural change during charging and discharging.[30,31] However, its low electrical conductivity (10^{-9} Scm^{-1}) results in initial capacity loss and poor rate capability. Efforts to ameliorate its rate capability include reductions of particle size, doping with small amounts of V^{+5}, Mn^{+4}, Ni^{2+}, or Cr^{3+}, and coating the $Li_4Ti_5O_{12}$ with conductive materials.[32–34]

Nanosized $Li_4Ti_5O_{12}$ is expected to exhibit improved rate capability because of the increase in the effective interfacial area between the metal oxide and electrolyte and the shorter lengths ions and electrons must travel during charging and discharging. Nanosized metal oxides tend to agglomerate, a process to suppress the agglomeration is important in order to enhance the electrical conductivity of an electrode and hence to achieve high-rate capability.[6] New methods of synthesizing nanosized materials under mild conditions are being developed using solution based synthesis. Microwave-assisted solution methods have distinct advantages over conventional solvothermal methods, such as lower synthesis temperatures, higher reaction rates, shorter reaction times, and smaller resulting particles. This chapter demonstrates $Li_4Ti_5O_{12}$/RGO as a hybrid electrode material, in which nanosized $Li_4Ti_5O_{12}$ is heterogeneously precipitated on to a RGO template, with excellent high-rate charge–discharge capability and cycling performance.

Scheme 10.2 shows a proposed route for the preparation of the $Li_4Ti_5O_{12}$/RGO hybrid. GO nanosheets were employed as a high surface area template for the selective heterogeneous nucleation and growth of TiO_2 nanoparticles. TiO_2/RGO hybrids were then prepared using microwave-assisted solvothermal synthesis[35] during which GO was partially reduced to RGO in a polyol solution,[36] leading to the formation of TiO_2/RGO. This TiO_2/RGO was then allowed to react with LiOH solution whereupon it was converted to $Li_4Ti_5O_{12}$ precursor (denoted as Li-Ti-O)/RGO nanohybrid under microwave-assisted solvothermal conditions. Finally, the Li-Ti-O/RGO nanohybrid was heat treated at 700°C for 5 h under a 5 wt% H_2/Ar gas atmosphere in order to prepare a nanohybrid material composed of phase-pure and highly crystalline $Li_4Ti_5O_{12}$

nanoplatelets and highly conductive RGO. In this synthesis, GO nanosheets were chosen as templates for the precipitation of TiO_2 nanoparticles as they can be readily exfoliated in solution to yield a stable dispersion mainly consisting of single layered sheets.[37] They also have a higher density of oxygen functional groups than RGO, such as hydroxyl, carboxyl and epoxy groups that interact strongly with metal oxide nanoparticles.[37]

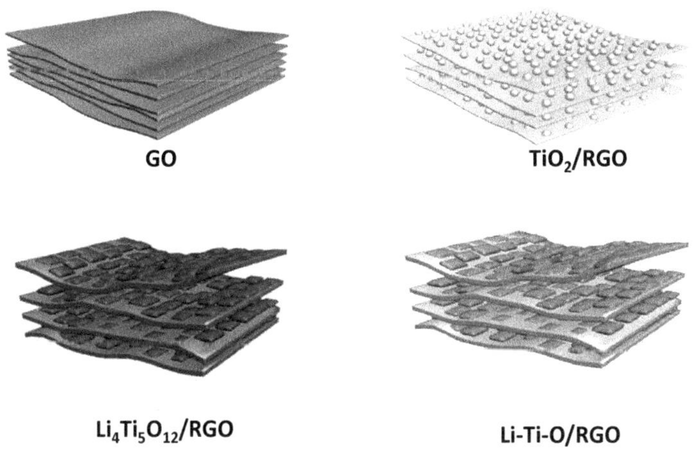

Scheme 10.2 Schematic illustration of the synthesis process for the $Li_4Ti_5O_{12}$/RGO hybrid. Reprinted with permission from Ref. 14. Copyright 2010, Elsevier.

Figure 10.1 shows X-ray diffraction (XRD) patterns and TEM images of the $Li_4Ti_5O_{12}$/RGO hybrid. The formation of phase-pure $Li_4Ti_5O_{12}$ was confirmed by Raman spectroscopy (data not shown). The XRD patterns of Fig. 10.1a can be indexed to phase-pure and highly crystalline spinel $Li_4Ti_5O_{12}$. A small and broad (002) diffraction peak of the disorderedly stacked reduced GO or graphene sheets appeared at 2θ of 25~27°.[38] The lattice constant of the spinel $Li_4Ti_5O_{12}$ nanoplatelets was calculated to be 8.364 Å from the XRD data, which is in good agreement with other reported values.[30,31]

The TEM images in Fig. 10.1b,c show the uniform dispersion of $Li_4Ti_5O_{12}$ nanoplatelets with sizes of 10–20 nm on RGO nanosheets. As in Scheme 10.1, the Li-Ti-O/RGO hybrid was formed via the microwave-assisted solvothermal treatment of TiO_2/RGO hybrid in LiOH solution. It should be noted that the anatase TiO_2 nanoparticles in the hybrid were of semi-spherical shape with particle sizes of 3–5 nm (data not shown). The loading amount of TiO_2 in the TiO_2/RGO hybrid was evaluated to be 69 wt%. Drastic changes in morphology from TiO_2 nanoparticles to the Li-Ti-O nanoplatelets in the hybrid suggest that the lithiation reaction might proceed via dissolution and reprecipitation on the RGO surface.[39] Further investigation is needed to elucidate the reaction mechanism of TiO_2/RGO to Li-Ti-O/RGO. Post heat treatment did not affect the sizes and shapes of the Li-Ti-O particles, but resulted in the formation of phase-pure and highly crystalline $Li_4Ti_5O_{12}$. The high-magnification TEM (HR-TEM) image (Fig. 10.1d) clearly shows a well-crystallized structure of $Li_4Ti_5O_{12}$ with lattice fringes of ca. 0.48 nm corresponding to (111) inter-planar spacing of $Li_4Ti_5O_{12}$.[40] The loading amount of $Li_4Ti_5O_{12}$ in the $Li_4Ti_5O_{12}$/RGO hybrid was evaluated to be 72 wt% from thermo-gravimetric analysis. In the synthesis of a $Li_4Ti_5O_{12}$/RGO hybrid electrode, the electrical conductivity of RGO is of the

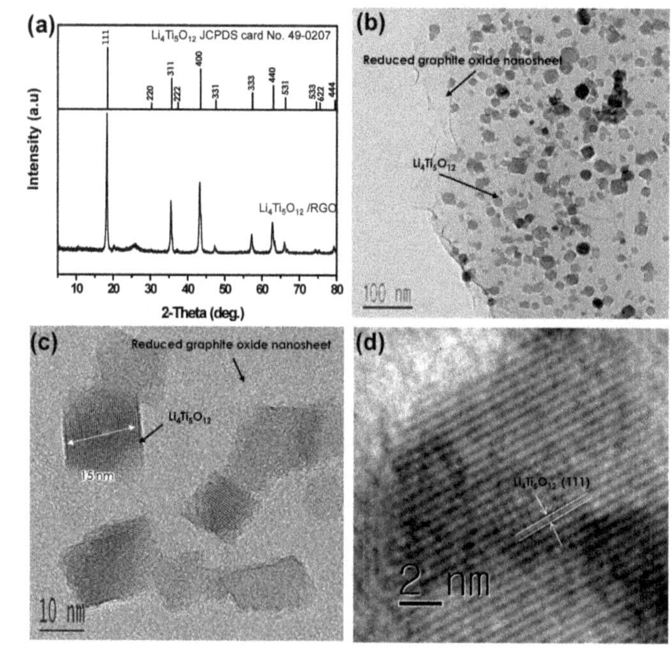

Figure 10.1 (a) XRD patterns of $Li_4Ti_5O_{12}$/RGO hybrid; spinel $Li_4Ti_5O_{12}$ (JCPDS No. 49-0207), (b) Transmission electron microscopy (TEM) image of $Li_4Ti_5O_{12}$/RGO hybrid and (c–d) HR-TEM images of $Li_4Ti_5O_{12}$/RGO hybrid. Reprinted with permission from Ref. 14. Copyright 2010, Elsevier.

utmost importance in order to achieve high-rate charge–discharge capability. Removal of residual oxygen introduced during the synthesis of GO is a key factor in improving this.[41] For a quantitative analysis of the reduction of RGO, the atomic ratio of carbon to oxygen (C/O ratio) was evaluated by XPS measurement (data not shown). Post heat treatment of the Li-Ti-O/RGO hybrid caused the C/O ratio of RGO to increase from 5.28 to 26.9 for the $Li_4Ti_5O_{12}$/RGO hybrid.

Figure 10.2a shows cyclic voltammograms (CVs) of the $Li_4Ti_5O_{12}$/RGO hybrid measured with potential scan rates of 0.2, 1, 2, 5 mVs^{-1} in the potential window of 1.0–3.0 V in 1 M $LiClO_4$ in PC. The CV at 0.2 mVs^{-1} clearly shows one set of well-defined current peaks at 1.53 and 1.6 V, which correspond to those of phase-pure $Li_4Ti_5O_{12}$.[37] Interestingly, the general shape of the CV was well maintained with increasing potential scan rates up to 5 mVs^{-1}, suggesting good high-rate capability of the $Li_4Ti_5O_{12}$/RGO hybrid. Conducting agent (≥15 wt%) is commonly used in fabricating electrodes of metal oxide nanoparticles in order to achieve electrical contact between the particles and current collector and among the particles themselves.[42] In this study no additional conducting agent was used to fabricate an electrode of $Li_4Ti_5O_{12}$/RGO hybrid.

Figure 10.2b shows the results of galvanostatic charge and discharge tests which investigate the high-rate capability of the hybrid electrode with an increase in the C-rate from 1 to 100 C-rate within the potential window of 1 2.5 V, where each set of discharge and charge curves was measured at the same C-rate. It shows typical charge and discharge behavior of the phase-pure spinel $Li_4Ti_5O_{12}$ with potential plateaus at 1.571 V (1 C) during charge and 1.543 V (1 C) during discharge.[31] The specific discharge capacity of the $Li_4Ti_5O_{12}$ in the $Li_4Ti_5O_{12}$/RGO hybrid was 154 $mAhg^{-1}$ at 1 C-rate. The $Li_4Ti_5O_{12}$ in the hybrid could deliver a discharge capacity of 128 $mAhg^{-1}$ of $Li_4Ti_5O_{12}$ at 50 C-rate (83% of 1 C discharge capacity) and 101 $mAhg^{-1}$ of $Li_4Ti_5O_{12}$ at 100 Crate (65% of 1 C discharge capacity) without the addition of any conducting agent in the electrode, which clearly demonstrates the excellent high-rate capability of the $Li_4Ti_5O_{12}$/RGO hybrid. It is noteworthy that the $Li_4Ti_5O_{12}$/RGO hybrid showed small increase in electrode polarization with increasing C-rate from 1 to 100 C-rate during charge/discharge and that this is responsible for the excellent high-rate capability of the hybrid compared with those reported previously.[43,44] $Li_4Ti_5O_{12}$/RGO hybrid exhibited the excellent capacity retention of 95 % at 1 C-rate and 96 % at 10 C-rate over 100 cycles, as shown in Fig. 10.2c.

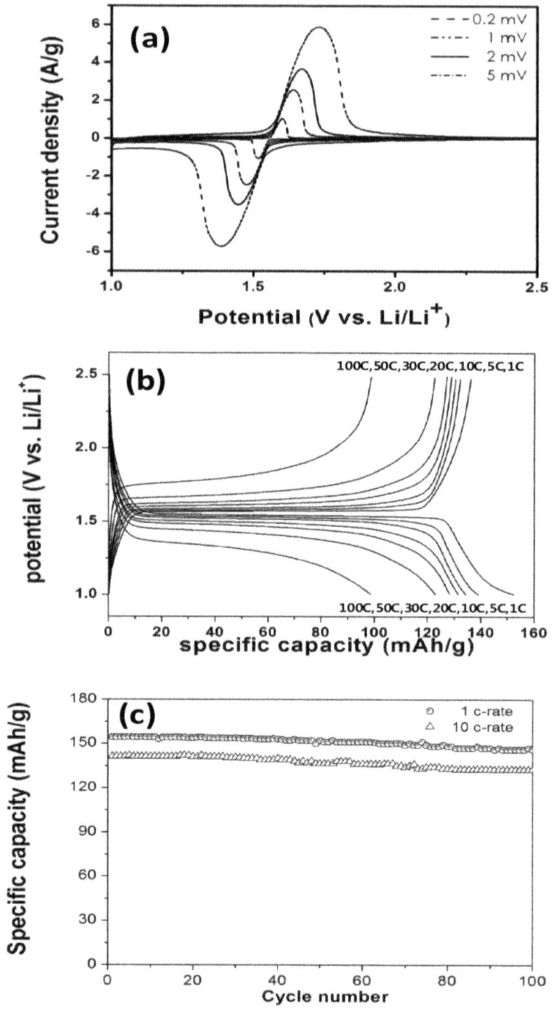

Figure 10.2 (a) CV of $Li_4Ti_5O_{12}$/RGO hybrid at scan rate of 0.2, 1, 2, 5 mVs^{-1}. (b) Charge–discharge curves at different C-rates of the $Li_4Ti_5O_{12}$/RGO hybrid between 1 and 2.5 V (*vs.* Li/Li^+): 1, 5, 10, 20, 30, 50, 100 C-rate. (c) Specific capacity of the $Li_4Ti_5O_{12}$/RGO hybrid during cycling between 1 and 2.5 V (*vs.* Li/Li^+) at 1 C-rate and 10 C-rate. Reprinted with permission from Ref. 14. Copyright 2010, Elsevier.

10.5 $LiMn_2O_4$/Reduced Graphene Oxide Hybrid Material[15]

Scheme 10.3 shows our two-step approach for the synthesis of the $LiMn_2O_4$/RGO hybrid. RGO was produced through chemical exfoliation of natural graphite and hydrazine reduction.[21] A MnO_2/RGO hybrid was first prepared using in situ redox deposition of MnO_2 on RGO in a solution of $KMnO_4$ and RGO in the first step. The synthesis mechanism of the MnO_2/RGO hybrid is similar to that of the MnO_2/CNT hybrid described elsewhere.[45-47] RGO has a work function of ~4.2 eV, which is equivalent to −0.2 V against a standard hydrogen electrode (SHE), and its electrode potential is well above the reduction potential of MnO_4^- ions, +1.692 V (*vs.* SHE).[48,49] Therefore, the spontaneous formation of MnO_2 on the RGO template might be due to the difference in the reduction potential between the RGO and MnO_4^- ions. Then, this MnO_2/RGO hybrid was allowed to react with LiOH aqueous solution for conversion to the $LiMn_2O_4$/RGO hybrid under microwave-assisted

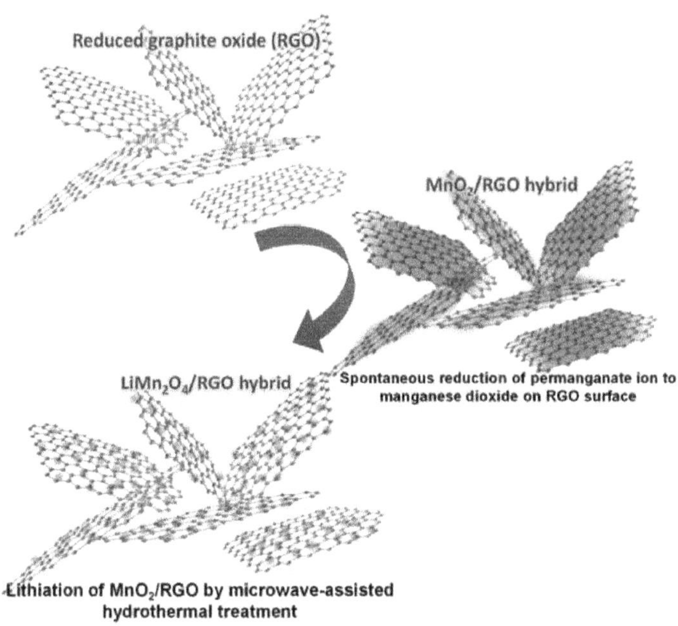

Scheme 10.3 Schematic illustration of the synthesis process for the $LiMn_2O_4$/RGO hybrid material. Reprinted with permission from Ref. 15. Copyright 2011, Royal Society of Chemistry.

hydrothermal (MAH) conditions. Previously, a binary oxide $NiCo_2O_4$/RGO hybrid was prepared using a two-step synthesis, which involved a deposition of binary NiCo hydroxide on GO nanosheets and its conversion to the binary oxide $NiCo_2O_4$/RGO hybrid through microwave-assisted solvothermal treatment.[50] This synthesis approach offers a general route to prepare a lithium transition metal oxide/RGO hybrid material from a transition metal oxide/RGO hybrid precursor.

Figure 10.3 shows the XRD pattern of the $LiMn_2O_4$/RGO hybrid (73 wt% of $LiMn_2O_4$ in the hybrid). All peaks could be indexed based on the cubic spinel $LiMn_2O_4$ structure (space group: *Fd-3m*). Rietveld refinement result using this structure indicates the formation of a well-crystallized spinel $LiMn_2O_4$ phase without any impurity phase on the RGO template. The refined lattice constant a (8.2134 ± 0.0002 Å) was slightly smaller than that of bulky stoichiometric $LiMn_2O_4$ (a = ~8.235 Å), which might be attributed to either the slightly lithium-rich stoichiometry or the nanosize effect.[51] The (002) peak of RGO was very weak, implying that the $LiMn_2O_4$ nanoparticles deposited on the RGO surface might prevent the restacking of the RGO sheets.[52]

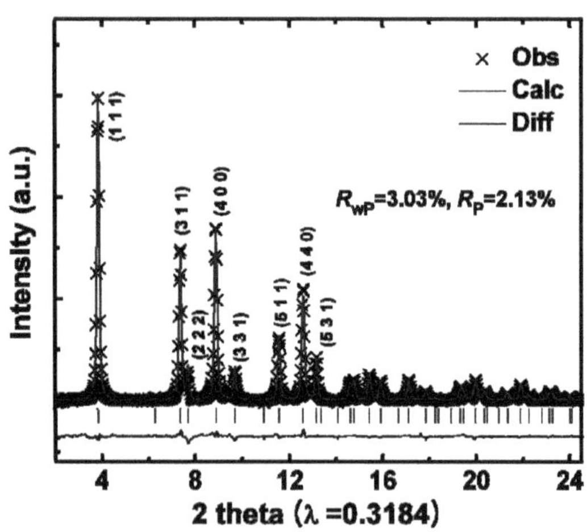

Figure 10.3 Rietveld refinement of the spinel $LiMn_2O_4$/RGO hybrid material. Reprinted with permission from Ref. 15. Copyright 2011, Royal Society of Chemistry.

TEM was used to investigate the morphological change during synthesis of the $LiMn_2O_4$/RGO hybrid from pristine RGO nanosheets. Fig. 10.4a shows the morphology characterization of pristine RGO nanosheets, which demonstrates its wrinkled paper-like structure. After MnO_2 coating process, the entire surface of RGO nanosheets was covered with thin layered MnO_2 at the nanometer scale, as shown in Fig. 10.4b. And Fig. 10.4c clearly shows the good dispersion and uniform coverage on the RGO nanosheets of the $LiMn_2O_4$ nanoparticles with 10–40 nm sizes via microwave-assisted hydrothermal treatment of the MnO_2/RGO hybrid in LiOH solution. A notable change in morphology from the thin layer of the precursor MnO_2 in Fig. 10.4b to the nanoparticles of spinel $LiMn_2O_4$ in the hybrid materials suggests the dissolution and reprecipitation mechanism of the formation of the $LiMn_2O_4$ phase on the RGO surface under the MAH conditions.[53] The clear lattice fringes observed in the high-resolution TEM image in Fig. 10.4d demonstrate the well-crystallized structure of the

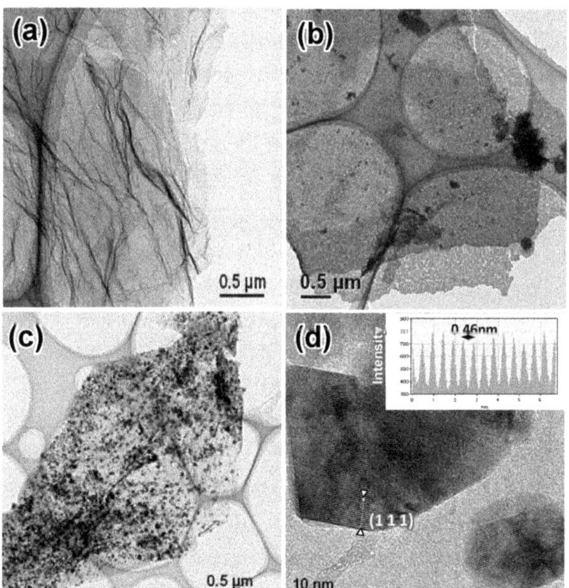

Figure 10.4 TEM image of (a) RGO nanosheets, (b) MnO_2/RGO hybrid precursor, and (c) spinel $LiMn_2O_4$/RGO hybrid. (d) HR-TEM image of an individual spinel $LiMn_2O_4$ nanoparticles on RGO nanosheets (the inset shows the direct line scanning analysis profile of the spinel $LiMn_2O_4$ nanoparticle). Reprinted with permission from Ref. 15. Copyright 2011, Royal Society of Chemistry.

$LiMn_2O_4$ nanoparticles in the hybrid, which is consistent with the XRD result. The 0.46 nm regular spacing of the lattice planes obtained from the direct line scanning analysis in the inset of Fig. 10.4d agrees very well with the d spacing of the (111) planes of $LiMn_2O_4$. Graphene can act as an effective microwave susceptor.[54] A higher temperature, therefore, is anticipated on the surface of the RGO than that of the surrounding aqueous media under microwave irradiation. This selective heating of RGO under microwave irradiation might lead to the precipitation of well-crystallized $LiMn_2O_4$ nanoparticles on the RGO surface through the selective heterogeneous nucleation despitethe low synthesizing temperature (200°C) and short reaction time (30 min).

To investigate the local structure around Mn atoms, Mn Kedge extended X-ray absorption fine structure (EXAFS) spectroscopy of the $LiMn_2O_4$/RGO hybrid was performed in comparison with the bulky sized $LiMn_2O_4$ reference (spinel, Sigma Aldrich). As shown in Fig. 10.5a background subtracted k^3-weighted Mn K-edge EXAFS spectrum (dotted line with circles) of the $LiMn_2O_4$/RGO hybrid is quite similar to that of the bulky sized reference (black line) except for a slight reduction in its amplitude, suggesting a very similar local structure around Mn in the $LiMn_2O_4$/RGO hybrid to that of the $LiMn_2O_4$ reference. Corresponding Fourier-transformed (FT) EXAFS spectra are plotted in Fig. 10.5b in order to distinguish the structural information of each shell at different bond lengths from the core Mn atom. The FT-EXAFS spectra in Fig. 10.5b display two major peaks. The first peak at R = ~1.5 Å is due to contribution from the nearest $Mn–O_6$ octahedra while the second peak at R = ~2.5 Å is attributed to $Mn–Mn_6$ octahedra in the second coordination shell. Note that the FT-EXAFS spectra were not phase corrected so that the actual bond lengths are approximately 0.2–0.4 Å longer. While the first Mn–O peak of the $LiMn_2O_4$/RGO hybrid is almost identical to that of the $LiMn_2O_4$ reference in its amplitude and position, the second Mn–Mn peak shows a little reduction in the amplitude compared with that of the $LiMn_2O_4$ reference revealing the reduced coordination number of the Mn–Mn bond in the $LiMn_2O_4$/RGO hybrid. Since the amplitude is proportional to the coordination number, this can be explained on the basis of the termination effects of very small particle sizes of the $LiMn_2O_4$ particles on the RGO template.[55]

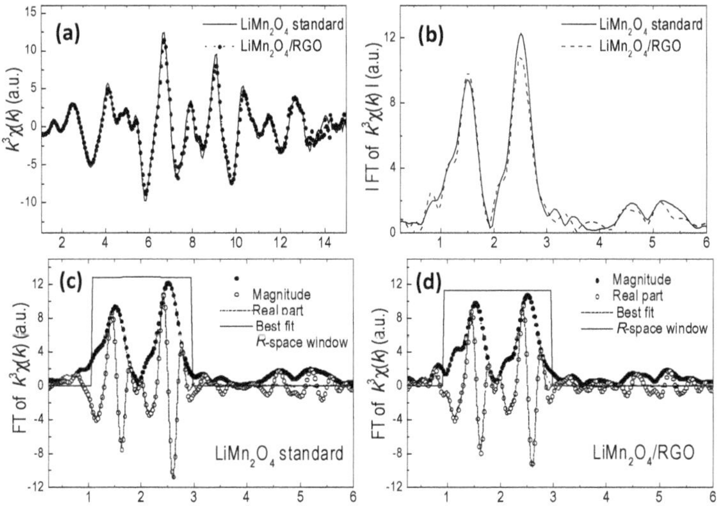

Figure 10.5 (a) k^3-weighted Mn K-edge EXAFS spectra and (b) corresponding k^3-weighted Fourier transform magnitude of the $LiMn_2O_4$ standard (solid line) and the $LiMn_2O_4$/RGO hybrid (dash line). Least-square fits of calculated FT-EXAFS phase and amplitude functions (dash line) to the experimental EXAFS spectra (solid and open circles) for (c) $LiMn_2O_4$ standard and (d) the $LiMn_2O_4$/RGO hybrid. Phase-shift was not corrected in the FT magnitude of EXAFS spectra. Reprinted with permission from Ref. 15. Copyright 2011, Royal Society of Chemistry.

In order to get quantitative structural parameters, analysis of the FT-EXAFS spectra using least-square fits of phase and amplitude functions generated by the FEFF 6.0 code was carried out. The fitted (solid lines) and experimental (solid and open circles) FT-EXAFS spectra for the $LiMn_2O_4$ reference and $LiMn_2O_4$/RGO hybrid are presented in Fig. 10.5c,d indicating excellent fitting result. Table 10.1 summarizes the structural parameters. It should be noted that the fitted coordination numbers for the nearest neighboring Mn–O bond in both samples show significantly smaller values of ~4.3 than six based on the crystallographic value. This is due to the well-known Jahn–Teller effect of Mn^{3+} ions in the $LiMn_2O_4$ structure which induces distortion of the $Mn–O_6$ octahedra. The slightly smaller Mn–Mn coordination number of 5.5 ± 0.6 for the $LiMn_2O_4$/RGO hybrid than 6.0 ± 0.3 for the $LiMn_2O_4$ reference confirms the nanosize effect.[55] Interatomic distances and Debye–Waller factors (i.e., degree of disorder) for the first two bonds of Mn–O and Mn–

Mn in Table 10.1 for these two samples are also very close, clearly demonstrating the well-ordered local atomic structure around the Mn atoms in the $LiMn_2O_4$/RGO hybrid, just as that in the bulky sized $LiMn_2O_4$. Interestingly, the Debye–Waller factor of the Mn–O bond in the $LiMn_2O_4$/RGO hybrid was even slightly smaller than that in the $LiMn_2O_4$ reference and this may suggest reduced Jahn–Teller distortion in the hybrid due to its unique hybrid structure.

Table 10.1 Mn K-edge EXAFS structure parameter of the $LiMn_2O_4$ standards and the $LiMn_2O_4$/RGO hybrid material

Sample	Path	N	ΔE/eV	r/Å	σ^2 ($\times 10^{-3}$Å^2)
$LiMn_2O_4$ standards	Mn–O	4.3 ± 0.3	2.44 ± 0.51	1.90(8) ± 0.003	3.60 ± 0.46
	Mn–Mn	6.3 ± 0.3	2.44 ± 0.51	2.90(9) ± 0.003	5.89 ± 0.35
$LiMn_2O_4$/ RGO hybrid	Mn–O	4.2 ± 0.3	−1.87 ± 0.78	1.90(5) ± 0.004	2.91 ± 0.58
	Mn–Mn	5.5 ± 0.6	−1.87 ± 0.78	2.90(4) ± 0.005	5.88 ± 0.66

Source: Reprinted with permission from Ref. 15. Copyright 2011, Royal Society of Chemistry.

Note: In both the samples, the EXAFS R factor, representing the absolute misfit between the model and data, was estimated to be less than 0.1%.

Abbreviations: N, Coordination number; ΔE, inner shell potential shift; r, bond length; σ^2, Debye–Waller factor

Nanoparticular $LiMn_2O_4$ materials synthesized at low temperatures often show low crystallinity and a high-temperature post-heat treatment is usually required in order to improve their crystallinity, causing particle agglomeration.[56] In contrast, the $LiMn_2O_4$ nanoparticles synthesized by the MAH method on the RGO template exhibit the unique structural properties of high crystallinity (by XRD) and well-ordered local atomic structure (by EXAFS), of the low synthesizing temperature and short synthesis time.

X-ray photoelectron spectroscopy (XPS) was also performed to identify the oxidation state of Mn in the $LiMn_2O_4$/RGO hybrid. The 3s core levels of the 3d transition metals are known to exhibit exchange splitting due to the interaction between the 3s core hole created in an XPS process and the 3d electrons.[57] From Fig. 10.6a, the binding energy of the 3s level of $LiMn_2O_4$ in the hybrid is 84.01 eV and the splitting energy is about 4.95 eV. Using a linear interpolate of the known splitting energies for Mn^{3+} (5.5 eV) and Mn^{4+} (4.5 eV) against the oxidation number, a formal Mn oxidation state in the $LiMn_2O_4$/RGO hybrid is determined to be +3.55 (Fig. 10.6b).

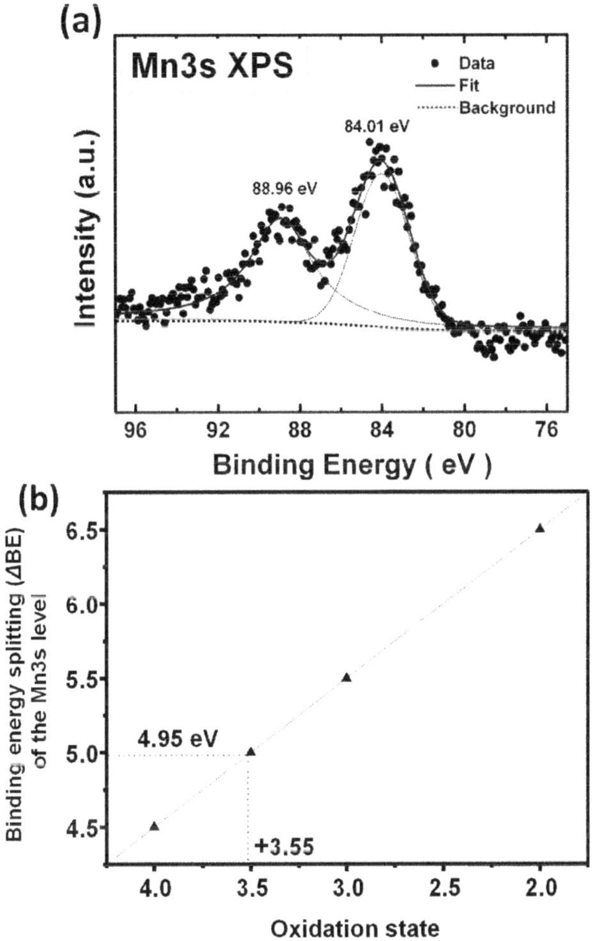

Figure 10.6 (a) Mn 3s core level XPS spectrum of $LiMn_2O_4$/RGO hybrid (b) oxidation number versus binding energy multiple splitting (ΔBE) of the Mn 3s lines. Reprinted with permission from Ref. 15. Copyright 2011, Royal Society of Chemistry.

Figure 10.7a shows the charge–discharge curves of the $LiMn_2O_4$/RGO hybrid electrode in an EC/DEC electrolyte of 1 M $LiPF_6$ at different C-rates. The electrodes were fabricated by slurry casting of 85 wt% active material ($LiMn_2O_4$/RGO hybrid), 10 wt% acetylene black, and 5 wt% PVDF binder dissolved in NMP solvent. Two plateaus centered at around 4 V, which are typical for stoichiometric $LiMn_2O_4$, were still distinguishable in the charge–discharge curves at the rate as high as 50 C. The specific capacity of the $LiMn_2O_4$ in the $LiMn_2O_4$/RGO hybrid was plotted with various C-rates in Fig. 10.7b. The rate performance of $LiMn_2O_4$/RGO was excellent, showing 85% capacity at 50 C-rate and 74% capacity at 100 C-rate with respect to the specific capacity of 137 mA h g^{-1} at 1 C-rate. The $LiMn_2O_4$/RGO hybrid also exhibited the excellent capacity retention of 90% at 1 C-rate and 96% at 10 C-rate over 100 cycles, as shown in Fig. 10.7c.

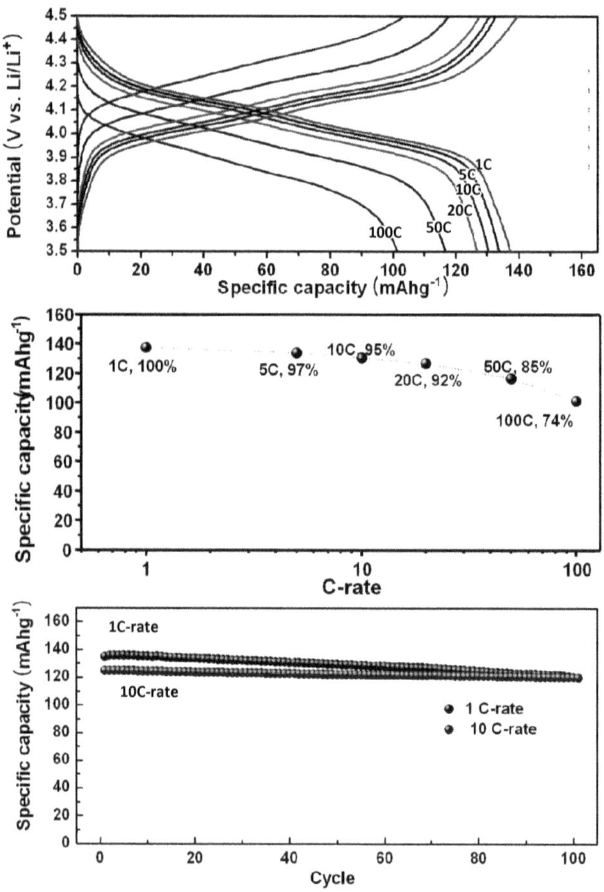

Figure 10.7 (a) Charge–discharge curves of the $LiMn_2O_4$/RGO hybrid electrode at various C-rates from 3.5 to 4.5 V (*vs.* Li/Li^+). (b) Specific capacity as a function of the C-rate and (c) specific capacity of the $LiMn_2O_4$ in the $LiMn_2O_4$/RGO hybrid electrode during cycling at 1 C-rate and 10 C-rate. Reprinted with permission from Ref. 15. Copyright 2011, Royal Society of Chemistry.

The superior electrochemical performance of the $LiMn_2O_4$/RGO hybrid electrode in terms of capacity and rate capability can be explained as follows: (1) The $LiMn_2O_4$ nanoparticles are well dispersed on the RGO nanosheets template, which prevents agglomeration of individual $LiMn_2O_4$ nanoparticles and consequently maintains their inherent highly active surface area. In contrast, the nanosized $LiMn_2O_4$ particles often agglomerated during the post-heat-treatment and hence show a significant capacity loss at high C-rates.[58] (2) The nanosized $LiMn_2O_4$ particles with high crystallinity and well-ordered local atomic structure are fully accessible for charge storage, which leads to a high capacity in contrast with the nanosized $LiMn_2O_4$ particles often losing their capacity at high-rate due to the lack of three-dimensional network for Li-ion transport caused by their low crystallinity or disordered local atomic structure.[59] (3) The highly conductive RGO sheets in the hybrid serve as electronic conductive channels for the $LiMn_2O_4$ nanoparticles, which decrease the resistance of the hybrid and electrode and contributes to the excellent high-rate capability.[60]

10.6 Capacitive Charge Storage Behavior of $LiMn_2O_4$/Reduced Graphene Oxide Hybrid[15]

In order to further understand its excellent high-rate performance, we also consider the capacitive effect of the $LiMn_2O_4$/RGO hybrid electrode which led to faster charge–discharge rates, by using the methodology suggested by T. Brezesinski *et al.*[61] The total stored charge can be separated into three components: the faradaic contribution from the Li-ion insertion process, the faradaic contribution from the charge transfer process with surface atoms (referred to as pseudocapacitance),[62] and the nonfaradaic contribution from the electric double-layer effect. The latter two capacitive components cannot be separated in the following analysis. A more detailed procedure was presented elsewhere.[63]

Using the methodology suggested by T. Brezesinski *et al.*, we can express the current response at a fixed potential as being the combination of two separate mechanisms, namely the surface reaction from capacitive effects and the bulk reaction from diffusion-controlled Li-ion insertion:[64]

$$i(v) = k_1 v + k_2 v^{1/2} \tag{10.1}$$

For analytical purposes, it can be rearranged to

$$\frac{i(v)}{v^{1/2}} = k_1 v^{1/2} + k_2 \tag{10.2}$$

In Eq. 10.1, $k_1 v$ and $k_2 v^{1/2}$ correspond to the current contributions from the surface capacitive effects and the bulk diffusion controlled insertion process, respectively. Plotting the potential scan rate dependence of the current allows us to determine k_1 and k_2 from the slope and the y-axis intercept point of a straight line, respectively, at each fixed potential using Eq. 10.2. This methodology thus enables us to distinguish between the currents arising from Li-ion insertion and those from capacitive effects (details are described elsewhere).[65] Since the capacitive behavior is responsible for the charge storage at the surface area which increases with a decrease in particle size, capacitive behavior is expected to become more pronounced as a particle size decreases.

Figure 10.8a,b show the CVs for the $LiMn_2O_4$/RGO hybrid electrode at potential scan rates of 1 mV s^{-1} and 10 mV s^{-1}, respectively. The total current (solid line) is obtained experimentally and the capacitive current (shaded region) is determined using Eq. 10.2. The CV data in Fig. 10.8a,b enable us to determine the total stored charge and the relative contributions associated with the bulk reaction from diffusion controlled Li-ion insertion and the surface reaction from capacitive effect.

By comparing the shaded area from the surface reaction with the total stored charge, we could find that the surface reaction contributed 59% and 71% of the total charge storage for the $LiMn_2O_4$/RGO hybrid at potential scan rates of 1 mVs^{-1} and 10 mVs^{-1}, respectively. We surmise that a large surface capacitive contribution of the $LiMn_2O_4$/RGO hybrid electrode is mainly due to the high surface area of small particle size of $LiMn_2O_4$ (10–40 nm) of the $LiMn_2O_4$/RGO hybrid. The fact that a large fraction of the current is capacitive in nature is in line with the study reported by J. Wang *et al.*[65] They discussed the capacitive charge storage of titanium oxide (TiO_2) nanoparticles as an electrochemical energy storage material. In the light of the above argument, it is recognized that the large surface capacitive contribution was responsible for the faster charge–discharge processes of the $LiMn_2O_4$/RGO hybrid electrode.

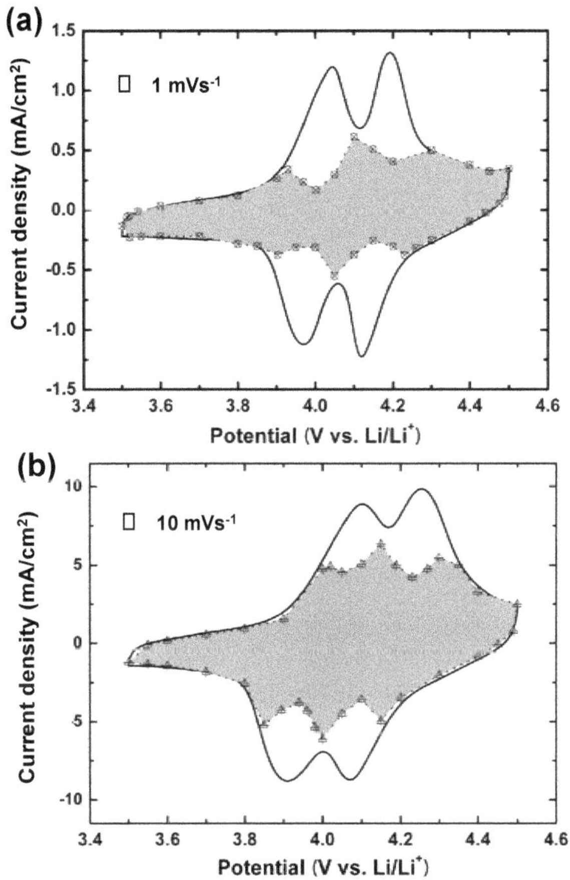

Figure 10.8 CV for the $LiMn_2O_4$/RGO hybrid electrode at the potential scan rate of (a) 1 $mV s^{-1}$ and (b) 10 $mV s^{-1}$. The total current (solid line) is obtained experimentally and the capacitive current (dotted line) is determined from Eq. 10.2. Shaded region shows the capacitive contribution to the total charge storage. Reprinted with permission from Ref. 15. Copyright 2011, Royal Society of Chemistry.

Figure 10.9 compares the contribution of the surface capacitive charge and the diffusion controlled insertion charge calculated from CVs in Fig. 10.8 to the total charge (gravimetrically normalized). It is evident that the capacitive charge storage is certainly unaffected by the increase in potential scan rate. The invariance of the charge storage from the surface capacitive effects could be ascribed to the fast kinetics of both double-layer and pseudocapacitive processes.In contrast,the insertion capacity decreased by 29% with an increase in potential scan rate from 1 mVs^{-1} to 10 mVs^{-1}. It is thought that less time for the Li-ion to diffuse into the $LiMn_2O_4$ in the hybrid might lead to the reduced insertion capacity with increasing potential scan rate. It is evident from this result that the high surface area of the nanosized $LiMn_2O_4$ in the $LiMn_2O_4$/RGO hybrid plays an important role in achieving the high-rate charging/discharging capability.

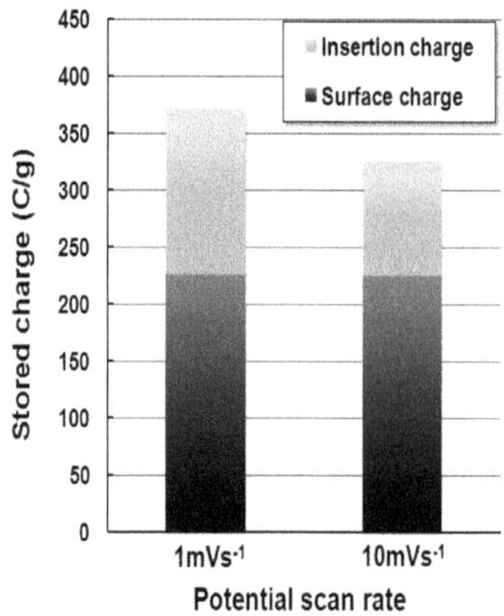

Figure 10.9 Comparison of the total charge stored in the $LiMn_2O_4$/RGO hybrid electrode at potential scan rate of 1 mVs^{-1} and 10 mVs^{-1}. The stored charge is divided into insertion charge and surface charge. Reprinted with permission from Ref. 15. Copyright 2011, Royal Society of Chemistry.

10.7 Concluding Remarks

In this chapter, we demonstrated the synthesis and electrochemical characterization of RGO and metal oxide/RGO hybrids for high-rate Li-ion battery applications. As outlined introduction, this study was performed with an aim to focus on the rational design and synthesis of the metal oxide/RGO hybrid material with the excellent high-rate charge–discharge capability for high-rate lithium-ion batteries. Our results provide compelling evidence that the well-crystallized and nanosized metal oxide (i.e., $Li_4Ti_5O_{12}$ and $LiMn_2O_4$)/RGO hybrid could achieve the high specific capacity and could deliver a remarkable high discharge capacity even at a very fast charge–discharge rate of 50 C-rate and 100 C-rate. These results demonstrate the promising potential of the metal oxide/RGO hybrid as an electrode material in Li-ion batteries with excellent rate capability. Furthermore, we considered the contribution of the surface capacitive charge and the diffusion controlled insertion charge for the $LiMn_2O_4$/RGO hybrid electrode material. It revealed that the high surface area of the $LiMn_2O_4$/RGO hybrid plays an important role in achieving the high-rate charging/discharging capability. To understand the unique charge storage mechanism of metal oxide/RGO hybrid materials completely, further study is required to get a better understanding of the Li-ion dynamics in hybrid materials, especially near the surface.

Acknowledgments

This work supported by the National Research Laboratory Program through the National National Research Foundation of Korea (NRF) grant funded by the Ministry of Education, Science and Technology (MEST) (Grant No. 2007-0055835). The work done at BNL (XRD and EXAFS work) was supported by the Assistant Secretary for Energy Efficiency and Renewable Energy, Office of Vehicle Technologies of the U.S. Department of Energy under Contract Number DEAC02-98CH10886.

References

1. K. S. Kang, Y. S. Meng, J. Berger, C. P. Grey, G. Ceder, *Science*, **311**, 977 (2006).
2. Y. Wang, G. Z. Cao, *Adv. Mater.*, **20**, 2251 (2008).
3. B. Kang, G. Ceder, *Nature*, **458**, 190 (2009).
4. P. G. Bruce, B. Scrosati, J. M. Tarascon, *Angew. Chem., Int. Ed.*, **47**, 2930 (2008).
5. A. S. Arico, P. Bruce, B. Scrosati, J. M. Tarascon, W. Van Schalkwijk, *Nat. Mater.*, **4**, 366 (2005).

6. H. S. Zhou, C. Jiang, M. Ichihara, I. Honma, *Electrochim. Acta*, **52**, 6470 (2007).
7. J. M. Tarascon, M. Armand, *Nature*, **414**, 359 (2001).
8. S. B. Ma, K. W. Nam, W. S. Yoon, S. M. Bak, X. Q. Yang, B. W. Cho, K. B. Kim, *Electrochem. Commun.*, **11**, 1575 (2009).
9. A. Magasinski, P. Dixon, B. Hertzberg, A. Kvit, J. Ayala, G. Yushin, *Nat. Mater.*, **9**, 461 (2010).
10. H. J. Dai, H. L. Wang, H. S. Casalongue, Y. Y. Liang, *J. Am. Chem. Soc.*, **132**, 7472 (2010).
11. K. H. Chang, Y. F. Lee, C. C. Hu, C. I. Chang, C. L. Liu, Y. L. Yang, *Chem. Commun* , **46**, 7957 (2010).
12. W. C. Ren, Z. S. Wu, L. Wen, L. B. Gao, J. P. Zhao, Z. P. Chen, G. M. Zhou, F. Li, H. M. Cheng, *ACS Nano*, **4**, 3187 (2010).
13. I. Honma, S. M. Paek, E. Yoo, *Nano Lett.*, **9**, 72 (2009).
14. H. K. Kim, S. M. Bak, K. B. Kim, *Electrochem. Commun.*, **12**, 1768 (2010).
15. S. M. Bak, K. W. Nam, C. W. Lee, K. H. Kim, H. C. Jung, X. Q. Yang, K. B. Kim, *J. Mater. Chem.*, **21**, 17309 (2011).
16. K. S. Novoselov, A. K. Geim, S. V. Morozov, D. Jiang, Y. Zhang, S. V. Dubonos, I. V. Grigorieva, A. A. Firsov, *Science*, **306**, 666 (2004).
17. C. Berger, Z. M. Song, X. B. Li, X. S. Wu, N. Brown, C. Naud, D. Mayou, T. B. Li, J. Hass, A. N. Marchenkov, E. H. Conrad, P. N. First, W. A., Heer *Science*, **312**, 1191 (2006).
18. M. Choucair, P. Thordarson, J. A. Stride, *Nat. Nanotechnol.*, **4**, 30 (2009).
19. M. J. McAllister, J. L. Li, D. H. Adamson, H. C. Schniepp, A. A. Abdala, J. Liu, M. Herrara-Alonso, D. L. Milius, R. Car, R. K. Prud'homme, I. A. Aksay, *Chem. Mater.*, **19**, 4396 (2007).
20. B. X. Fan, W. Peng, Y. Li, X. Li, S. Wang, G. Zhang, F. Zhang, *Adv. Mater.*, **20**, 1 (2008).
21. S. Stankovich, D. A. Dikin, R. D. Piner, K. M. Kohlhaas, A. Kleinhammes, Y. Jia, Y. Wu, S. T. Nguyen, R. S. Ruoff, *Carbon*, **45**, 1558 (2007).
22. D. Li, M. B. Muller, S. Gilje, R. B. Kaner, G. G. Wallace, *Nat. Nanotechnol.*, **3**, 101 (2008).
23. S. Pei, H. M. Cheng, *Carbon*, **50**, 3210 (2012).
24. K. P. Loh, Q. Bao, G. Eda, M. Chhowalla, *Nature Chemistry*, **2**, 581 (2010).

25. S. Dubin, S. Gilje, K. Wang, V. C. Tung, K. Cha, A. S. Hall, J. Farrar, R. Varshneya, Y. Yang, R. B. Kaner, *ACS Nano*, **4**, 3845 (2010).

26. J. Krupka, W. Strupinski, *Appl. Phys. Lett.*, **96**, 082101 (2010).

27. D. Wang, R. Kou, D. Choi, Z. Yang, Z. Nie, J. Li, L. V. Saraf, D. Hu, J. Zhang, G. L. Graff, J. Liu, M. A. Pope, I. A. Aksay, *ACS Nano*, **4**, 1587 (2010).

28. S. Stankovich, D. A. Dikin, G. H. B. Dommett, K. M. Kohlhass, E. J. Zimney, E. A. Stach, R. D. Piner, S. T. Nguyen, R. S. Ruoff, *Nature*, **442**, 282 (2006).

29. B. S. Kong, J. X. Geng, H. T. Jung, *Chem. Commun.* (2009), 2174–2176. Doi: 10.1039/B821920F.

30. K. S. Park, A. Benayad, D. J. Kang, S. G. Doo, *J. Am. Chem. Soc.*, **130**, 14930 (2008).

31. T. Ohzuku, A. Ueda, N. Yamamoto, *J. Electrochem. Soc.*, **142**, 1431 (1995).

32. K. Naoi, S. Ishimoto, Y. Isobe, S. Aoyagia, *J. Power Sources*, **195**, 6250 (2010).

33. J. Huang, Z. Jiang, *Electrochim. Acta*, **53**, 7756 (2008).

34. Z. Lin, X. Hu, Y. Huai, L. Liu, Z. Deng, J. Suo, *Solid State Ionics*, **18**, 412 (2010).

35. J. Widoniak, S. Eiden-Assmann, G. Maret, *Colloids Surf. A*, 270–271, **329** (2005).

36. Y. Zhou, D. Bao, L. A. L. Tang, Y. Zhong, K. P. Loh, *Chem. Mater.*, **21**, 2950 (2009).

37. H. Wang, H. S. Casalongue, Y. Liang, H. J. Dai, *J. Am. Chem. Soc.*, **132**, 7472 (2010).

38. Z. Du, X. Yin, M. Zhang, Q. Hao, Y. Wang, T. Wang, *Mater. Lett.*, **64**, 2076 (2010).

39. D. Fattakhova, P. Krtil, *Mater. Res. Soc. Symp. Proc.*, **703**, (2002) V3.26.1.

40. J. Li, Z. Tang, Z. Zhang, *Electrochem. Comm.*, **7**, 894 (2005).

41. Y. Liang, D. Wu, X. Feng, K. Mullen, *Adv. Mater.*, **21**, 1679 (2009).

42. A. Odani, V. G. Pol, S. V. Pol, M. Koltypin, A. Gedanken, D. Aurbach, *Adv. Mater.*, **18**, 1431 (2006).

43. A. S. Prakash, P. Manikandan, K. Ramesha, M. Sathiya, J. M. Tarascon, A. K. Shukla, *Chem. Mater.*, **22**, 2857 (2010).

44. G. Wang, K. Yan, Z. Yu, M. Qu, *J. Appl. Electrochem.*, **40**, 821 (2010).

45. Z. J. Fan, J. Yan, T. Wei, W. Z. Qian, M. L. Zhang, F. Wei, *Carbon*, **48**, 3825 (2010).
46. S. B. Ma, K. Y. Ahn, E. S. Lee, K. H. Oh, K. B. Kim. *Carbon*, **45**, 375 (2007).
47. X. Jin, W. Zhou, S. Zhang, G. Z. *Chen. Small*, **3**, 1513 (2007).
48. A. Benayad, H. J. Shin, H. K. Park, S. M. Yoon, K. K. Kim, M. H. Jin, H. K. Jeong, J. C. Lee, J. Y. Lee, *Chem. Phys. Lett.*, **475**, 91 (2009).
49. S. Trasatti, *Pure Appl. Chem.*, **58**, 955 (1986).
50. K. H. Chang, Y. F. Lee, C. C. Hu, C. I. Chang, C. L. Liu, Y. L. Yang, *Chem. Commun.*, **46**, 7957 (2010).
51. I. Honma, M. Okubo, Y. Mizuno, H. Yamada, J. Kim, E. Hosono, H. S. Zhou, T. Kudo, *ACS Nano*, **4**, 741 (2010).
52. X. Wang, C. Xu, J. W. Zhu, *J. Phys. Chem. C*, **112**, 19841 (2008).
53. Q. Feng, H. Kanoh, Y. Miyai, K. Ooi, *Chem. Mater.*, **7**, 1226 (1995).
54. A. H. Windle, K. R. Paton, *Carbon*, **46**, 1935 (2008).
55. S. Calvin, M. M. Miller, R. Goswami, S. F. Cheng, S. P. Mulvaney, L. J. Whitman, V. G. Harris, *J. Appl. Phys.*, **94**, 778 (2003).
56. H. S. Zhou, D. L. Li, M. Hibino, I. Honma, *Angew. Chem. Int. Ed.*, **44**, 797 (2005).
57. V. A. M. Brabers, F. M. Vansetten, P. S. A. Knapen, *J. Solid State Chem.*, **49**, 93 (1983).
58. C. H. Jiang, S. X. Dou, H. K. Liu, M. Ichihara, H. S. Zhou, *J. Power Sources*, **172**, 410 (2007).
59. S. B. Tang, M. O. Lai, L. Lu, S. Tripathy, *J. Solid State Chem.*, **179**, 3831 (2006).
60. C. H. Jiang, E. Hosono, H. S. Zhou, *Nano Today*, **1**, 28 (2006).
61. S. H. Tolbert, T. Brezesinski, J. Wang, B. Dunn, *Nat. Mater.*, **9**, 146 (2010).
62. B. E. Conway, V. Birss, J. Wojtowicz, *J. Power Sources*, **1**, 66 (1997).
63. T. Brezesinski, J. Wang, J. Polleux, B. Dunn, S. H. Tolbert, *J. Am. Chem. Soc.*, **131**, 1802 (2009).
64. T. C. Liu, W. G. Pell, B. E. Conway, S. L. Roberson, *J. Electrochem. Soc.*, **145**, 1882 (1998).
65. J. Wang, J. Polleux, J. Lim, B. Dunn, *J. Phys. Chem. C*, **111**, 14925 (2007).

Chapter 11

High-Energy and High-Power Li-Ion Cells: Practical Interest/Limitation of Nanomaterials and Nanostructuration

S. Jouanneau, S. Patoux, Y. Reynier, and S. Martinet

CEA LITEN, 17 Rue des Martyrs, F-38054 Grenoble Cedex 9, France

sebastien.martinet@cea.fr

11.1 Introduction

In this chapter, we deal with the practical interest of nanomaterials and also of nanostructuration on the enhancement of properties of Li-ion cells, either for high power or high energy.

The first two parts of the chapter are dedicated to positive and negative materials, respectively, with a broad overview of the status of the R&D specially oriented on nanomaterials optimization and nanostructuration effect. Then we will switch to some examples of practical performances and behavior of Li-ion cells using these new materials or nanostructured electrodes.

Nanomaterials for Lithium-Ion Batteries: Fundamentals and Applications
Edited by Rachid Yazami

ISBN 978-981-4316-40-8 (Hardcover), 978-981-4364-23-2 (eBook)
www.panstanford.com

11.2 Positive Electrode Materials

More and more research programs are focusing on nanomaterials for lithium (Li) batteries. It is clearly a benefit from having small particles for power application (diffusion is faster in smaller particles). At the opposite, the interest in "nano" for energy density is not so obvious. For a given insertion material, there is no direct relation between "nano" and specific capacity and energy. Only insertion materials that have low practical over theoretical capacity ratio might be improved at the nanoscale. This is, for instance, the situation of the well-known $LiFePO_4$ compound (~165 mAh/g at 3.45 V *vs.* Li^+/Li), mostly dedicated to automotive applications.* For the last 10 years, practical capacity (and energy) of this material has been enhanced by working on the synthesis aspect. Due to limitation in conductivity, it was shown that the electrical performances benefit from size reduction and presence of smart carbon coating at the nanoscale. In terms of volumetric energy, the situation is even worse. Nanomaterials have low density, thus reducing the capacity and energy by unit of volume.

In this section on positive electrode materials, we will first discuss the above-mentioned $LiFePO_4$ together with some notions on various polyanionic structure materials. Then, we will present the last developments on high-voltage spinel oxides and high-capacity layered oxides, considered as very promising candidates for future high-energy Li-ion batteries. For both families, the upper voltage limits are above the stability window of conventional electrolytes. For this reason, nanosize particles are naturally not appropriate to prevent rapid degradation of the electrolyte.

11.2.1 Olivine-Type Materials: Improving the $LiFePO_4$ Properties Using "Nano"

In the late nineties, at the University of Texas, J. B. Goodenough and his team depicted a new promising positive electrode material based on a robust three-dimensional polyanionic framework. For the first time, they highlighted the electrochemical capability of

*Due to an energy ranging from 110 to 160 Wh/kg at the cell level, the $LiFePO_4$/graphite redox system cannot be considered as an alternative to the conventional Li-ion systems for high-energy applications.

$LiFePO_4$ (olivine-type structure) while studying the substitution of $(PO_4)^{3-}$ and $(SO_4)^{2-}$ for O^{2-}, and making innovative works on numerous polyanionic structure materials. The concept of insertion/extraction of Li (and sodium) to/from the positive electrode material was previously extended by Goodenough's,[1,2] Delmas's,[3-5] and Quarton's[6,7] groups, to the three-dimensional structures framed with XO_n^{m-}polyanionic entities (X = P, S, Mo, W, and so on) in the eighties. The lithium iron phosphate studied at this time became very popular and is now considered as one the best candidate to be used at the positive electrode of Li-ion cells for automotive application. $LiFePO_4$ is safe, robust, inexpensive when compared with $LiCoO_2$, and it may sustains high-rate capability with good practical capacity. However, $LiFePO_4$ suffers from a lower potential (3.45 V *vs.* Li^+/Li) and lower material density (3.6 $g \cdot cm^{-3}$) than common spinel and layered oxides (~4 V *vs.* Li^+/Li; 4.5 and 5.1 $g \cdot cm^{-3}$; respectively). It means that $LiFePO_4$ is not suitable for small-size portable application where high gravimetric and volumetric energy densities are required. This is why, the research of polyanionic structure materials is continuing on materials such as silicates, fluorosphosphates, and $LiMPO_4$ counterparts (M = Mn, Co, Ni).

Like natural triphylite, synthetic $LiFePO_4$ adopts the olivine structure-type, built on oxygen hexagonal packing with Li and Fe atoms occupying half of the octahedral sites, and P atoms occupying 1/8th of the tetrahedral sites. The peculiar distribution of Li^+ and Fe^{2+} within the octahedral sites generates $[TeOc_2]$ layers. This connectivity has a special impact on both electronic and ionic conductivity in $LiFePO_4$. FeO_6 octahedra share corners between each other and electronic delocalization is made difficult. Li conductivity is not three-dimensional and proceeds mostly along $[001]_{Pnma}$. The theoretical capacity of 170 $mAh \cdot g^{-1}$ at a flat potential plateau at 3.45 V *vs.* Li^+/Li (Fig. 11.1), the non toxicity, and the cost of iron makes it a very good candidate to be used in industrial Li batteries, particularly for hybrid vehicles. For these reasons, the olivine $LiFePO_4$ has gained much interest since the pioneering discovery by Padhi[8] in 1996 to finally be commercialized by several companies today. To this date, there are more than 1000 articles in the literature dealing with the physico-chemical and electrical properties of $LiFePO_4$, as well as many methods of preparation. We invite the reader to refer to a recent book published in 2008 that summarize the literature data on this topic.[9] Apart from the abundantly discussed, and sometimes controverted, redox mechanism (essentially a two-phase process at room temperature),[10] which is affected by the presence of doping element,[11] temperature and particle size,[12] we want to introduce some works concerning the particle size, the carbon coating, and the doping that were made to improve the electrical properties of $LiFePO_4$.

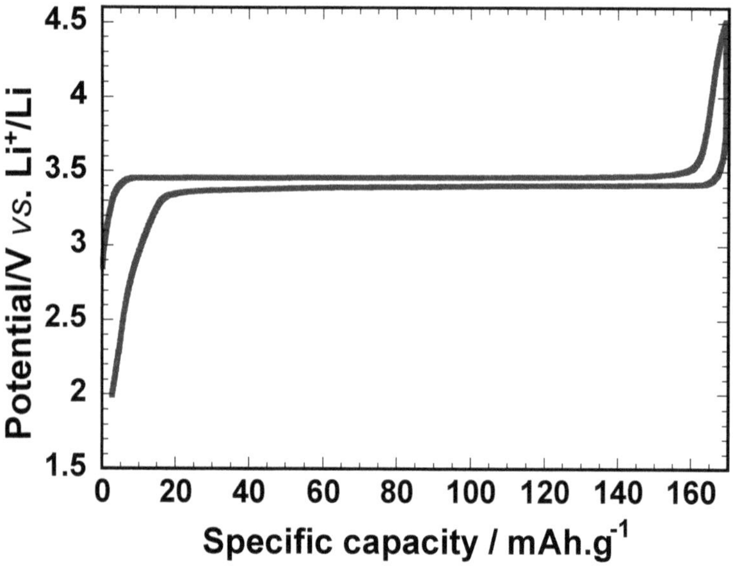

Figure 11.1 Potential *versus* capacity profile for a $LiFePO_4$/Li coin cell at room temperature.

First, it is important to carefully control the synthesis parameters of $LiFePO_4$ in order to prevent iron (II) oxidation, concomitant with the formation of stable Fe(III) phosphates such as $LiFeP_2O_7$, $Li_3Fe_2(PO_4)_3$, and eventually amorphous phases for low-temperature experiments. At the opposite, formation of iron phosphides, such as Fe_2P, has been observed at high temperature under reducing conditions.[13] As recalled above, particle size is a critical parameter for power application. Since the first syntheses made by Padhi at the University of Texas by solid state reaction (~50 μm particle size), numerous methods have been employed (hydrothermal, aqueous co-precipitation, sol–gel, emulsion-drying, microwave heating, polyol, sprays-pyrolysis, and so on). In all the cases, variants are possible. Lowering the temperature from 800°C to 500°C, and below, helps to reduce the particle size. For example, synthesis of $LiFePO_4$ nanoparticles in polyol medium at 335°C allows to obtain low-size monodispersed particles (20 nm wide and 50 nm long) with uniform morphology.[14] In this communication, charge/discharge capacities are excellent from low rate to 5 or 10 C rates (100% at 0.94 C, 85% at 7.5 C, 55% at 30 C), but electrodes are prepared using excessive carbon black additive (30% by weight). The Tarascon's group succeeded in preparing $LiFePO_4$ with controlled particle size and morphology at tem-perature of only 250°C in ionic liquid media.[15] This innovative low-temperature solvothermal–hydrothermal method is very different from conventional routes, energetically favorable, and may allow to tailor on-demand material with primary particle size in the order of a few hundreds of nanometers. However, rate capability has to be improved in the future.

Together with the particle-size reduction by modifying the synthesis method and associated parameters, significant increase in the electric conductivity has been obtained by *in situ* carbon coating of the $LiFePO_4$ particles. Many different recipes are possible and carbon sources are unlimited (from dry peanuts, citric acid, or sucrose, to carbon black or carbon nanotubes, for example). They have been well-documented in the literature since the public announcement of Ravet in 1999 about the possibility to reach the theoretical capacity.[16] At that time, the cycling was performed at 80°C (polymer cell), which favor the conductivity of materials. Carbon coating is a very important notion for polyanionic structure materials to circumvent the low electronic conductivity (polyanion groups are usually insulating and depreciate the diffusion of electron from a transition metal element to another). When operated *in situ* by thermal decomposition of an organic precursor during the synthesis of the material itself, it may also prevent agglomeration of particles and/or particle growth. It is difficult to manage a total control on the carbon coating (uniformity, thickness, and distribution at the surface of $LiFePO_4$) all the more that only a small change can drastically affect the electrical performances. This is one of the reasons why so many works have been performed over the last 15 years. Usually, the thickness of carbon is estimated by transmission electronic microscopy, and it varies between 5 and 10 nm for 1–2% carbon by weight of the $LiFePO_4$/C composite.[17] When an organic source is used, such as acrylonitrile, introduced in the initial aqueous solution with all the precursors needed for its polymerization, it is necessary to achieve the reaction by thermal treatment at temperature above 500°C to remove all H and O atoms under inert or reducing atmosphere. Note that the presence of carbon favored a local reducing atmosphere as it can react with oxygen to form CO and CO_2. In particular, the carbothermal synthesis of $LiFePO_4$ makes use of this process.[18] The electrochemical performance of "$LiFePO_4$" is often directly correlated with the structural factors of incorporated carbon, such as sp^2/sp^3 and disordered/graphene ratios.[19] Depending on the precursors, additives (e.g., ferrocene, pyromellitic acid, etc.), and temperatures, it is possible to adjust these ratios and to finally modify the charge/discharge capacities and rate capabilities.

Another approach to improve the electrical performances of "$LiFePO_4$" consists in adding foreign elements, such as Zr, Nb, Cu, Ag, Mg, Al, Ti, etc., with the advantage over carbon coating not to lower the energy density if properly operated.

Numerous and opened discussions have been made on this topic. The conclusions are still confusing as contradictory information has sometimes been reported in the literature, particularly about the conductivity of such materials. It has to be pointed out that high conductivity does not necessarily lead to improvement in the electrochemical properties. The simple situation should be that the doping is either made on the Li site ($Li_{1-x}M_xFePO_4$) or the iron site ($LiFe_{1-x}M_xPO_4$). In some cases, doping is not even always established as a two-phase composite may form, based on a $LiFePO_4$ bulk. Unexpected residual carbon from the synthesis precursors may affect the conductivity by creating a thin carbon coating during the heat treatment.[20] Presence of small amount of conducting phosphides (formed under local reducing atmosphere, as said before), such as Fe_2P, may also modify the conductivity. In the same mind, we have to mention the possibility to make a Li-deficient sample ($Li_{1-x}FePO_4$) with the advantage, according to Kim *et al.*,[21] to enhance the electrical conductivity by a factor of 10^6 to 10^8.

Figure 11.2 shows the result of a combined boron-doping[22] and carbon-coating to exemplify the above-mentioned ways of improvement. It gives low-size "$LiFePO_4$" with excellent electrical performance.

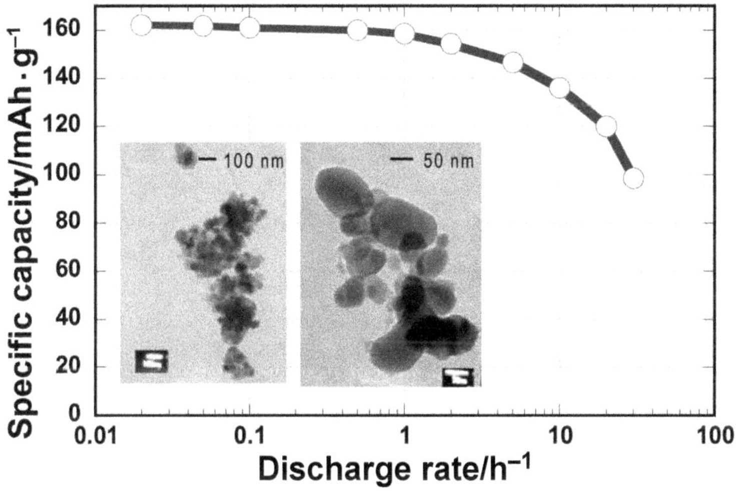

Figure 11.2 Rate capability of a boron-doped $LiFePO_4$ with carbon-coating/Li coin cell at room temperature. A transmission electronic microscopy image of $LiFePO_4$ is showed in insert.

In comparison to the considerable work on $LiFePO_4$, isotype compounds such as $LiMnPO_4$ and $LiCoPO_4$ (~no redox activity for $LiNiPO_4$)[23] have attracted moderate interest from now, probably due to the difficulty to demonstrate good cycling. From a theoretical point of view, these materials are far more interesting than $LiFePO_4$ because of similar capacity and higher potential (4.1 V *vs.* Li^+/Li for $LiMnPO_4/MnPO_4$ and 4.8 V for $LiCoPO_4/CoPO_4$).

The substitution of manganese for iron in $LiMPO_4$ would increase the voltage from 3.45 (pure $LiFePO_4$) to 4.1 V (pure $LiMnPO_4$) while the other attractive properties of olivines (stability, safety, price, etc.) remain intact. Unfortunately, the past research has identified several problems with $LiMnPO_4$, especially the extremely low intrinsic. As $LiMnPO_4$ did not present any practical capacity in the early days to the exception of a very controversial paper,[24] first communications reported electrochemical behaviour of the solid solution $LiFe_{1-x}Mn_xPO_4$. Yamada reinvestigated the $Li_{1-y}Fe_{1-x}Mn_xPO_4$ phase diagram in details.[25] For $x \leq 0.6$, performances are not very different from $LiFePO_4$,[26] and the average potential slightly increases in the proportion of the redox activity at 3.45 and 4.1 V *vs.* Li^+/Li (both phenomena are preserved, even if the Fe^{2+}/Fe^{3+} reaction occurs according to a single-phase reaction). But, at higher values, a drastic loss of capacity is observed, even at low current densities.

The origin of the different electrochemical activities of isostructural $LiFePO_4$ and $LiMnPO_4$ is still not well understood. Initially, Yamada[27] suggested that the low electrochemical activity of $LiMnPO_4$ was due to slow kinetics and internal friction during Li extraction/insertion. Later, Delacourt[28] succeeded in preparing small particles of $LiMnPO_4$ (100 nm size) by direct precipitation in aqueous solution. In this case, the reversible capacity reached 70 $mAhg^{-1}$ at C/20, whereas it is only 35 $mAhg^{-1}$ for 1 μm particles. In parallel, Yonemura[29] obtained 150 $mAhg^{-1}$ of discharge capacity at C/100 with small particles. From these studies, it seems that particle size is critical in determining useful capacity of $LiMnPO_4$. It was confirmed by the measure of the intrinsic conductivity of $LiMnPO_4$ (10^{-12} $S\cdot cm^{-1}$),[30] which is 1000 times lower than the intrinsic conductivity of $LiFePO_4$ (10^{-9} $S\cdot cm^{-1}$). Temperature is crucial to prepare small-size particles. Sol–gel synthesis, with temperature of 520°C led to particles with the average diameter of 130 nm and reversible capacities of 156 $mAhg^{-1}$ at C/100 and 135 $mAhg^{-1}$ at C/10.[31] Polyol synthesis, at a temperature below 300°C, allowed to prepare very small particles of $LiMnPO_4$ (20–30 nm thick platelets) with high reversible capacity retention.[32] Along the same line goes the proposal by S. Whittingham[33] and S. Franger[34] who finds the hydrothermal synthesis a viable synthesis method for preparation of olivines ($LiMPO_4$, M = Fe, Mn, Co, and Ni). $LiMnPO_4$ (no coating) was prepared at 300°C from an original method, initially used with success for $LiCoPO_4$, but electrochemical result were surprisingly not really modified in comparison with basic high-temperature solid-state synthesis. Like for $LiFePO_4$, presence of appropriate carbon coating is also probably crucial. At present, research to improve electrochemical performance of $LiMnPO_4$ is still on the way in several laboratories to make this compound a viable solution for automotive application, for instance. Development of $Li(Fe,Mn)PO_4$ materials, with low Fe cation content is probably more realistic from an application point of view.

Concerning $LiCoPO_4$, the situation is somewhat different. Charge/discharge reaction is usually not completed, but nominal capacity is often superior to the one observed for $LiMnPO_4$, and potential is 0.7 V above the Mn counterpart.[35-39] We partially optimized the synthesis method, and obtained good reversible capacity, low polarization, and acceptable rate capability (Fig. 11.3). Nevertheless, cycle life and rapid self-discharge were still observed as reported in the literature, even for small-size (~150 nm) and carbon-coated (~10 nm thick) particles of $LiCoPO_4$.[40] Challenge on this compound is high, and energy density of more than 250 Wh/kg is possible at cell level. The potential of this compound (4.8 V *vs.* Li^+/Li), giving more energy to the cell, is above the limit of stability of the conventional electrolyte. From this point, it is similar to the high-voltage spinel oxides discussed below. But, the latter have intrinsic better cycle life than $LiCoPO_4$. Among other limitations, cobalt element might have catalytic effect on the degradation of the electrolyte. A publication studied the thermal stability of charged $LiCoPO_4$. Synchrotron diffraction and differential thermal analyses revealed that the cobalt phosphate ($CoPO_4$) is probably less stable than the layered $Li_{0.5}CoO_2$.

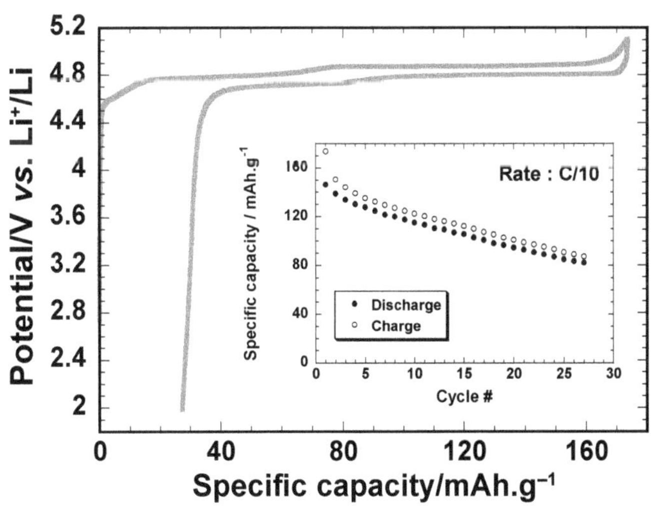

Figure 11.3 Potential *versus* capacity profile for a $LiCoPO_4$/Li coin cell at room temperature. The cycle life is reported in insert.

In the search for new positive-electrode materials for Li-ion batteries, silicates and fluorophosphates have also gained growing interest in the last few years. Among the silicates, Li_2FeSiO_4 has practical capacity reaching the theoretical value,[41] but the lower electronegativity of Si *vs.* P results in a lowering of the Fe^{3+}/Fe^{2+} redox couple. Discharge potential is about 2.8 V *vs.* Li^+/Li. It is 0.65 V below $LiFePO_4$ for similar capacity. Contrary to Li_2FeSiO_4, Li_2MSiO_4 with M = Mn and Co were expected to exchange two moles of electron (extraction of two Li),[42,43] which correspond in theory to twice the capacity of the iron counterpart (~330 mAh/g). Study of Li_2MnSiO_4 revealed the instability of the structure upon initial delithiation, with a strong tendency to amorphize.[44] However, it was concluded, by DFT calculations that the solid solution $Li_2Mn_xFe_{1-x}SiO_4$ could be used, and exchange of more than one electron is possible (in proportion of the Mn/Fe ratio). For instance, $Li_2Mn_{0.25}Fe_{0.75}SiO_4$ has capacity of 190 mAh/g in first discharge (140 mAh/after 10 cycles), with a potential of 2.7 V *vs.* Li^+/Li.[45] The existence of several polymorphs of Li_2CoSiO_4 was reported a long time ago when studying various Li_3PO_4-type compounds.[46] Their (poor) electrochemical performances were discussed in several papers.[47] Based on crystallographic considerations, it seems difficult to reversibly extract the two Li ions while maintaining the integrity of the structure.

Partial removal of Li can be operated from Li_2VOSiO_4.[48] The capacity of the $Li_2VOSiO_4/LiVOSiO_4$ redox couple is 155 mAh/g. Experimentally, 0.7 Li^+ per unit formula can been extracted after ball-milling of the sample in presence of carbon. Discharge capacity is 90–100 mAh/g, and the potential is 3.6 V *vs.* Li^+/Li. In the best case, Li_2VOSiO_4 would only approach the performance of $LiFePO_4$. In a computational investigation on fluorinated polyanionic compounds as positive electrode for Li batteries, a hypothetical $LiVSiO_4F$ compound, with structure based on the parent Li_2VOSiO_4 was recently presented.[49] It is not sure whether this fluorosilicate does really exist, and the corresponding capacity would still be close to the one of Li_2VOSiO_4, but, it has the merit to pave the way for new possible structures.

On a basic approach, fluorine has high Pauling's electronegativity that would enhance the redox potential for a given M^{n+}/M^{m+} couple. Another potential interest of fluorine is the ability to bring stability to the material, and to allow new structure to be expected. A review paper on fluoride-based electrode materials for advanced energy storage was published by G. Amatucci in 2007.[50]

Recently, fluorinated solvents for stable and safe electrolyte have become a hot topic. Concerning the electrode materials, fluorine (F^-) has been used for a long time as a doping element (replacing a small fraction of oxygen O^{2-}) for layered and spinel oxides with more or less success. For layered oxides, it was initially used for doping $LiNiO_2$. Global improvement of capacity retention upon cycling and safety have been noticed for this sample and attributed to various small changes, such as difference in phase transitions during charge and modification of the well-known cations mixing. Doping is surely interesting, but it should be considered as a way to optimize existing materials, something similar on this point to the coating of materials.

Apart from doping, fluorine may generate new materials, such as fluorophosphates. For instance, 4 V vanadium-based compounds, such as $LiVPO_4F$ (and $NaVPO_4F$)[51] with theoretical capacity of 156 mAh/g and $Li_5V(PO_4)_2F_2$[52] [and $Na_3V_2(PO_4)_2F_3$[53] and $Na_3V_2O_2(PO_4)_2F$[54]] with theoretical capacity of 170 mAh/g may be cited. Electrochemical results of these materials have already been presented in tens of papers with more or less success. It means that their practical use is feasible, but energy at cell level will be certainly not above 200 Wh/kg. A second example concerns Li_2MPO_4F materials, which have distinct interest. Li_2FePO_4F (and Na_2FePO_4F) has been recently prepared and characterized by Nazar's group.[55] Li_2FePO_4F is synthesised by ion exchange in a refluxing solution of LiBr using Na_2FePO_4F, previously obtained by solid-state reaction or sol–gel. Like Natrium Super Ionic CONductor (NASICON)-related materials, it is possible to use Z_2FePO_4F (Z = Li, Na) either in Li-ion or Na-ion cells, with potential cost reduction for sodium system but lower energy. In both case, the theoretical capacity is moderate (124–146 mAh/g) and the potential between 3.6 and 3.0 V *vs.* Li^+/Li. The other Li_2MPO_4F phases are more interesting in terms of energy density (possibility to have twice the capacity of the iron analogue and higher voltage), but is also more risky. There is no reference about Li_2MnPO_4F in the literature to our knowledge, even if crystal structure of Na_2MnPO_4F has been reported.[56] Synthesis and crystal structure of Li_2NiPO_4F has been presented a few years ago,[57] but no electrochemical response has been obtained for the moment (same situation as for $LiNiPO_4$). Finally, preliminary results by Okada *et al.*[58] confirmed the possibility to use Li_2CoPO_4F in Li-ion batteries. The extraction of Li (1 mole per unit formula only) occurs at ~5.0 V *vs.* Li^+/Li. Only one electrochemical curve with a single cycle obtained at low rate [galvanostatic intermittent titration technique (GITT) mode] was presented, and the reversibility at high voltage was only partial. With a theoretical capacity of 287 mAh/g at 5 V ($Li_2Co^{2+}PO_4F \rightarrow Co^{4+}PO_4F$), Li_2CoPO_4F would be a "perfect" candidate for very high-energy density Li-ion cells (>>250 Wh/kg in theory) at the condition that it is possible to reversibly extract two Li atoms.

11.2.2 Li-Rich Layered Oxides

Layered oxides have certainly been the most extensively studied topic in the large field of Li batteries. Thirty years ago, Goodenough's group[59] demonstrated the possibility to extract 0.5 mole of Li ions from the layered structure of $LiCoO_2$, leading to $Li_{0.5}CoO_2$ at 4.2 V *vs.* Li^+/Li. The common specific capacity of the $LiCoO_2/Li_{0.5}CoO_2$ redox couple is 140 mAh/g, half of the theoretical capacity corresponding to the complete extraction of Li ($LiCoO_2/CoO_2$). This compound is still used in today's commercial cells with capacity between 140 and 160 mAh/g. Similar study was performed on $LiNiO_2$, once again by J. B. Goodenough and coworkers at Oxford.[60] The possibility to convert layered oxide into spinel oxide, favored at high temperature, for the $Li_{0.5}NiO_2$ composition was noticed. A more recent article by John Reed and Gerbrand Ceder depicted and explained the structural relations between layered and spinel oxides.[61] The transformation does not require any oxygen rearrangement since the structures share a cubic closed-packed oxygen anion sublattice, with transition metal element (noted *M*) occupying octahedral interstitial positions while the Li occupy either octahedral or tetrahedral interstices. The nature of *M* controls the facility to operate the change toward more stable spinel formed in the order Mn >> Ni >> Co.

During the following years, a significant amount of researches have been directed at finding an optimal composition with layered $LiMO_2$ structure. The total replacement for manganese or iron of cobalt or nickel, leading to lamellar $LiMnO_2$ and $LiFeO_2$, has not been successful due to poor electrochemical performance. An alternative might lie in the partial substitution of Co- or Ni-layered oxides. Together with $LiCoO_2$, the $LiNi_{0.8}Co_{0,15}Al_{0,05}O_2$ composition was thus adopted in commercial cells. The former usually has higher rate performance and higher density, and the latter has higher practical capacity, lower cost, and higher stability in cycling, but safety is considered as a drawback for both compounds. The selection of $LiCoO_2$ and $LiNi_{0.8}Co_{0,15}Al_{0,05}O_2$ was driven by the portable market increase in the last 15 years (laptops, cell phones), and the development initiated for EV and HEV (for $LiNi_{0.8}Co_{0.15}Al_{0.05}O_2$). It is probable that their use will be prohibited for large-scale automotive application for cost and safety reasons.

Still looking for positive electrode materials with high capacity as well as low cost and low toxicity to replace Ni- and Co-based oxides, the series of materials $LiNi_{1-y-z}Co_yMn_zO_2$ emerged. Such compounds potentially combine the rate performance and density of $LiCoO_2$, the high practical capacity of $LiNiO_2$, and the structural stabilization imparted by the presence of Mn^{4+}.[62,63] Specifically, intensive effort has been directed toward development of $LiNi_{0.5}Mn_{0.5}O_2$,[64–66] $LiNi_{1/3}Co_{1/3}Mn_{1/3}O_2$,[67–69] and variants as possible replacements for $LiCoO_2$. Computational[70–72] and spectroscopic studies[73] show that, under normal cycling conditions, Mn remains in the +4 oxidation state, thus avoiding complications associated with the Jahn–Teller distortion of Mn^{3+} ions in six-coordinate sites (main problem of the $LiMn_2O_4$ spinel oxide, as discussed below). These materials exhibit slightly sloping discharge profiles. The capacities obtained are dependent upon composition and voltage limits, with 200 mAh/g reported for $LiNi_{1/3}Co_{1/3}Mn_{1/3}O_2$ when cells with Li anodes are charged to 4.6 V. Ion mixing[74] (i.e., Ni in Li sites and vice versa) can reduce the effective capacity and adversely affect rate capability, and is more likely to occur when Ni contents are relatively high. Cobalt is electro-active in these compounds and may replace some of the Ni, reducing this tendency[75] and improving rate capability.[76] In addition, $LiNi_{1/3}Co_{1/3}Mn_{1/3}O_2$ possesses increased thermal abuse tolerance compared with $LiNi_{0.8}Co_{0.15}Al_{0.05}O_2$.[77] $LiNi_{1/3}Co_{1/3}Mn_{1/3}O_2$, like most of the layered oxides, is usually synthesized by the mixed hydroxide method,[78] in which a $M(OH)_2$ (M = Ni, Co, Mn) precursor is precipitated from a metal nitrate solution and is then reacted with a Li salt at elevated temperature. This is not difficult in practice, but pH, temperature, flow conditions, concentrations, and other factors influence the crystal structure, purity, and physical properties of the products and must be carefully controlled for optimum electrochemical performance.[79]

Structure and electrochemistry of many substituted phases like $LiNi_{1/3}Co_{1/3-y}M_yMn_{1/3}O_2$ (M = Ti, Al, Fe, and so on)[80] positive electrode materials have been regarded. Electrochemical behavior in Li cells of electrodes based on these materials is substantially altered compared with the $LiNi_{1/3}Co_{1/3}Mn_{1/3}O_2$ parent. These differences can be directly attributed to changes in the structural characteristics induced by the substitutions. For instance, $LiNi_{1/3}Co_{1/4}Fe_{1/12}Mn_{1/3}O_2$ exhibits lower capacity and poorer rate capabilities due to kinetic limitations resulting from an increase in the antisite cation defect concentration as implied by a reduced $c/3a$ ratio. Phase pure $LiNi_{1/3}Co_{1/3-y}Al_yMn_{1/3}O_2$ compounds are formed for $0 \leq y \leq 1/4$ but a γ-$LiAlO_2$ impurity is observed for $y = 1/3$. Although Al substitution results in decreased capacity between 4.3 and 2.0 V when electrodes are discharged in Li cells, capacity retention and rate capability are substantially improved. This can be attributed to decreased antisite mixing and an increased Li slab dimension compared with $LiNi_{1/3}Co_{1/3}Mn_{1/3}O_2$.

Apart from finding the best material composition and synthesis conditions, coating of particles has also been discussed in the literature. In a review paper on cathode materials modified by surface coating for Li-ion batteries, the authors drew a list of the coating agents usually reported (MgO, Al_2O_3, SiO_2, TiO_2, ZnO, SnO_2, ZrO_2, $Li_2O \cdot 2B_2O_3$-glass, and other materials) and their interaction with active materials.[81] Even if the coating does play a major role in a first time when looking for a new material as it does not directly participate to the electrochemical process (no direct effect on the nominal capacity), it often helps to enhance the cycle life by preventing direct contact with the electrolyte solution.† In some cases, coating may also suppress phase transition, improve the structural stability, and decrease the disorder of cations in crystal sites.

Different from simple coating, we also have to mention here the "core-shell" materials. The principle consists in preparing a thick and stable shell (micrometer order) completely encapsulating a high-energy but unstable core. Synergetic (positive) effect of the two materials is expected. Formation process of microscale core-shell structures is tedious, and it requires, at least, three distinct steps (e.g., hydroxide co-precipitation, formation of core-shell

---

†Unless coating is operated on the electrode, the conductive carbon presented in the electrode is not protected against direct contact with the electrolyte.

hydroxide, formation of core-shell oxide). For example, a few papers investigated the electrochemical performance and stability of the core-shell structured $Li(Ni_{0.8}Mn_{0.1}Co_{0.1})_{1-x}(Ni_{0.5}Mn_{0.5})_x O_2$[82] and $Li(Co)_{1-x}(Ni_{0.5}Mn_{0.5})_xO_2$[83] in which the $LiNi_{0.8}Co_{0.1}Mn_{0.1}O_2$ shell contributes to the stabilization of the material, whereas the $LiNi_{0.8}Co_{0.1}Mn_{0.1}O_2$ and $LiCoO_2$ cores enhance the global energy. Due to the difficulty to obtain homogeneous powders (and possibly due to the lack of real improvement), particularly for large-scale production, as per our knowledge, this technique has not been generalized yet.

In all the cases, the practical capacity of the most-common positive commercial material ($LiCoO_2$, $LiNi_{0.8}Co_{0.15}Al_{0.05}O_2$, $LiNi_{1/3}Co_{1/3}Mn_{1/3}O_2$, and so on) ranges from 140 to 200 mAh/g according the cut-off voltage, giving energy density usually between 160 and 200 Wh/kg. Improved material capacities, mainly due to higher cut-off voltage and lower cycle life, allow reaching 220 or 240 Wh/kg at cell voltage as in the Sony's Apelion cell, for instance.

The particular case of layered manganese oxides has to be presented because of their high theoretical capacities and the potential for enhanced Li-ion diffusion through the two-dimensional van der Waals gaps.[84] However, orthorhombic $LiMnO_2$[85,86] which has a corrugated structure, and monoclinic $LiMnO_2$ prepared from α-$NaMnO_2$[87,88] convert to spinel upon cycling in Li cells. The oxygen arrays in these layered structures are nearly cubic close-packed (ccp) like that of spinel, requiring only rearrangement of cations for transformation to occur. First-principle calculations based on density functional theory (DFT) indicate that Mn^{2+} ions can diffuse readily through tetrahedral sites into the octahedral sites located in the Li layers.[89] Mn^{2+} ions are produced through charge disproportionation according to the equation:

$$2Mn^{3+}(\text{octahedral}) \rightarrow Mn^{2+}(\text{tetrahedral}) + Mn^{4+}(\text{octahedral})$$

Migration requires the simultaneous presence of Mn^{3+} ions and vacancies in the Li layers; phase conversion can be suppressed by partial substitution with ions subvalent to manganese (increasing the average oxidation state of the remaining manganese) or with multivalent ions more electronegative than Mn.[61,90] For example, $LiNi_{0.5}Mn_{0.5}O_2$,[64] $LiNi_{1/3}Mn_{1/3}Co_{1/3}O_2$,[66] and also $Li[Li_{0.2}Cr_{0.4}Mn_{0.4}]O_2$,[91] all cycle stably. In each case, as said before, Mn is in the +4 oxidation state in the as-made compounds and is not redox active under normal cycling conditions. Therefore, it may not be strictly correct to consider these materials substituted manganese oxides, but rather manganese-substituted Cr, Ni, or Ni–Co oxides. More importantly, drawbacks associated with Cr (toxicity), Ni (poor abuse tolerance), and Co (cost) are still present with these materials, albeit to a lesser degree than in the unsubstituted analogues.

Besides, Li-rich manganese-based layered oxides have been strongly studied as an alternative to the conventional layered oxides. Thackeray was the first to mention the redox activity of the Li_2MnO_3 compound in the beginning nineties.[92] Li_2MnO_3 has a layered rocksalt structure that can be described in conventional layered $LiMO_2$ notation as $Li[Li_{0.33}Mn_{0.67}]O_2$. The structure is made of layers of Li ions alternating with layers of Li and Mn ions in a 1:2 ratio, and separated from one another by layers of cubic close-packed oxygen planes. It is very similar to the ideal layered structures of $LiCoO_2$ and $LiNiO_2$, but with poor electrochemical response. It seems that Li_2MnO_3 cannot be used alone in Li cell, even if a recent communication tends to rehabilitate this material.[93] Electrochemical performance of Li_2MnO_3 is found to depend strongly on the synthesis conditions as lower particle size and increase in amount of stacking faults enable the material to be charged at lower potential to a higher capacity.

In 1997, Japanese researchers at the Mitsui Mining and Smelting Co. published pioneering work on $Li_2MnO_3 \cdot LiCoO_2$ mixed compounds.[94] Several other layered oxides including a Li_2MnO_3 component were subsequently investigated by distinct groups, such as Pacific Li in the beginning of noughties ($Li_2MnO_3 \cdot LiCrO_2$),[95] Dahn's group in Canada [$Li(Li,Mn,Co,Ni)O_2$ phases],[96,97] the Japanese National Institute of Advanced Industrial Science and Technology ($Li_2MnO_3 \cdot LiFeO_2$-type phases),[98] SAFT ($Li_{1+x}(Ni_{0.425}Mn_{0.425}Co_{0.15})_{1-x}$; $0 \leq x \leq 0.12$),[99,100] the University of Iwate in Japan,[101,102] and the Manthiram's group at the University of Texas.[103,104] In addition to these groups, the contributions of Thackeray's[105–107] and Amine's[108,109] groups at Argonne National Laboratory at Chicago are essential. From these studies, it is concluded that excess Li can be incorporated into the layered structure. Depending on the authors and on the compositions, it is considered as a "solid solution" of Li_2MnO_3 and $LiMO_2$ or a "composite structure" in which a Li_2MnO_3 (layered) component is structurally integrated with a layered $LiMO_2$. The general composition can thus be represented as $LiM_{1-y}[Li_{1/3}Mn_{2/3}]_yO_2$ or $yLi_2MnO_3\cdot(1-y)LiMO_2$ where M can be Cr, Mn, Fe, Co, Ni, or mixtures thereof. To distinguish from conventional-layered oxides, these materials might be labeled as layered-layered oxides, Li-rich layered oxides, or simply as high-capacity layered oxides. Very high practical capacity, from 220 to 280 mAh/g at room temperature, can be reached by careful control of composition, synthesis, and cycling conditions. It is thus accepted that the Li_2MnO_3 "component" provide improved structural and electrochemical stability. Cycling has to be per-formed up to 4.6 or 4.8 V *vs.* Li^+/Li that is at potential above the stability limit of the usual layered oxides themselves and electrolyte. Alternatively, Li_2MnO_3 can be replaced by related materials, such as Li_2TiO_3 and Li_2ZrO_3, but its performance is lowered (Ti^{4+} and Zr^{4+} are not reactive in these oxides).[110]

According to the "composite structure" notation,[111] the mechanism involved during charge/discharge may be summarized as follows:

During first charge

$$\text{(A)}\quad yLi_2MnO_3\bullet(1-y)LiMO_2 \rightarrow yLi_2MnO_3\bullet(1-y)MO_2 + (1-y)Li$$

$$\text{(B)}\quad yLi_2MnO_3\bullet(1-y)MO_2 \rightarrow yMnO_2\bullet(1-y)MO_2 + yLi_2O$$

During discharges

$$\text{(C)}\quad yMnO_2\bullet(1-y)MO_2 + Li \rightarrow yLiMnO_2\bullet(1-y)LiMO_2$$

During subsequent charges

$$\text{(D)}\quad yLiMnO_2\bullet(1-y)LiMO_2 \rightarrow yMnO_2\bullet(1-y)MO_2 + Li$$

Initial charge thus consists in two distinct phenomena (Fig. 11.4). First, M^{3+} ions are oxidized into M^{4+}, such as in conventional layered oxides. The gentle increase in the potential is characteristic of a solid solution. Second, the profile changes, and the potential tends to describe a plateau at high potential that may correspond to a loss of oxygen from the active material. Following charges and discharges show a single phenomenon, but the achieved reaction could not be so simple. For example, we observed changes in the profile and appearance of a new signal between 3 and 2 V *vs.* Li^+/Li during prolonged cycling. Although we suspect a partial transition from layered to spinel structure and a segragation between transition metal element (Ni and Mn for example) with a particle surface progressively Ni richer than the bulk, additional researches are need. However, the discharge capacity is not impeded by such a modification, contrarily to the voltage profile that follows a decay from 3.6V to 3.4V vs Li^+/Li over the first 50 cycles.

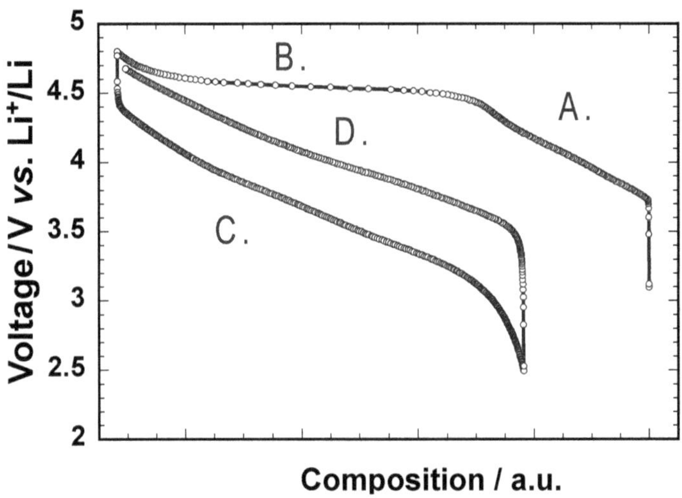

Figure 11.4 Electrochemical profile of a $y\text{Li}_2\text{MnO}_3\cdot(1-y)\text{LiMO}_2$/Li coin cell at room temperature.

Due to the possibility to extract Li from the Li_2MnO_3 component at high voltage (~4.8 V *vs.* Li^+/Li), very high capacities are thus expected as said before. For instance, values as high as 311 mAh/g in (first) discharge have been obtained on $0.7\text{Li}_2\text{MnO}_3\cdot0.3\text{LiMn}_{1/3}\text{Ni}_{1/3}\text{Co}_{1/3}\text{O}_2$.[112] After a few cycles, the capacity of the best materials is more or less stabilized at 220–250 mAh/g. The main limitation is possibly kinetics that generally requires cycling at high temperature (~50°C) at low rate (~C/30). Cycle life is also often altered in regard to the literature data. We obtained promising results using specific compositions with adapted synthesis conditions (Fig. 11.5). The latter lowers the first-cycle irreversible capacity. At room temperature, the discharge capacity is rather stable at about 250 mAh/g, and it is 300 mAh/g at 55°C. Such a result allows proposing very high-energy density at cell level, possibly 280 Wh/kg in combination with graphite or 280 to 350 Wh/kg in combination with silicon/carbon composites. Cost and safety are other potential advantages of these layered-layered oxides in comparison with conventional layered oxides.

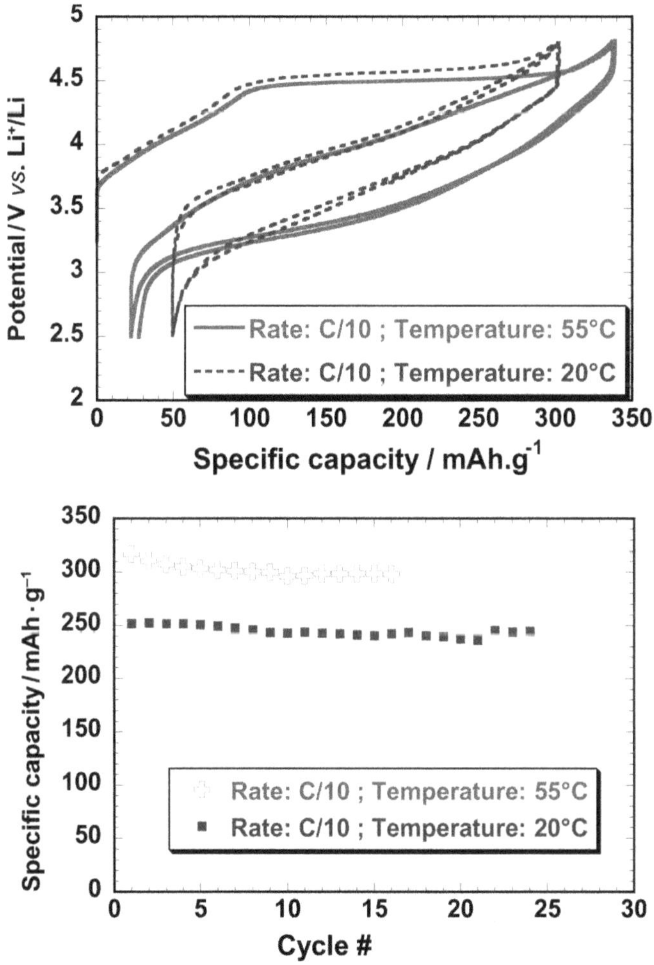

Figure 11.5 (Above) Potential *versus* capacity profiles for a recently patented (CEA-LITEN) positive electrode material with $y\text{Li}_2\text{MnO}_3\cdot(1-y)\ \text{LiMO}_2$ composition. (Below) Corresponding capacity *versus* cycle life plot.

Thanks to very high capacity, at an average discharge potential of about 3.5 V *vs.* Li^+/Li, the layered-layered oxides are possibly a promising solution for future high-energy density Li-ion batteries. Before, potential drawbacks have to be circumvented: a high first cycle irreversibility up to few tens of mAh/g, a low rate capability, limited low temperature performances and the partially misunderstood reaction processes and its evolution during cycling. Hopefully, we are confident that solutions exist to face all these issues. Some of these have already been proposed. Following the work of Thackeray on the same topic, Manthiram recently reported a way to eliminate the irreversible capacity loss by blending the layered-layered oxide with LiV_3O_8 and $Li_4Mn_5O_{12}$, both materials are able to reversibly insert Li ions.[113] A recent paper by Thackeray's group has discussed the rate capability limitation, showing a possible enhancement of the capacity at a given rate by fluorination of the material and/or Li-Ni-PO_4 treatment of the electrodes.[114]

11.2.3 Toward 5V Spinel Oxides

Simultaneously to the first researches on layered oxides, Thackeray and Goodenough[115,116] have presented the possibility to use oxides with three-dimensional spinel structure for Li batteries. In particular, all the Li ions can be reversibly extracted at 4.0 V *vs.* Li^+/Li from the $LiMn_2O_4$ spinel, leading to λ-MnO_2 at the end of charge, thereby achieving a specific capacity (and potential) close to $LiCoO_2$ and counterparts (140 mAh/g). The electrochemical performances of this compound, including its cycle life, strongly depend on the stoichiometry and the oxidation state of manganese.[117,118] Spinel $LiMn_2O_4$ was often showed as the most-serious alternative to replace standard layered oxides at the positive electrode because of its abundance, low-cost, and environment friendliness. Unfortunately, the use of $LiMn_2O_4$ was hindered essentially because of its electrochemical instability at high temperatures. In fact, the main problems of $LiMn_2O_4$ is related to the high content of Jahn–Teller Mn^{3+} ions, which have a tendency to dismute into Mn^{4+} and highly soluble Mn^{2+}. Then, dissolved manganese ions have a strong tendency to poison the surface of the (graphite) negative electrode to the detriment of the cycle life and the long-term storage, accelerated at high temperature. Note that the use of small (nanosize) particles tends to accelerate the undesirable dissolution of manganese. However, through a dual substitution of small amounts of aluminium for manganese and of fluorine for oxygen, it was showed to limit the Jahn–Teller instability (at room temperature, $LiMn_2O_4$ undergoes first-order structural transition from a single cubic phase to a mixed cubic and a disturbing tetragonal phase). Controlling the electrode/electrolyte interface through surface modification (e.g., coating with silica or borate glasses) is another important strategy to improve significantly the cycle life. Coating with zirconium dioxide or aluminium phosphate may also act as HF getter (in fluoride-based Li-salt such as $LiPF_6$), the agent responsible for accelerating the Mn^{2+} dissolution.

Despite a limited market share to 10% of the global Li-Ion battery market, $LiMn_2O_4$, with its advantages of safety and low cost, is a candidate to be used in positive electrode mixed with other oxides for larger scale commercialization such as automotive application.

The spinel oxides have theoretical capacity between 140 and 150 mAh/g. In the case of the 4V spinel oxides ($LiMn_2O_4$ and derivatives), the energy density at cell level is thus limited to 160–180 Wh/kg (carbon negative electrode) or probably slightly more with new negative electrode (e.g., Si/C composites). To fulfil applications requiring high-energy density, high-potential spinels are undoubtedly promising candidates. Initially, these materials have been considered as substituted $LiMn_2O_4$ materials (see above), because they were prepared to study the 4 V retaining capacity and reduce the Jahn–Teller distortion.[119,120] In 1997, K. Amine[121] and J. R. Dahn[122] envisioned the "5 V" component of $LiMn_{2-x}M_xO_4$ spinel oxides (M = Ni, Fe with x = 0.5, 1 and M = Ni with $0 < x < 0.5$). Subsequent researches have focused on simple or multi-substitutions (e.g., iron, copper, titanium, chromium, cobalt, nickel) for manganese to enhance the capacity at high voltage. Only a few compounds that do not incorporate manganese at all were also studied.[123] The most famous is certainly $LiNiVO_4$, which adopt an inverse spinel structure (Li in octahedral sites, not in tetrahedral sites). Even before the Mn-based spinels, it was shown that $LiNiVO_4$ sustains high redox activity.[124] This sample has poor general electrical performances and, to our knowledge, it is no longer studied now. For the last six or seven years, intensive efforts have been devoted to the development of $LiNi_{0.5}Mn_{1.5}O_4$ and variants as positive electrode materials for high-energy density Li-ion batteries. Compared with $LiMn_2O_4$, Mn remains essentially in the +4 oxidation state, under normal cycling conditions, avoiding complications associated with the Jahn–Teller distortion of Mn^{3+} ions in six-coordinate sites. The theoretical capacity of the $LiNi_{0.5}{}^{II}Mn_{1.5}{}^{IV}O_4/Ni_{0.5}{}^{IV}Mn_{1.5}{}^{IV}O_4$ redox system is 146.7 $mAh\cdot g^{-1}$. It corresponds to the oxidation of nickel (II) at 4.7–4.75 V *vs.* Li^+/Li (700 $Wh\cdot kg^{-1}$ of active material, referring to Li metal).

High-voltage spinel oxides have been studied at CEA-LITEN for the last 10 years, giving much care to the nickel-based compositions ($LiNi_{0.5}Mn_{1.5}O_4$ and $LiNi_{0.4}Mn_{1.6}O_4$).[125,126] Generic $LiNi_{0.5}Mn_{1.5}O_4$ can either be crystallized in $P4_332$ or Fd-3m space group, depending on the synthesis conditions.[127,128] It is well known that direct spinel oxides usually adopt the Fd-3m space group (e.g., $LiMn_2O_4$) with Li ions located in tetrahedral 8*a* sites and transition metal elements filling the 16*d* octahedral sites. The particular Ni/Mn ratio (1/3), under certain synthesis conditions, allowed a cationic ordering over two distinct octahedral positions, thus nesting the nickel ions in the 4*a* sites and manganese ions in the 12*d* sites (S. G.:$P4_332$). Note that the disordered phase is the most familiar, and the one usually studied in the literature due to better electrochemical performance than the ordered phase.[129] It is often considered as non-stoichiometric, and occasionally expressed as $LiNi_{0.5}Mn_{1.5}O_{4-\delta}$. Oxygen vacancies have been related to the minor redox activity noticed at 4 V *vs.* Li^+/Li, which is probably due to the oxidation of Mn^{3+} into Mn^{4+}. Presence of $Li_xNi_{1-x}O$ is generally observed as impurity of the non-stoichiometric spinel oxide.

We have optimized the electrochemical performances of high-voltage spinel oxides and selected the $LiNi_{0.4}Mn_{1.6}O_4$ and $LiNi_{0.5}Mn_{1.5}O_4$ compositions for future developments. Note that we have also regarded, as many authors, single and multiple substitutions of Co, Cu, Cr, and Ti for Ni and Mn. Such compounds did not have any particular advantage, possibly except the chromium-doped materials, already reported by Rojo *et al.*[130] However, we believe that the presence of chromium is not justified in terms of performances (similar), and detrimental in terms of cost and toxicity. All the other substituted compounds (Co, Cu, Ti) have poorer electrochemical performances than $LiNi_{0.4}Mn_{1.6}O_4$.

In a recent paper, we depicted the advantage of using high-voltage nickel-based spinel oxides through the presentation of optimized electrochemical performances *versus* Li, graphite, and $Li_4Ti_5O_{12}$.[125] The Figures 11.6 and 11.7 show the electrochemical profile, the rate capability, and the cycle life of the $LiNi_{0.4}Mn_{1.6}O_4$/Li system in coin cell configuration with conventional electrolyte (in excess). Although the usual relative electrolyte is instable at high potential, the results we have obtained are already very promising for "5 V" high-energy density cells (*vs.* graphite) and for "3 V" high-power batteries cells (*vs.* $Li_4Ti_5O_{12}$). However, the self-discharge is far-above acceptable value (up to 80%/month), which prevents commercial use for the moment. We believe that the material itself cannot be further optimized significantly. Improvement of the stability of the electrolyte is expected to introduce high-voltage materials on the market.

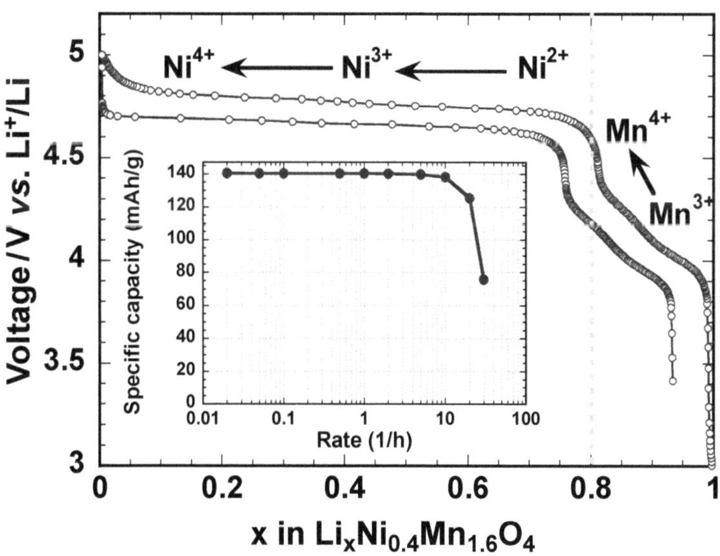

Figure 11.6 Potential-composition profile of the $LiNi_{0.4}Mn_{1.6}O_4$ spinel oxide (C/5 rate@20°C). Specific capacity *versus* discharge rate (inset).

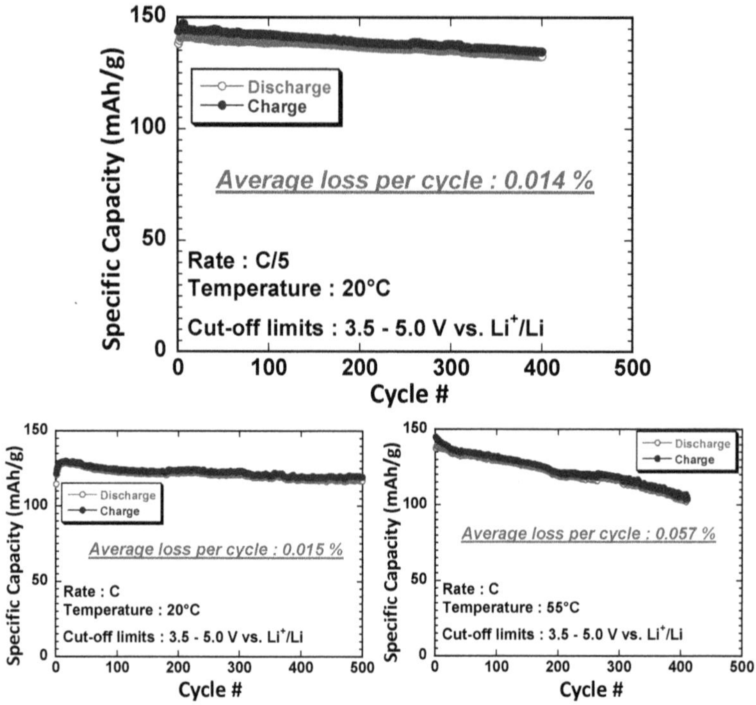

Figure 11.7 Specific capacity *versus* cycle number for the $LiNi_{0.4}Mn_{1.6}O_4$/Li system at 20 and 55°C under C and C/5 rate (conventional electrolyte).

Ab initio calculation methods[131] predicted that the current solvents for the electrolytes used in Li-ion batteries often have oxidation stability to 5.5 V *vs.* Li^+/Li (propylene carbonate, ethylene carbonate, dimethyl carbonate). In practice, the potential limits are far below as illustrated by the work of Kanamura *et al.*[132,133] For instance, the stability of PC-based electrolyte varied from 5 V (calculation) to 4.2 V *vs.* Li^+/Li (experiment). The origin of this difference may be multiple: "catalytic" effect of the insertion materials, traces of water in the electrochemical system, influence of Li salts. In some cases, the decomposition of the electrolyte is accompanied by the formation of a solid film on the surface of the positive electrode.[134,135]

In the literature, the solutions explored usually consist in coating the active material by electro-inactive inorganic species (Al_2O_3, $AlPO_4$, TiO_2, ZrO_2). It is probable that the protective layer is efficient to prevent the oxygen loss from layered oxides at high-state of charges. Insofar as this coating is carried out on the particles before the realization of the composite electrode, this technique may only protect the active particles with respect to the electrolyte oxidation, not the electrode. The other components of the electrode (carbon, binder, metallic current collector) are still directly in contact with the electrolyte. In other words, *ex situ* method cannot be used to ensure an effective protection of the electrode.

Since the nineties, it is well established that the introduction of a small amount of vinylidene carbonate (VC) in the electrolyte favored the formation of a controlled and homogeneous protective layer with good Li^+ conductivity at the surface of graphite. Following the work done on the negative electrode, we seek to achieve a stable solid electrolyte interface on the positive electrode side by *in situ* oxidation/polymerization of an additive. This approach, not widespread until now for the positive electrode, is being followed by Ube Industries in Japan.[136] In an exploratory approach, we have initiated in 2007 a first evaluation of the impact of various additives in conventional electrolyte on the self-discharge. In the best case, a reduction by a factor of three to four has been observed (Fig. 11.8), thus giving interest to this original approach for further development. The selected additives need to have lower oxidation potentials than the electrolyte solvents to form, during first charge, an insoluble protective layer on the positive electrode surfaces. The surface film has to be Li^+ conducting, and the additive must not show any detrimental influence on the negative electrode. Some of the properties can be anticipated such as the solvation ability *versus* Li^+, the electronic conductivity, but others might be more difficult to predict such as the formation of a compact and stable film in a complex medium. Several additive families are reported in the literature for various issues and systems.[137] For instance, trimethyl phosphate,[138] phenyladamantane, cyclohexylbenzene, thiophene and biphenyl or triphenyl are used to improve the cyclability of $LiCoO_2$/graphite system. The most studied compound is the biphenyl,[139] which oxidizes at 4.45 V *vs.* Li^+/Li. The amount of biphenyl in the electrolyte must be controlled in order to obtain a thin layer, and prevent Li^+ blocking. The developments and future studies to be realized on the high-voltage spinel oxides have to be dedicated to the minimization of the surface reactivity of the compound toward the electrolyte, while maintaining high kinetic properties and good cycle life.

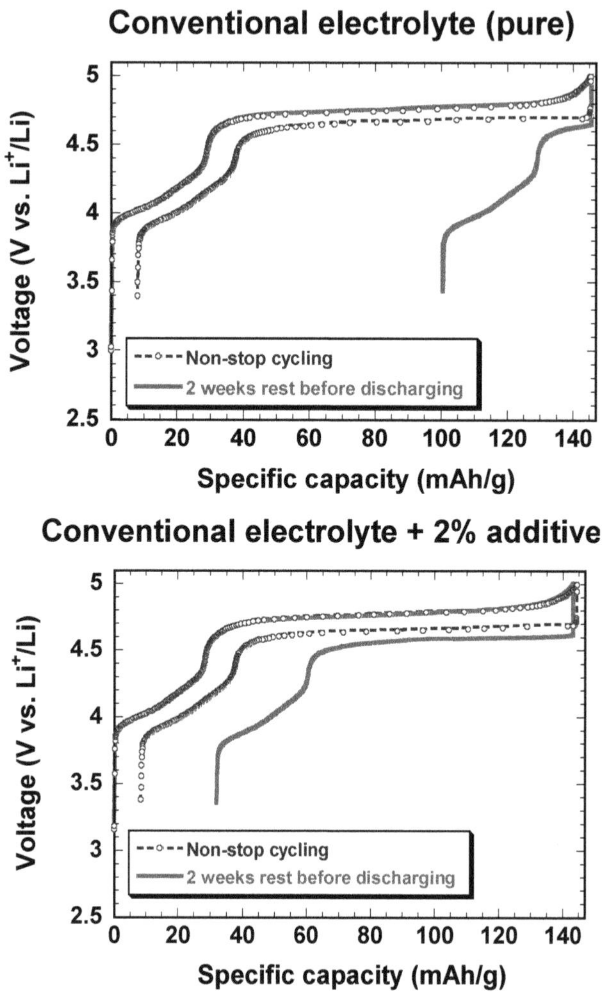

Figure 11.8 Effect of a specific additive tested by CEA to reduce the self-discharge at high voltage for the $LiNi_{0.4}Mn_{1.6}O_4$/Li system.

Another approach to obtain good cyclability and low self-discharge may consist in the replacement of conventional liquid carbonates with new fluorinated solvents. Due to the inductive effect of fluorine atom (electronegativity), these solvents may present a good stability in oxidation. Some fluorinated solvents were synthesized by direct fluorination using F_2, electrochemical fluorination or chemistry of fluorinated compound.[140] The advantages of fluorinated solvents are (*ii*) the non-flammability; i.e., high flash point, (*i*) a dielectric constant higher than those for non-fluorinated linear solvent such as carbonates, (*iii*) the high performance obtained at low temperatures as electrolyte due to low melting point and (*iv*) a good electrochemical stability in oxidation due to the withdrawing electron effect of fluorine atoms. The direct fluorination is generally a non-selective reaction[141] giving several fluorinated compounds that are difficult to separate. The anodic monofluorination of lactones, carbonates, cyclic ethers, or esters using ionic liquid fluoride salt is the most promising and permits to obtain several fluorinated solvents with a good field and selectivity.[142–144] A large variety of compounds are obtained, but the electrochemical stability is usually not presented. Kabayashi *et al.* performed the direct fluorination of ethylene carbonate.[145] The best electrochemical properties are obtained with the 4,4 fluoro 1,3-dioxolan-2-one. This compound presents lower melting point and boiling point, and a smaller dielectric constant than non-fluorinated analogue compounds. A decrease in the viscosity and an increase in the conductivity are obtained. On the other hand, the incorporation of several fluorine molecules in the ethylene carbonate induces an increase in the viscosity and a decrease in the electrochemical performance and conductivity. High electrochemical performance was also obtained for $LiCoO_2/Li_4Ti_5O_{12}$ high-power Li-ion batteries using LiTFSI in 3-(2,2,2-trifluoro)ethoxypropionitrile electrolyte.[146] The thermal stability of fluorinated ester electrolytes was studied by Yamaki *et al.*[147] A good thermal stability *vs.* metallic Li was obtained. The cycling capacity of the graphite anode in EC/DMC+ MFA (Methyl difluoroAcetate) has been evaluated by Nakajima *et al.*[148] MFA-based electrolyte gives a better capacity than other fluoroester solvents. The reduction of MFA was observed at 0.87 V *vs.* Li^+/Li, which is close to the reduction of EC. The reduction of MFA induces the formation of a stable passive layer on the graphite electrode.

We have also evaluated the interest in room temperature ionic liquids (for 5 V systems) without success at that stage, and are presently investigating fluorine-based solvents. Whatever the solutions tried to intend to limit the reactivity of high-potential positive electrodes/electrolyte interfaces above 4.5 V *vs.* Li^+/Li, the selected electrolyte must be chosen in order to obtain, also, a good electrochemical stability *versus* the negative electrode.

Besides the problems of self-discharge and cycle life, energy density as high as 220–240 Wh/kg at cell level is possible with high-voltage spinel oxides ($LiNi_{0.4}Mn_{1.6}O_4$ *vs.* graphite). It is expected to go beyond 250 Wh/kg by using new high-performance negative electrode.

11.3 Negative Electrodes

11.3.1 Overview

The carbon graphite material has become the favorite to replace the Li metal[149] from the sale of a system containing a negative electrode based on graphite by Sony in 1991.[150,151] This is known as Li-ion battery, also known as rocking-chair as opposed to Li metal, Li is in the electrode materials in ionic form. The hexagonal structure of graphite is a stack of graphene plans to facilitate the intercalation of Li ions between the planes. The success is mainly due to a low cost, natural abundance, a good reversibility with a potential close to 0 V (*vs.* Li^+/Li) (except for the first cycle) and the formation of a stable passivation film on the surface of the electrode. Despite the popularity of graphite, some limitations motivate research to other alternative materials. The main limitation is its capacity now reaches the theoretical capacity is 372 $mAh \cdot g^{-1}$ while markets were still seeking more energy density.

However, it must be remembered that the limiting electrode is the positive electrode. This means that having a high capacity for the negative will not increase the overall capacity of the battery very significantly. Knowing the capacities of positive and negative electrodes, we can calculate the capacity per mass of the Li-Ion cell.

According to Fig. 11.9, setting the capacity of positive electrode 140 and 200 $mAh \cdot g^{-1}$, the evolution of the gain of the overall capacity is similar, approximately 14%. System capacity increases up to a capacity of 1000 $mAh \cdot g^{-1}$ of Anode. Beyond this capacity, the system does not seem to benefit from a strong increase in capacity. Much research is conducted on the families of negative electrode materials with better properties, including improved ability mass.

From 1992 to 2004, successive improvements have helped to double the energy density mass Li-ion batteries, which rose from 90 to 200–220 Wh/kg.

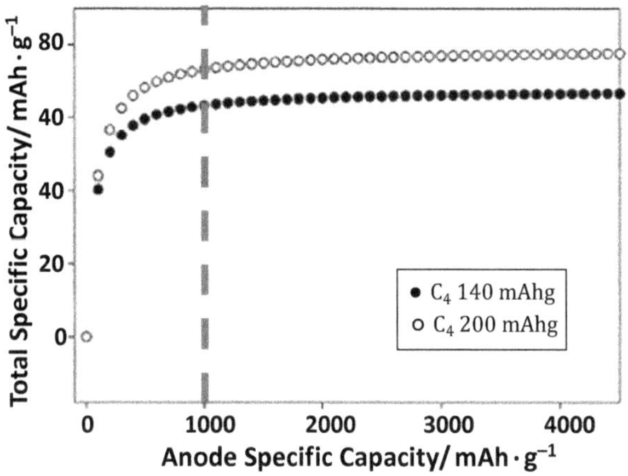

Figure 11.9 Capacity per total mass of two Li-ion cells based on the negative electrode capacity. The total mass capabilities include the masses of active materials and other compounds accumulator constituents. The capacity of the positive electrode are considered 140 mAh·g^{-1} and 200 mAh·g^{-1}. Reprinted from Ref. 152. Copyright 2007, Elsevier.

Among these families of negative electrode materials, the phosphides of transition elements occur recently as promising materials with a capacity between 800 and 1000 mAh·g^{-1} like Cu_3P and CoP_3[153 155] but their development is limited because of their bad performances in cycling and their toxicity.

Oxides of transition metals are also of interest (e.g., CoO, Co_3O_4, NiO, Cu_2O, MnO) by reacting with Li following a conversion reaction (reversible destruction of the original structure in metallic nanoparticles) with a capacity of 600–800 mAh·g^{-1} with potential between 1 and 1.5 V.[156] Their development is currently underway but the sharp polarization between the charge and discharge is a major hindrance to any application in the short/medium term.

Another family of particular interest concerns the elements (Al, Si, Sn, Ge, Pb, Sb, Bi) which can form binary alloys electrochemically with Li.[157–159] In 1997, Fujifilm[160] proposed negative electrodes based on tin oxide [tin composite oxide (TCO): $Sn_xAl_yB_zP_pO_n$]. The presence of oxygen in the compound neverthe less led to an irreversible capacity equivalent to the reversible capacity due to SnO_2.[161] Research has continued to be conducted on this material since 2005, when Sony marketed a new Li-ion batteries, Nexelion, provided with a tin anode (Sn–Co–C).[162,163] The nanocomposites of tin and cobalt are included in a carbon matrix. This configuration allows the battery to reach a capacity of more than 30% of that of traditional Li-ion batteries.

In Fig. 11.10, the silicon appears as the most promising material as the active material. Indeed, its specific capacity can reach 10 times that of graphite or 3579 $mAh \cdot g^{-1}$ at room temperature. During the insertion of Li, Si is converted into a Li–Si alloy, which can cause a change in volume of 280%. This significant change in volume repeated over successive charge and discharge lead to a deterioration of the morphology of the Si electrode. The material is then weakened to its spray then leads to a loss of contact with the rest of the electrode. These adverse events are reflected immediately on the electrochemical performance of Si (drastic loss of capacity). It seems that the use of Si in the form of thin films[164,165] or nanometric powders[166] can mitigate the extent of these phenomena and achieve an excellent reversible specific capacity close to the the oretical value.

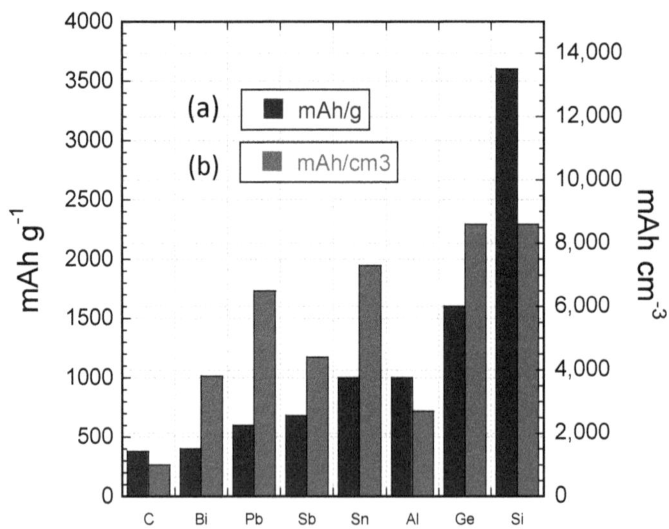

Figure 11.10 The gravimetric capacity (b) and volume (a) for some alternative materials. The capacity of graphite is given as references.

Finally, oxides based on titanium as a negative electrode material, although of less importance in terms of energy density are of interest in terms of safety. Unlike graphite, the intercalation of Li is performed around 1.5–1.6 V, thus avoiding the deposition of Li metal on the surface of the electrode[167] and this for high current densities. Thus, these titanium oxides are of interest in terms of power density.

In the case of $Li_4Ti_5O_{12}$ theoretical capacity is 175 $mAh\cdot g^{-1}$ (reversible insertion of three moles of Li^+ ions in the structure of $Li_4Ti_5O_{12}$ leading to a reduction of part of Ti^{4+} ions in Ti^{3+} ions in $Li_7Ti_5O_{12}$)[168,169] with good stability over the cycles and a high-power response. $Li_4Ti_5O_{12}$ has a spinel structure type $Li_{tétra}$ $[Li_{1/3}Ti_{5/3}]_{octa}O_4$ with cubic symmetry Fd-3m. Its structure has tunnels in the three dimensions of space provides a good diffusion of Li^+ ions in the material.[170] During the initial insertion of Li, then following cycles of charge/discharge, the expansion or contraction phenomena are minimal. Indeed, the oxidized phase ($Li_4Ti_5O_{12}$) and the reduced phase ($Li_7Ti_5O_{12}$) have a volume substantially identical ($\Delta V = \pm\ 0.07\%$). For this reason, $Li_4Ti_5O_{12}$ is considered as a material "zero strain."[171] An excellent cohesion within the electrode and between electrodes, separators and current collectors is expected throughout the cycling. This avoids the appearance of cracks and therefore the matrix damages may cause a loss of capacity and performance. In practice, the capacity is recovered close to the theoretical value for slow rate cycling. Depending on the material synthesis and morphology of grains, performance, and power (capacity recovered at high rates) are more or less interesting.

The potential for integration of Li ions being above the reduction potential of the electrolyte, almost no passivation layer is forming at the interface electrode/electrolyte (the electrolyte does not deteriorate). In addition, this material presents a high thermal and chemical stability. It could be used for cycling at high speed and/or low temperature. Moreover, because of its high potential for insertion with respect to the potential deposit of Li, there is no risk of short circuits at high current densities.[172] Finally, working on this potential, it is possible to use a current collector of aluminum, lighter than a copper collector (used for graphite electrodes). However this relatively high potential (negative electrode) does not achieve energy densities as high as for graphite.

The titanium oxide $Li_4Ti_5O_{12}$ constitutes an alternative to the graphite negative electrode for applications in power, long life and enhanced safety. Currently $Li_4Ti_5O_{12}$ is in advanced development phase and will soon be marketed for Li-ion batteries for fast charging. Applications, such as portable equipment, smart cards, or smart electric traction are already envisaged for this material.

Theoretical capacity (175 mAh·g^{-1}) of Li titanate is still relatively low and leads the scientific community to interest in the titanium oxide TiO_2. Some structural forms of TiO_2 (B or H, etc.),[173–175] less compact than the anatase TiO_2 have the same advantages as $Li_4Ti_5O_{12}$ compared with carbon graphite. The form of TiO_2 "Bronze" (noted TiO_2-B) is a negative electrode material interesting because it presents a theoretical capacity significantly higher than $Li_4Ti_5O_{12}$ (335 mAh·g^{-1}) or the couple TiO_2/Li TiO_2). To date, only few researches on this compound were conducted in comparison with those on $Li_4Ti_5O_{12}$. The interest of the scientific community created by its electrochemical properties is recent. At least eight varieties of allotropic TiO_2 are reported in the literature. TiO_2-B is one of two forms less compact, leading to the insertion of Li. As the structure of TiO_2-B is not dense with large interstitial spaces compared to the radius of Li^+ ions (see Fig. 11.11), the insertion/extraction of Li ions in the structure does not produce significant distortion of the crystal unit. TiO_2-B is based on a structure of layers of octahedrons TiO_6 bound by their tops. The sheets are connected by oxygen atoms forming a three-dimensional network. The structure

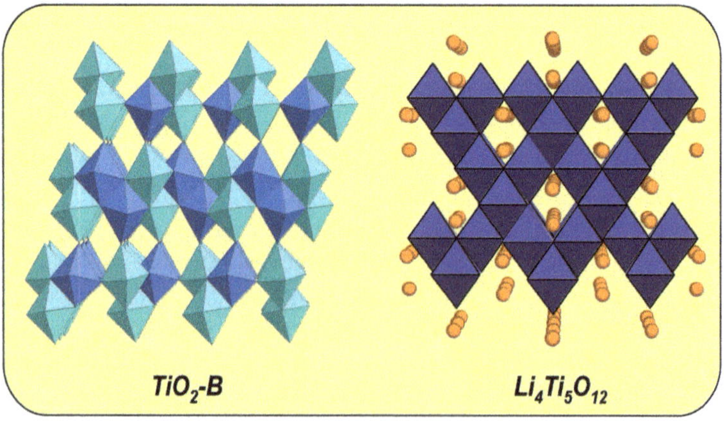

Figure 11.11 Illustration of the crystal structures of titanium-oxide negative electrode.

of TiO_2-B comes from a distortion of an ideal model composed of regular octahedrons type MO_6. TiO_2-B is a metastable phase of titanium dioxide TiO_2. That is why only a few electrochemical studies have focused on this material since its discovery more than 25 years ago.[173]

Titanium oxides present good characteristics for high-power applications. These properties will be optimized by going to nanoparticles and/or nanostructures to promote a better ionic diffusion and thus a better response to high current density.

Many current work on the active electrode materials for Li battery designed especially to seek compositions and physico-chemical characteristics to overcome the most current limitations especially in terms of energy density and power, for example by the synthesis of nanoparticules (reducing diffusion path of Li^+). The difficulty for the application will be to stay in a size/specific surface area domain with a maximum practical electrode density and a feasible process. The composite Si/C on the one hand, introducing a potential energy gain and Li titanates on another hand due to their attractive properties in terms of power, will be detailed in this chapter. In both cases we will show the impact and interest, if any, of their nano (metric and/or structured) aspect on their performance as negative electrode for Li-ion battery.

11.3.2 Si/C Composite Nanostructuration

Many studies concern, for example, nanoparticles of Si, tin, or metal alloys to replace the currently used graphite. In this way, it is possible to move from 350 $mAh \cdot g^{-1}$ stored in the graphite used in Li-ion battery business to more than 1000 $mAh \cdot g^{-1}$ (3800 $mAh \cdot g^{-1}$ theoretical). Various materials including nanosilicon in the form of thin films or particles, deposited or incorporated into a conductive carbon matrix are on the way to achieve this performance. However, the optimization of the electrodes and of the complete battery system incorporating these compounds still require several years of study. Although the operating voltage of Si/C is around 0.5 $V/Li^+/Li$, remaining very close to that of graphite, Si/C can achieve very large energy, limitations in power remain comparable to those of graphite (problems of dendritic growth Li involving significant security issues).

11.3.2.1 Li–Si system

To understand the interest to move toward o, nanoparticles and nanostructure of silicon, it is important to look at the Li–Si system.

From the phase diagram established in 1976 by Sharma[176] and actualized by Okamoto[177] and coulometric titration of stability areas of potential for each intermetallic crystalline Li–Si compound at 415°C established in 1981 by Huggins *et al.*[32,33] it is possible to make the graph presented in Fig. 11.12 which gives the relationship between the volume expansion and the Li quantity in the structure. The theoretical volume expansion increases gradually from the metastable structure Li-Si. Contrary to the intercalation of Li between the graphene plans that show a moderate expansion close to 10%,[178] this sharp increase in volume will destabilize the structure of the electrode and lead to a real electrochemical grinding provided the lithiation follows the diagram binary Li–Si at room temperature. For $Li_{21}Si_5$, the volume returned to a silicon atom is 4 times larger than that of pure silicon.

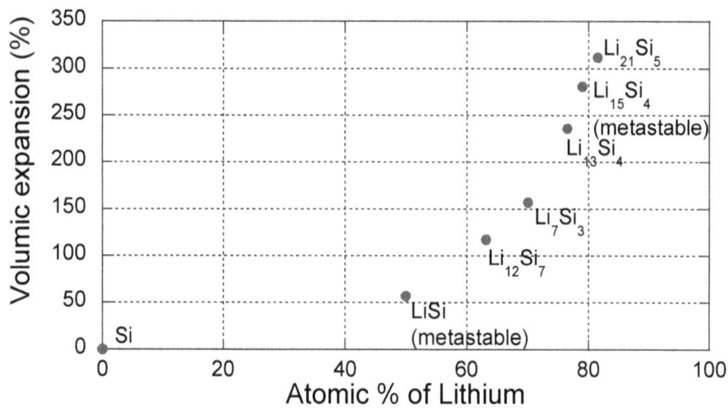

Figure 11.12 Theoretical volumic density expansion (%) for various alloys of binary diagram Li–Si.

At 415°C, the electrochemical formation of each intermetallic crystalline silicon compound results on galvanostatic curves with a plateau. At room temperature, only one plateau is observed in the insertion of Li, as shown in Fig. 11.13. The compound is metastable, $Li_{15}Si_4$, observed by X-ray diffraction (XRD)[179], Raman spectroscopy,[180] and electron microscopy at high-resolution (HREM)[181] at the end of discharge. At a temperature of 25°C, the formation mechanism of the alloy $Li_{15}Si_4$ by electrochemical route go through an amorphization step at the solid state. According to the *in situ* XRD study, the first part of the lithiation results in a biphasic reaction with the coexistence of particles of silicon and an amorphous phase composition defined by $Li_{3,5}Si$. This result confirms the analysis made by Limthongkul. Obrovac[182] showed that under a potential landfill less than 50 mV (*vs.* Li^+/Li), the amorphous alloy $Li_{3,5}Si$ recrystallized in a crytalline phase $Li_{15}Si_4$ not present in the diagram binary Li–Si and not identified by the electrochemical lithiation at 415°C.

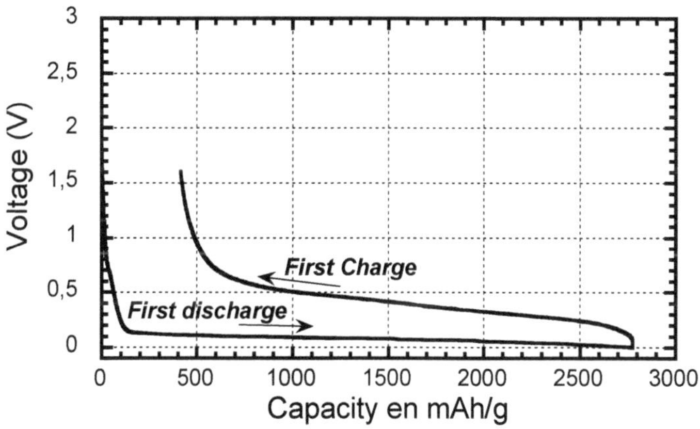

Figure 11.13 Discharge/charge curve at the first cycle for a silicon-based electrode at 25°C and C/10 rate.

The most electrochemically lithiated phase at room temperature is therefore not $Li_{21}Si_5$ but $Li_{15}Si_4$. The maximum capacity reachable is not 4008 $mAh \cdot g^{-1}$ but 3579 $mAh \cdot g^{-1}$. $Li_{15}Si_4$ formation is accompanied by a volume expansion of 281%.

The phenomena of insertion-deinsertion of Li in the structure of silicon is complex. For a better understanding of the process, Fig. 11.14 shows the stages of amorphization and crystallization of the silicon structure. This scheme was developed based on the work of Obrovac[182,183] and the work of Dahn team.[179] The diffusion of Li from the surface to the core is very slow at room temperature, 10–13 $cm^2 \cdot s^{-1}$ according to calculations by impedance.[184] For silicon particles larger than 10 microns, given the value of the diffusion coefficient at room temperature, the simple calculation (Lithiation time × Diffusion coefficient = Lithiated thickness of the particle) give that silicon particles will not be fully affected by the

Li insertion during charge/discharge at C/10. We can imagine a border of amorphous non-crystalline Si–Li alloy including the non lithiated crystalline silicon, as shown in Fig. 11.6. During the first reduction, silicon coexists with an amorphous alloy of Li_ySi with a defined composition $y = 3.5 \pm 0.2$, then we see the crystallization in $Li_{15}Si_4$ at the end of the lithiation below 60 mV (*vs.* Li^+/Li).[185] This stage of crystallization occurs after the formation of an amorphous phase of variable composition Li_xSi with $3.5 < x < 3{,}75$. The following reaction is a biphasic transformation, between the crystalline silicon, Cr–Si, and the amorphous alloy, a-Li_ySi, during the first lithiation to an amorphous alloy.

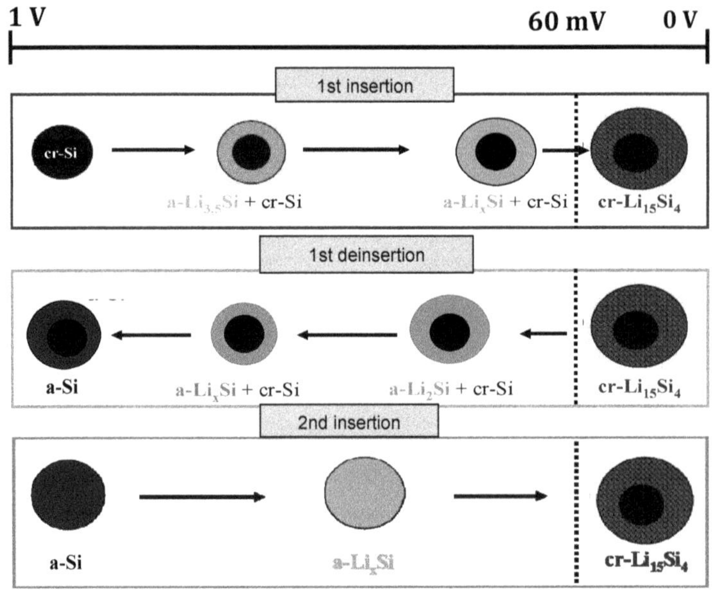

Cr–Si = Crystalline silicon, a-Si = Amorphous silicon, a-Li_xSi = Amorphous alloy

Figure 11.14 Microstructural changes during the galvanostatic cycling with variable Li composition according to Dahn. Adapted from Ref. 179.

At lower potential, the alloy structure is arranged into $Li_{15}Si_4$ (or $Li_{3,75}Si$). During the extraction of Li, the crystalline alloy is back into a disorderly a-Li_zSi, where $z = 2$, then amorphous alloy a-Li_xSi (x variable) and finally to an amorphous silicon.

Depending on the particle size and scanning speed, the insertion of Li may not be complete at the end of the first discharge. This leads to a gradual lithiation of silicon in the core of the particle. After a few cycles, the insertion rate increases and the particles, initially crystallized, become amorphous, probably due to progressive fracturing of the original particles. The volume expansion shown in Fig. 11.12 and the coexistence of the phases described above are combined with constraints at the material level. The particles of silicon become weak until a spray, which leads to loss of electrical contact and a brutal capacity fall.

To limit the deep degradation of the material, a partial lithiation until a-$Li_{3,5}Si$ was suggested by Obrovac[183] by setting the lower limit of potential. Obrovac thinks that the main degradation of the particles is caused by the lithiation of the heart of silicon particles.

Indeed, the more deeply the Li diffuses, more micro-cracks created by the variation of volume grow up until of the material cracking from the lithiated areas formed. Moreover, the phenomenon of cracking is pronounced during delithiation because the presence of cracks in the SEI promotes the penetration of cracks during the second swelling. By XRD, the crystalline compound $Li_{15}Si_4$ is formed at 60 mV (*vs.* Li^+/Li). Obrovac sought to maintain a biphasic structure without reaching the material core in the case of silicon particles of 45 μm. To identify a potential limit for end of discharge, several electrodes have been tested in galvanostatic mode with a current 150 $mA \cdot g^{-1}$. The electrode is limited for the first cycle at 1200 $mAh \cdot g^{-1}$, and then cycled between 0.9 V and the potential limit between 120 and 190 mV. The potential is limited and imposed until measuring a current of less than 30 $mA \cdot g^{-1}$. In these cycling conditions, the potential limit is determined at 170 mV. The silicon is only lithiated on a ring on the outside boarder of the particle and has not undergone the transformation into crystalline $Li_{15}Si_4$. This study shows that it is possible to mitigate the adverse events that occur during the structural changes by limiting the lower bound of potential.

Instead of defining a potential for the lower limit, Jung[185] limits the specific capacity. Programs controlling the cycling capacity exchanged during charge and discharge provide better cycling performance. To illustrate this behavior, Fig. 11.15 shows thin films electrode of silicon layers capacities *versus* cycle numbers starting at 4200, 3000, and 1300 $mAh \cdot g^{-1}$. Despite a capacity above the design capacity, with a limit of 4200 $mAh \cdot g^{-1}$, the sample behaves as a material without any particular limitation. The capacity of the electrode falls deeply from the first cycles to be zero after 40 cycles while a limitation at 3000 mAh·g^{-1} indicates that the capacity is stable for 100 cycles. The best performance is attributed to the electrode limited to 1300 mAh·g^{-1} with a number of stable cycles twice as large as the previous sample. But this development is accompanied by an increase in the window of charge/discharge during cycles which may not be feasible in a complete Li-ion cell.

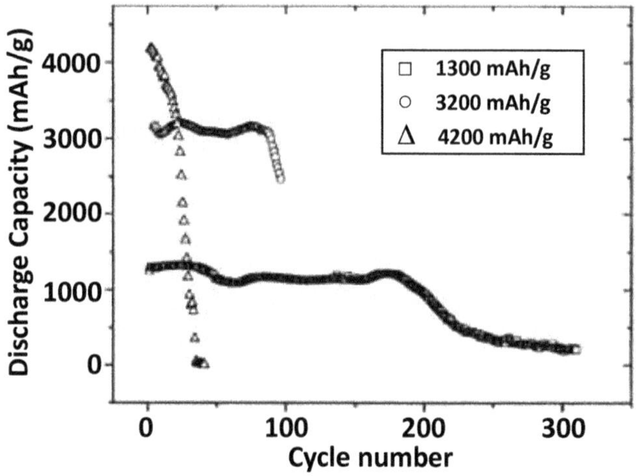

Figure 11.15 Electrochemical performance of thin layer of silicon electrodes. Limiting capacity: 4200 mAh·g^{-1} (Δ), 3000 mAh·g^{-1} (○), 1300 mAh·g^{-1} (□). Reprinted from Ref. 185. Copyright 2003, Elsevier.

11.3.2.2 Particule size and nanostructuration effects

11.3.2.2.1 *Positive impact*

One way to increase the capacity of the electrode based on silicon while enhancing the holding in cycling is to reduce the particles size of the active material. However, the decrease in particle size can in no way remove the mechanical stresses incurred during the cycles of insertion/disinsertion Li. According to theoretical calculations produced by Wolfenstine,[186] size limit to remove the mechanical stress is lower by an order of magnitude to the size of the crystalline unit of silicon. This does not mean that the reduction in size has no effect on the electrochemical performance of electrodes with silicon. The electrochemical behavior of silicon as negative electrode material may be different depending on the particle size especially for a silicon nano (average diameter less than 1 μm). Liu *et al.*[187] have shown the beneficial effect of reducing particle size of silicon from 20 to 2 μm. Their study indicates that this reduction facilitates the kinetics of charge/discharge of the electrode. The transition to a submicron size[166] allows to further increase the capacity and cyclability of the silicon electrode. Various studies have been conducted to reduce the size of silicon particles from several methods of synthesis.[187,188] A journal article on the subject[152] lists these methods and associated performance.

Several strategies are possible to develop nano-Si/C. One such strategy is to consider the carbon as conductive substrate on which silicon nanoparticles are deposited using method of vapor deposition (CVD). The second approach is to imprison the silicon in a carbon matrix by atomization. Both techniques are mature and can be chosen for a possible industrialization. In proposing a composite carbon-silicon, the main advantages are the ability to combine the high specific capacity of silicon and the electronic conductivity of carbon. This combination needs to be feasible and economically viable on an industrial scale. In the case of carbon graphite, is also added a good stability upon cycling of this material. The different configurations of carbon/silicon should allow good cohesion between the two materials.

In a first approach the carbon as graphite substrate can serve as conductor. The nanosized silicon is then deposited on the surface of carbon by the CVD method from silane (SiH_4). Silicon will be in contact with graphite, which can provide additional capacity. Novak[189–191] studied this type of process for negative electrode of Li-ion battery. The deposition of silicon nanocarbon by gas: CVD (chemical vapor deposition) is particularly interesting. The main idea is to deposit silicon thin film on a substrate to improve performance. Deposition of silicon thin films on submicronic substrate have been tried by different researchers.[187,192–195] The results have been good with a capacity retention over 500 cycles, but for relatively low amounts of active material corresponding to films of a few hundred nanometers thick. Another possibility is to join the silicon particles on a material that is both electronic conductor and electrochemically active as graphite. A study on the MesoCarbon MicroBeads (MCMB) mésocarbone by Xie[196] shows that small amounts of silicon deposited on the surface by

CVD improve the capacity of the material (462 $mAh\cdot g^{-1}$) with a composition of 3.6% of Si instead of 290 $mAh\cdot g^{-1}$ for single MCMB). But the capacity decreases by 45% during the first cycles. This loss is attributed to the passive film formation and to the loss of active material during the cycling. Timcal graphite KS-6 was also tested as a substrate for the deposition of silicon nanometer by TVD (thermal vapor deposition).[192,193,199] The final compound shows good electrochemical properties, with a good specific capacity that is not affected by strong rates (500 $mAh\cdot g^{-1}$) and a loss of 0.15% of the capacity per cycle. The reduction of particle size to a nanometer size influence the electrochemical behavior, since the effect of mechanical stress is limited. In this context, the method of CVD on graphite offers the synthesis of composites with actual profits of 30% in specific capacity.[189–191,196,197]

The other approach is to imprison the nanometer silicon in a carbon matrix. Initially, a polymer is loaded with nanoparticles of silicon that after pyrolysis, are converted into carbon with an electrical conductivity increased. The layer of carbon will also buffer changes in volume of silicon to avoid the spray of particles and loss of contact with the current collector.[192,198,199] In this case, capacity of 1300 $mAh\cdot g^{-1}$ have been obtained during 60 cycles by limiting the capacity[192] (combining with Obrovac method). This routes seems thus to be interesting for high-energy composites.

Another way is the mechanical grinding synthesis consisting in grinding of micrometric powders of several compounds in order to achieve better homogeneity of the mixture. Through this process, the particle size is reduced in the nanometer range which will help the kinetic diffusion discharge.[187] Another advantage of this technique is the ability to synthesize the composite at high scale. Zuo postpones encouraging capacity for a composite graphite/silicon, the order of 400–500 $mAh\cdot g^{-1}$ depending on the silicon rate but also a large irreversible capacity of 30–40%.[200] Heat treatment at 200°C shows an increase of 15% of capacity for the composite, however, the irreversible capacity of the first cycle remains high. Several carbons were tested as natural graphite,[187,200–206] the graphite MCMB[207–209] or carbon black,[187,206] with this process. Carbon black is not an active material but a conductor additive. The composite of silicon and carbon prepared by mécanosynthesis has good electrochemical performance. The technique allows an intimate mixture of materials, 30% of carbon black contributes to the stability on cycling (>96%) over 50 cycles with a capacity of 600 $mAh \cdot g^{-1}$. The sufficient carbon rate reduces the possibility of loss of contact between particles of silicon and carbon. It also avoids the possibility of agglomeration of nanoparticles of silicon. But this type of composite is below the target of more than 1000 $mAh \cdot g^{-1}$. Khomenko studied three types of carbon: graphite, hard carbon, and carbon black graphitized.[206] Graphite is the most promising materials because with the same rate of silicon in the composite, the electrode reaches 830 $mAh \cdot g^{-1}$ reversible capacity with graphite, 625 $mAh \cdot g^{-1}$ with carbon black and 620 $mAh \cdot g^{-1}$ with the carbon disk. In addition, for applications in Li-ion battery, avoid too much irreversibility capacity mainly due to the rate of silicon. In this study, the optimal rate of silicon was given around 8–10% by mass.

Besides the aspects related to the particle size of silicon, efforts have been focused on the morphology. Indeed, to be reduced to a distortion in two dimensions as in the case of thin films[210] instead of three dimensions for the electrodes usual, it is necessary to synthesize silicon nanoparticles having one or two directions of growth privileged.

In the case of silicon nanowires[211,212] spatial location of the volume expansion in one direction provide to corresponding electrode both a very good cyclability and a large capacity close to theoretical capacity even when high current densities are used. Chan[211] reported a capacity of 3500 $mAh \cdot g^{-1}$ after 20 cycles for a C/5. This capacity decreases to 2100 $mAh \cdot g^{-1}$ when the charge/discharge rate is increased to 1 C. Similar results are obtained by Laïka *et al.*[212]

11.3.2.2.2 *Counterpart*

The silicon stored in open air is covered with a layer of oxide, SiO_2, which is formed spontaneously. During the first reduction, the SiO_x is reduced first to form Li_2O and Si. Then the reaction can be continued by the insertion of Li in silicon. This reaction is assumed irreversible. The reduction of native oxide does not appear at the same time as the decomposition of the electrolyte resulting in the formation of the SEI on the surface of electrode material (active material + electronic conductor). According to a study on the oxide SiO_2,[213] the curve of differential capacity (dQ/dV *vs.* V) shows a peak at 0.25 V (*vs.* Li^+/Li) during the first reduction that does not appear anymore in subsequent cycles. We can therefore assume that this potential of 0.25 V corresponds to the reduction of the oxide. Morita has shown that the size of the native oxide on the surface of silicon particles is 1.5 nm whatever is the particle size.[214] Thus considering a layer of oxide of 1.5 nm, the irreversibility due to the reduction of SiO_2 can be calculated based on the particle size (see Fig. 11.16). The size of silicon particles strongly influence the irreversible capacity compared to the total capacity. Indeed, the contribution of irreversibility in relation to the total capacity is equal to 47% for particles of 10 nm in diameter to 4% for 100 nm and become negligible for micrometric particles.

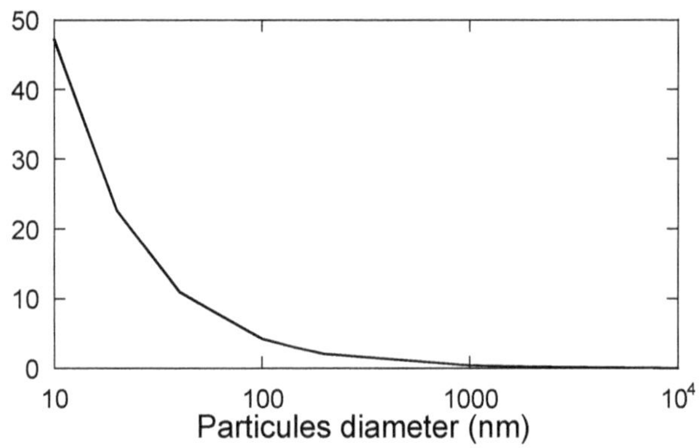

Figure 11.16 Evolution of irreversibility capacity (% in relation with total capacity) as a function of the particle size of silicon.

In the presence of water, alkyl Li(RIOCO$_2$Li) carbonates react to give $2ROCO_2Li + H_2O \rightarrow Li_2CO_3 + 2\ ROH + CO_2$. The decomposition of anions $LiPF_6^-$ produces crystallites of LiF majority component in the passivation layer at the end of the first cycle according to XPS analysis[215] and PF_5. The $LiPF_6$ and PF_5 can react with the presence of traces of water in electrolyte solutions by producing HF. HF is known to be very aggressive facing silicon. Reacting with the silicon, it breaks the Si–Si bonds in the network, resulting in cracks in the material. LiF makes the passivation layer less homogeneous and less permeable to the electrolyte resulting in deterioration of the electrodes.[216] The cracks induced by this weakening will increase during the contraction of the silicon particles during the extraction of Li. A fresh new surface will be in contact with the electrolyte. It is clear that the interaction between silicon and the electrolyte will depend on the specific surface area provided by the electrode, which may change during the cycling. This phenomenon will be pronounced for compounds finely divided.

So, if nanosize can be a good option to have a good compromise between capacity and cyclability on a theoretical point of view, the counterpart is more SiO_2 and more electrolyte decomposition that could be a break to further practical developments.

These works also highlight the need to adapt the electrode matrix (binder, electronic conductor additive, porosity) to the morphology and size of particles of active material. In particular, carbon black is a material non-electroactive *versus* Li, the total capacity of the electrode is influenced depending on the composition of carbon black. The capacity of the composite is limited given the proportion of silicon rarely exceeding 50%. Significant research efforts are conducted on the methods of synthesis of the composite Si/C but cannot be dissociated from the environment within the composite electrode. The thin film deposition on a current collector offers a limited area but the use of nanoscale powders are quickly confronted with problems of maintaining the electronic percolation within the electrode.

11.3.2.3 From material to battery

Beyond the specific capacity in $mAh \cdot g^{-1}$ of the composite, an important criterion at the level of the battery will be the specific capacity per surface area. This quantity has a direct impact on the overall energy density of the battery. For varying surface area capacity of composites with Si/C, it is possible to play on the composite ratio Si/C and/or on coating slit and dry when coating the ink on the current collector. Increasing the ratio Si/C exacerbates the problems associated with the expansion of Si and increase the coating slit arouses the problem of conduction within the electrode (optimum in electronic percolation/porosity, homogeneity on the thickness, proportions of components, and so on) and the impact on volume changes is more important that the one on thin coatings. Optimization of the formulation and processing of electrodes will have a crucial role on the final performance.

In particular, the binder should allow good cohesion between the particles of the electrode and provide good adhesion to the current collector. The common binder in the formulation of the electrode is polyvinylidene fluoride (PVDF) in the solvent *N*-methyl pyrolidone (NMP). Yet it is a thermoplastic with an average elongation that does not contain the significant changes in volume of silicon subjected to charge/discharge cycles. Of all the possible binders, an elastomer, styrene-butadiene rubber (SBR) was proposed in the formulation of electrode with silicon.[209,217,218] Compared with PVDF, the SBR is more flexible due to its low modulus of Young[218] which enables it to withstand changes in volume of the electrode with a better adherence of the electrode on the current collector. The SBR is soluble in water, giving it an additional advantage from an environmental and economic point of view. Carboxymethylcellulose sodium (Na-CMC) is added to the aqueous emulsion as thixotropic agent. A quantity of 1–2% of this polymer can insure sufficient stability of the electrode while the standard formulation contains an average of 10% PVDF. Beyond 5–6%, the binder forms an impermeable barrier to Li ions. The diffusion of ions is longer, the battery may not deliver a high capacity. Electrodes based only on CMC show better performance than the mixture SBR/CMC.[219–221] Convalents bridges between the CMC and SiO_2 on the surface of silicon particles are formed during the condensation reaction. This layer then protects the surface of silicon against reactions with the electrolyte. As a result, the irreversible capacity is therefore reduced in the first few cycles.

We must therefore find a compromise in all of this to stabilize on the largest possible number of cycles the largest capacity per surface area as possible.

Finally, the balancing of the two electrodes taking into account the capacity and the irreversible formation of the passivation layer, the management charge and discharge of the battery will be some of the many parameters to refine.

11.3.3 High-Rate Titanates: Interest of Nanosize?

Titanium oxides of different structures can be obtained in different nanoforms depending on synthesis conditions: nanoparticles (~50–100 nm), nanowires.[222–224] Due to their morphological characteristics (nanomaterials, high specific surface area, and so on), these compounds are indeed of interest for Li-ion batteries application but they will also be materials of choice for the development of an asymmetric electrode system with $Li_4Ti_5O_{12}$[225] or TiO_2-B,[226] between batteries and supercapacitors.

11.3.3.1 Properties of lithium titanates

Figure 11.17 shows the results of comparative electrochemical TiO_2-B, $Li_4Ti_5O_{12}$, and graphite. The first two compounds were prepared respectively according[227,228] with no specific sizes of nanometric particles.

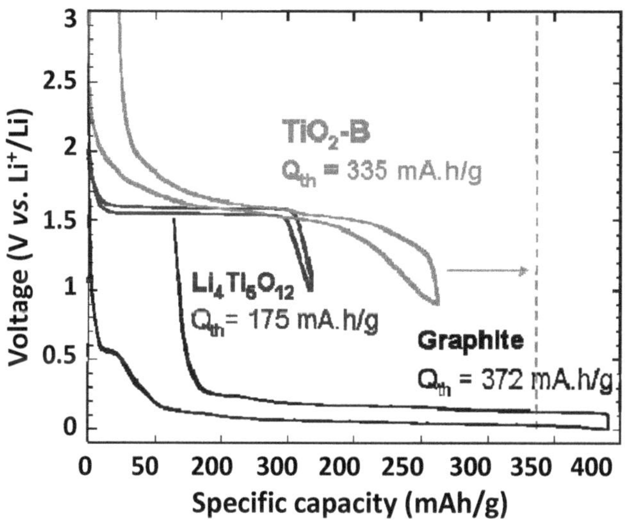

Figure 11.17 Potential *versus* capacity curve for three negative electrode materials for Li-ion battery.

The $Li_4Ti_5O_{12}$ was prepared according to patented synthesis protocol.[227] The different stages of the process involve alternate periods of mechanical grinding and thermal treatment under air. The powder obtained has a specific surface area of about 10 m^2/g. The specific surface area is correlated to particle size. In this case, the material is bi-modal (a component around 400 nm and around 50 μm). The commercial electrode materials, such as $LiCoO_2$ have a specific surface area below 1 m^2/g.

The TiO_2-B was prepared in three steps according to.[228] The first reaction is a in solid-state between anatase TiO_2 and KNO_3 at 1000°C for 48 h to form $K_2Ti_4O_9$. This compound is then hydrolyzed in aqueous HNO_3 to exchange K^+ ions by protons. Dehydration of $H_2Ti_4O_9$, H_2O to 500°C for several hours, allow to obtain a pure microcrystalline powder of TiO_2-B. The specific surface area of powders prepared are between 10 and 45 m^2/g.

At low rate (Fig. 11.17), TiO_2-B has a specific capacity greater than $Li_4Ti_5O_{12}$. In identical cycling conditions, it appears that the two products deliver at high rate (5 C and 10 C) similar capacities (Fig. 11.18). Thus, the gap between the capacities of TiO_2-B and $Li_4Ti_5O_{12}$ is reduced as the current rate is increased. For more extreme rates (several tens of "C"), $Li_4Ti_5O_{12}$ would be the most interesting. In a basic configuration of Li-metal, high-speed tests at 10 C (charge or discharge in 6 minutes) are affected by limitations

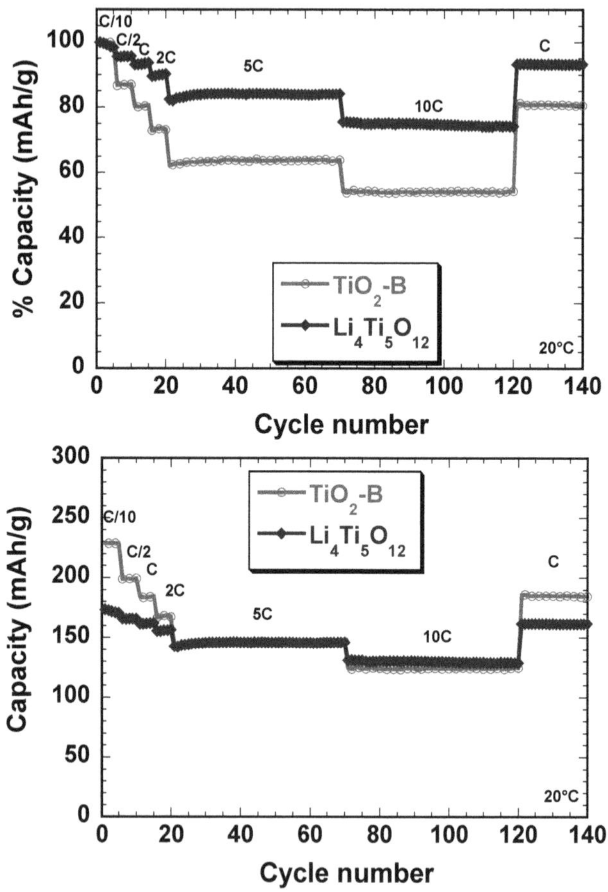

Figure 11.18 Capacity (high) and percentage of capacity recovered (bottom) based on the number of cycles for TiO_2-B and $Li_4Ti_5O_{12}$ at various charge/discharge rate (*vs.* Li-metal).

in the Li electrode. $Li_4Ti_5O_{12}$ powder is more dense than that of TiO_2-B, allowing a gain in volumetric energy density. The advantage of $Li_4Ti_5O_{12}$ appears very significant if we report the curve representing the percentage of capacity depending on the number of cycle and the rate (100% corresponding to the initial capacity at slow rate). $Li_4Ti_5O_{12}$ allow to charge or restore 75% of its maximum capacity in 6 minutes (about 55% TiO_2-B). It should be noted that the tests represent the worst scenario because the successive charges and discharges are made to the same rate. For applications requiring fast return of energy (battery discharge corresponding to a fast charge of the negative electrode), a moderate charge of the battery (discharge of the negative electrode) rate could be maintained.

In these conditions, $Li_4Ti_5O_{12}$ is therefore more interesting than TiO_2-B for high-power applications (pulses of a few seconds or minutes). The titanium oxide $Li_4Ti_5O_{12}$ is an alternative to the graphite negative electrode for applications in power, long life and enhanced safety.

11.3.3.2 Particule size effects

11.3.3.2.1 *$Li_4Ti_5O_{12}$*

Various works of literature aim to prepare titanium oxide nano-sized or nanostructured to improve performance in power.

Nanomyte™ $Li_4Ti_5O_{12}$[229] is a stable electrode material and exhibits excellent cyclability and high capacity retention (130–155 mAh/g) at fast charge and discharge rates (≥5 C). These nanostructured $Li_4Ti_5O_{12}$ particles have a spinel type crystal structure and are composed of crystals that are less than 100 nm in size; the collection of particles having an average diameter of less than about 300 nm.

$Li_4Ti_5O_{12}$ Spinel Nanowires have been prepared and tested for high-rate Li-ion intercalation electrodes.[230] A comparison with fine $Li_4Ti_5O_{12}$ particles average particle size (0.7 μm) is given Figs. 11.19 and 11.20. The reversible first discharge capacity is 165 mAh/g, and no capacity fading was observed out to 30 cycles. Figure 11.19 shows rate capabilities of the sample between 1 and 2.5 V at rates of 0.5, 1, 5, and 10 C with a fixed charge rate of 0.1 C. The first capacity at 0.5 C is identical to that at 0.1 C, but there is a slow capacity fading with increasing current. At 5 and 10 C = 1600 mA/g rates, the capacity retention is 95 and 93%, showing 157 and 155 mAh·g^{-1}, respectively. Unlike the wires, the particles exhibited rapid capacity fading at higher C-rates, with only 77 and 51 mAh/g at 5 and 10 C rates, respectively. The result is far inferior to that of the nanorods, and the capacity retention of the ultrafine particles decreased by 49% and 59% at 5 and 10 C rates, respectively, compared to the nanowires. A comparison with results obtained by[230,232] give less differences between fine particules, even nano-sized and nanowires performances. According to Kim and Cho, the electrode density of the spinel nanorods is 2.8 g/cm^3 while that of the ultrafine particles is 1.5 g/cm^3, and therefore the volumetric density of the nanorods and the ultrafine particles are estimated to be 464 and 250 mAh/cm^3, respectively, meaning it is easier to densify nanowires instead of fine particle of 0.7 μm.[233] On the other end, according to Jouanneau *et al.*,[230] the tapping density of their $Li_4Ti_5O_{12}$ (bi-modal 400 nm and 50 μm) is twice the one obtained on nanosized (<300 nm) so that volumetric electrode density. The counterpart of the spherical nanosized effect on the electrode/battery power density can be explained with an excessively high Brunauer–Emmett–Teller (BET) surface area, since over 20 wt% binder and 20 wt% carbon black are required to get an electrode. This meant a large decrease in electrode density, and only 1 g/cm^3 was obtained, which is 2.8 times smaller than the one of the nanowires. With nanowires, it is possible to reach high volumetric density of the electrodes.

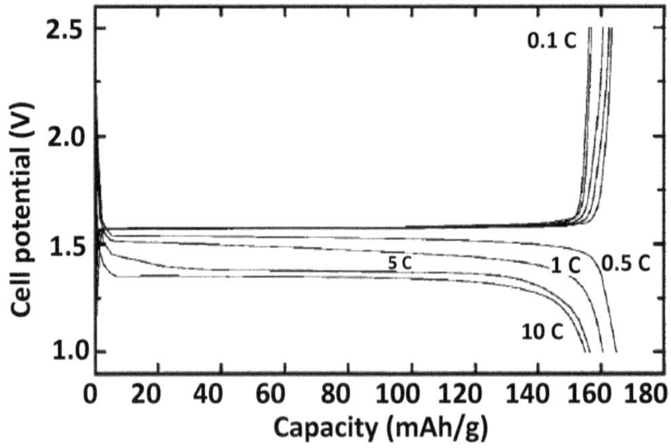

Figure 11.19 Voltage profiles at different rate for $Li_4Ti_5O_{12}$ nanowires at different rates of 0.5, 1, 5, and 10 C). The charge rate was fixed at 0.1 C = 16 mA/g.

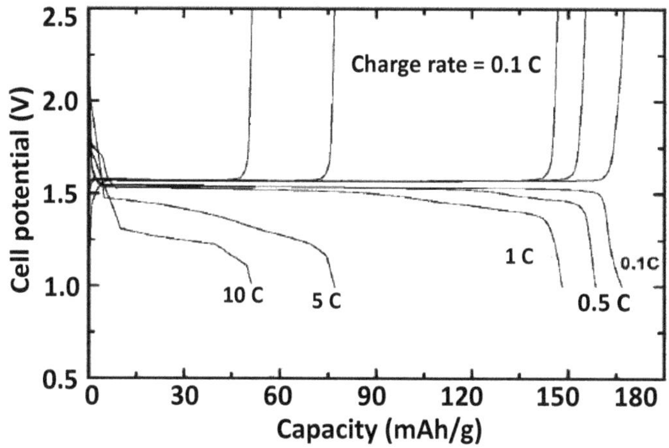

Figure 11.20 Rate capability test of the ultrafine $Li_4Ti_5O_{12}$ particles at different C rates (0.5, 1, 5, and 10 C). The charge rate was fixed at 0.1 C = 16 mA/g.

It seems then two options are available to reach high-power density electrode on negative side for $Li_4Ti_5O_{12}$: nanowires or bi-modal distribution of nano- and micro-sized compounds.

11.3.3.2.2 *TiO_2*

In practice, the different structures of TiO_2 (anatase, rutile, TiO_2-B), the TiO_2-B in the form of nanowires may issue up to 300 mAh·g^{-1} at slow rate in 30 hours and 200 mAh·g^{-1} at faster rates.[231,232] For low capacity electrode surface (<0.5 mAh/cm^2), the maximum capacity obtained at a very slow rate (C/100) is 260 mAh·g^{-1}. Recent studies allow obtaining in practice around 60% of this capacity to the slow rate (C/10) and very recently some authors[222] obtained 75% of theoretical capacity from TiO_2 nanowires up to 10 C. The size, morphology, specific surface area and microstructure appear as key parameters for performance improvements.[233–238] The recent discovery of the synthesis of TiO_2-B in the form of nanotubes and nanowires and electrochemical performance associated has contributed to a real passion for this material. Other authors show that TiO_2/C core-shell nano-composites[239] as anode materials for Li-ion batteries present a better cycle performance and a higher diffusion coefficient of Li ions compared with the virginal TiO_2 nanoparticles. This should induce better power performance.

Porous rutile TiO_2 nanorod micospheres[240] performances are illustrated Fig. 11.21.

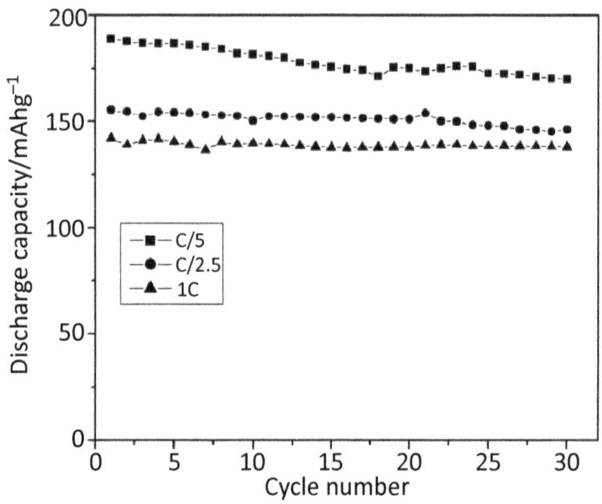

Figure 11.21 Discharge capacity with respect to cycle number for the hierarchical porous rutile nanorod microsphere electrodes at different charge/discharge rate.

The discharge specific capacities of 189, 155, 142 mAh g^{-1} are obtained at C/5, C/2.5, and 1 C rates, respectively. These hierarchical porous rutile TiO_2 nanorod microspheres can be inserted up to 0.73 Li ($Li_{0.73}$ TiO_2, 246 mAh·g^{-1}) during the first discharge at a C/10 charge/discharge rate. After 30 cycles, their stabilized capacity is as high as 192 mAh g^{-1}.

Nevertheless, all these data are provided on electrodes with low specific capacities per surface area and lots of work is needed to obtain similar results on denser electrode. We can suspect the same behaviour as the one for $Li_4Ti_5O_{12}$ meaning nanowires or nanorods will be preferable than spherical nanosized particules.

11.3.3.3 Main synthesis route for nanoscale lithium titanate

11.3.3.3.1 *$Li_4Ti_5O_{12}$*

A method of producing nanostructured $Li_4Ti_5O_{12}$ particles,[241] comprising the following steps: (a) dissolving a Li containing salt in an organic solvent to form a liquid solution, (b) adding a dispersion of nanoparticles of TiO_2 to the liquid solution, said TiO_2

nanoparticles having an average primary particle size of less than 100 nm; (c) heating the liquid solution to facilitate diffusion of Li ions into the nanoparticles; (d) separating the solids from the liquid solution; and (e) heat treating the solids to form a crystal structure, wherein the crystals are up to 100 nm in size.

Nanosized lithium titanium oxide $Li_4Ti_5O_{12}$ powder can also be prepared using LiCl as a high-temperature flux.[242] XRD, scanning electron microscope, and transmission electron microscope measurements indicate that the obtained $Li_4Ti_5O_{12}$ particles are uniform and the particle size of $Li_4Ti_5O_{12}$ powder can be controlled by flux content and heating time. Under the optimal synthetic condition, the particle-size distribution is narrow 100 nm.

A large scale production of one-dimensional $Li_4Ti_5O_{12}$ nanowires (Fig. 11.22) is achievable by firing a mixture of Li acetate and TiO_2 nanowires.[230]

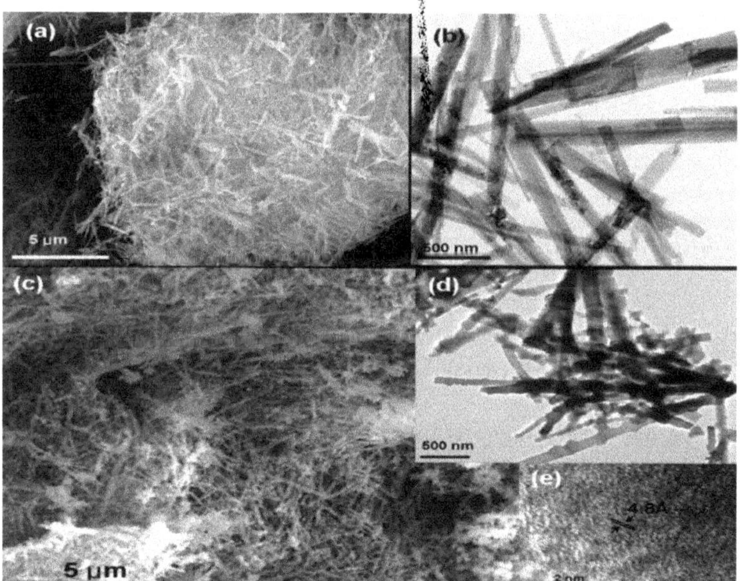

Figure 11.22 TEM images of $Li_4Ti_5O_{12}$ nanowires.

11.3.3.3.2 *TiO_2*

Synthesis of mesoporous TiO_2 particles using ionic liquid (IL)-assisted sol–gel method was investigated by Yoo *et al.*[243] and they obtained very good structural properties including high surface area, stable crystal structure, controlled porosity and tailor-designed pore size distribution. Due to the unique properties of the ionic liquid, including zero vapor pressure and water-immiscibility, they obtained very well nanostructured TiO_2 particles (see Fig. 11.23) and they show that the length of alkyl group on the cation part of ionic liquid could determine the pore size and volume of TiO_2 particles.

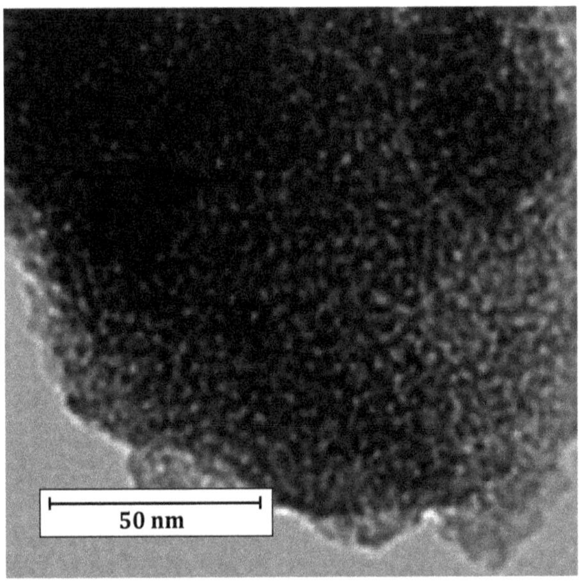

Figure 11.23 TEM image of sample prepared using [Bmin][PF6] with heat treatment at 100°C.

So there are many prospects for improvement by playing more on the morphology and nanostructuring via original synthesis conditions to become to a capacity close to that of graphite.

Another synthesis route[244] consists on a step requiring extreme conditions to solubilize TiO_2-anatase in concentrated soda. The hydrothermal synthesis conditions allow to obtain a hydrogenated sodium titanate of formula $Na_yH_{2-y}Ti_nO_{2n-1}\cdot xH_2O$. After an exchange of Na^+ ions by protons and dehydration of the total material, TiO_2-B is obtained in the form of nanotubes or nanowires. The difficulty of characterization of the product and the conditions required for the hydrothermal synthesis route can be a hindrance for the application.

A fast cooling step can help to limit the size of the particles. For high-power applications, small particle size is generally preferred (diffusion paths of Li shortcuts). The specific surfaces area of powders are between 10 and 45 m^2/g.

11.3.3.4 From material to battery

One of the counterparts of the hydrothermal route, for example, is the fact that the materials may have very high specific surface (>60 m^2/g) making their processing in composite electrodes very sensitive if we want to ensure simultaneously a good electronic percolation, a good diffusion within the electrode and a good cohesion of the composite with adherence to the current collector. This would significantly increase the quantities of binders and conductive additives or optimize contacts and interfaces between the constituents of a composite by a process well monitored and optimized to minimize additives that will inevitably lead to a loss in energy density. Moreover, the optimization of active environment within the composite electrode has a major role in highlighting the intrinsic performances of these remarkable compounds. The objective is to reproduce these performances on most loaded electrodes (surface capacity greater or equal to 1 mAh/cm^2) in order to lead to high energy and specific power densities. The aim is the validation as an active material for negative electrode for Li-ion battery having properties in terms of energy density and power very competitive (with 100–120 Wh/Kg up to 4700 W/kg in pulse). The purpose is more specifically the validation of certain allotropic forms of TiO_2 as negative electrode material property to charge very fast (less than 3 minutes) and replace $Li_4Ti_5O_{12}$.

11.4 Integration in Prototypes

While an abundant literature about nanomaterials elaboration and characterization is available (see previous chapters for positive and negative electrodes), very few articles are devoted to their processing and integration into Li-ion prototypes.

Indeed, cautious analysis of materials is better carried out with thin electrodes containing large amount of electronic percolator (carbon black mainly), in order to avoid ionic and electronic limitation due to electrode processing. That is why many academic or more fundamental studies do not take into account processing. A good example of this fact can be seen in a recent article written by Ceder and Kang[245] in which very fast charging capabilities are claimed for $LiFePO_4$ electrodes, but at the expense of a considerable amount of percolator (e.g., 65%) and thin electrode thus rendering it impractical for most applications.

On the other hand these parameters must be carefully optimized to successfully incorporate nanomaterials into commercial batteries.

11.4.1 Electrode Processing: Impact of Material Size

The first challenge when dealing with nanomaterials is their extremely large surface area, typically in excess of 10 m^2/g when conventional materials such as $LiCoO_2$ are less than 1 m^2/g. Smaller particles tend to agglomerate and if not properly dispersed lead to aggregates which impedes the improvements anticipated by nanosize materials.[246] This problem is even more prominent when using aqueous processing. In that case hydrophobic electronic percolator such as carbon black with more than 50 m^2/g must be dispersed in water. Carboxymethyl cellulose[247] (CMC) used primarily as a thickener and binder in the electrode can also serve as a surfactant agent, but other additives are often mandatory to get a stable dispersion in water.

The amount of binder has also to be adjusted according to this large surface area to obtain the desired flexibility and electrode cohesion. More binder is necessary and can result in lower energy density.

When dealing with nanomaterials one has also to pay a particular attention to moisture, since these compounds can be very hygroscopic. Moisture is known to react with the electrolyte and form HF which can seriously lower battery life.[248] It is also more difficult to remove moisture from nanopowders when compared to micron sized particles. For instance it has been demonstrated that two days of drying at 120°C under vacuum is necessary to obtain optimal impedance of acethylene black and $LiFePO_4$ when only half a day under the same condition is enough for micron sized nickel cobalt aluminium (NCA) cathode.[249]

Finally more surface area means more passivation layer on the electrodes. Indeed the solid electrolyte interphase (SEI) due to electrolyte decomposition is a surface reaction and the irreversible capacity thus grows with decreasing particle size. The capacity lost during the first battery charge is a critical parameter because it means a part of the often costly positive electrode becomes inactive during the process due to lack of remaining Li in the system. This is one of the reasons why it is difficult to consider nanotubes as an anode active material for instance.

11.4.2 The Uncommon Case of Nanosilicon Electrode

Because of its very large theoretical capacity (3850 mAh/g) silicon has been extensively studied as a graphite anode replacement in Li-ion batteries.

However this extraordinary capacity comes at the price of a nearly 300% volume expansion during Li insertion leading to $Li_{3,75}Si$. Using nanosilicon allows better cycle life since cracks are less likely to propagate in the smaller particles.

An electrode incorporating this active material in a reasonable proportion to have an excess capacity compared to graphite (372 mAh/g) will undergo some radical expansion.

For instance a typical positive electrode loading in Li-ion batteries is of the order of 3 mAh/cm^2, which translates in a 25 μm thick electrode for the silicon negative electrode.

As a matter of fact, the huge capacity of silicon is mitigated by the limited positive one, and although a 150 μm negative can be easily processed, it would lead to a 18 mAh/cm^2 loading, which would translate in an impractical thickness for the matched positive.

Since silicon volume expansion is isotropic, the electrode thickness will increase during the first charge by as much as 300%, or 100 μm for the 25 μm electrode. This expansion could be bearable if a single electrode stack was used. However, commercial batteries have several stacked electrodes, or they are wound several times around a core, and the cumulative effect of the negative electrodes can cause more than a mm expansion. Besides, swelling hinders the reversibility of the electrode by cracking and electronically disconnecting particles. If the battery is wound the volume increase will create tensions inside the jelly roll which can cause short circuits.

For all these reasons, volume expansion must be contained as much as possible. A possible solution is to limit silicon capacity use, and thus swelling in some proportion.[250]

Another workaround is to use only a small proportion of silicon, enough to increase the capacity compared to graphite but not too much to keep overall swelling low.[251] As can be seen on the simulation of Fig. 11.24, the benefit of increased negative electrode capacity is indeed more visible between 300 and 1500 mAh/g and the energy density levels off above this value due to positive electrode limitations. Also one can note that the gain of capacity compared to conventional Li-ion batteries is large when the energy density increase is less pronounced. This is because Silicon mean charge voltage (~0.45 V *vs.* Li^+/Li) is higher than that of graphite. Of course this simulation does not plot the volumetric energy density which is likely to be variable if silicon expansion is not contained.

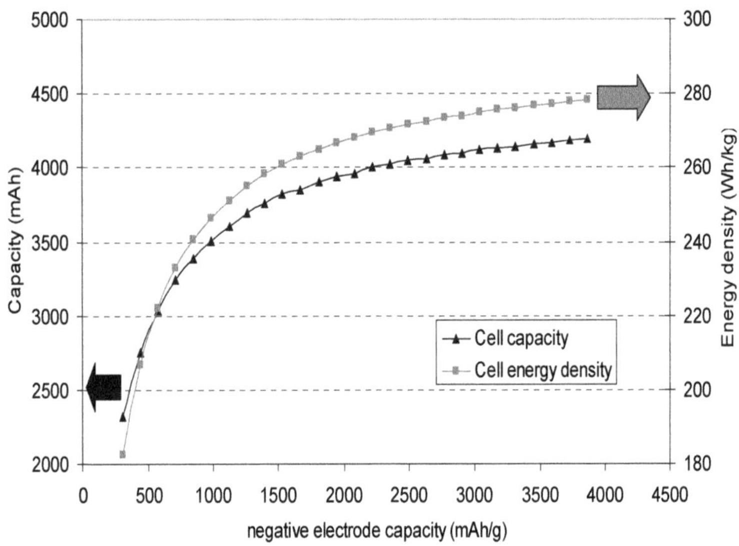

Figure 11.24 Simulation of a 18650 class cell capacity and energy density as a function of negative electrode capacity (NCA positive with a 3.5 mAh/cm^2 loading).

11.4.3 Nanomaterials as Additive in Electrodes of Lithium Batteries

One of the first use of nanomaterials in Li-ion batteries was as electronic percolator, since carbon black is a cheap and easy way to increase electrode conductivity. Many different carbon types have been test and used since the birth of Li-ion batteries, such as acetylene black, graphite, carbon fibers and nanotubes.[252]

Fibers can have sub micron diameter with several micron length, helping long range conductivity in the bulk of the electrode, while small carbon black aggregates percolate voids between active material particles. Since a limited passive layer is formed on the positive electrode, it can contain large amount of theses percolators. On the other hand putting large surface area additive in negative electrodes with potentials lower than 0.8 V *vs.* Li^+/Li is not recommended since it will cause a dramatic increase of the irreversible capacity.[253]

The amount of additive has to be kept at the lowest necessary (usually a few percents) to obtain percolation, since it does not contribute to the battery capacity, and can be costly in the case of carbon fibers. The advantage of using nanomaterials in this particular case is to benefit from a maximum of electrical contacts with a minimum of weight and volume. Vapor-grown carbon fibers (VGCF) is a typical example of this use because the sub micron diameter fibers can connect particles over a long range in the electrode for a minimal added weight.[254] Nanotubes could have a similar use but are today still too expensive.

11.4.4 Examples of Performance in Full Cells

11.4.4.1 High energy: toward 300 Wh/kg

Coupling a high-energy positive electrode, such as Li-rich layered oxide, with a high-energy silicon electrode makes it possible to reach fairly high-energy densities. Nanomaterials as additives and active materials in this case enable optimal use of the capacity. Using $y Li_2MnO_3$-$(1-y)LiMO_2$ cathode material synthesized in our laboratory, preliminary testing at CEA on 1.2 Ah prototypes shows promising results, with 243 Wh/kg reached at 55°C and low cycling rates, compared to 200 Wh/kg with the same design using conventional materials (NCA *vs.* graphite). Figure 11.25 shows the third cycle of the stacked cell, using a laminated pouch design, conventional electrolyte (carbonate mix with $LiPF_6$ salt) and a Celgard separator. Still, many optimizations are needed to fix the swelling and cycle life issues, but this kind of system could already be viable for some niche applications, where very large energy densities are needed for a relatively small cycle life.

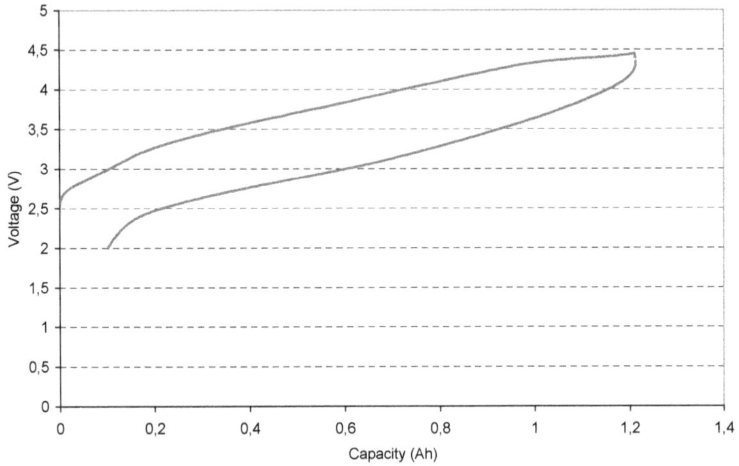

Figure 11.25 Third C/10 cycle at 55°C for a high energy 1.2 Ah pouch cell containing yLi_2MnO_3-$(1-y)LiMO_2$ positive and Si-negative electrode (total mass = 14.4 g).

Based on this result larger energy densities can be anticipated on bigger cells (Fig. 11.26). Indeed large scale batteries maximize the amount of active materials *vs.* inactive parts (can, current collector, tabs...). The energy density levels off well above 300 Wh/kg which shows the potential of this electrochemical couple.

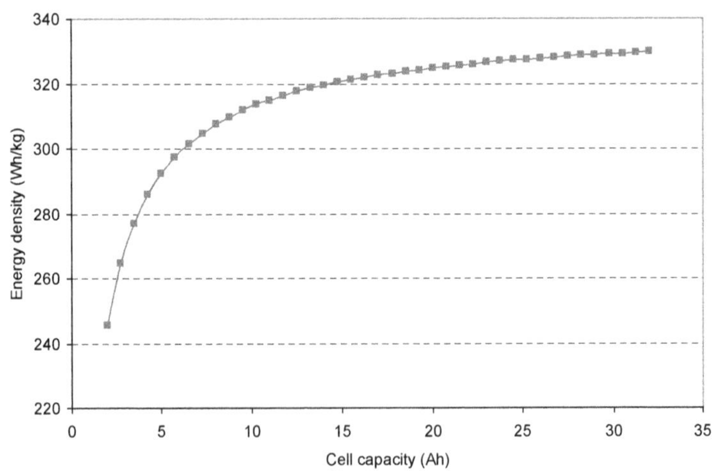

Figure 11.26 Simulation of the energy density evolution for a 12 cm high-wound cell with the same characteristics as the prototype seen in Fig. 11.25 as a function of cell capacity.

Experiments have also been carried out on the same cell format with a $LiFePO_4$ electrode using nanopowders synthesized at CEA and a silicon anode. An energy density of 145 Wh/kg on the second cycle has been reached for a 800 mAh prototype. This couple could be useful for application where both energy density and safety are critical issues, like for instance electric vehicle use. The discharge curve of the cell is reported on Fig. 11.27.

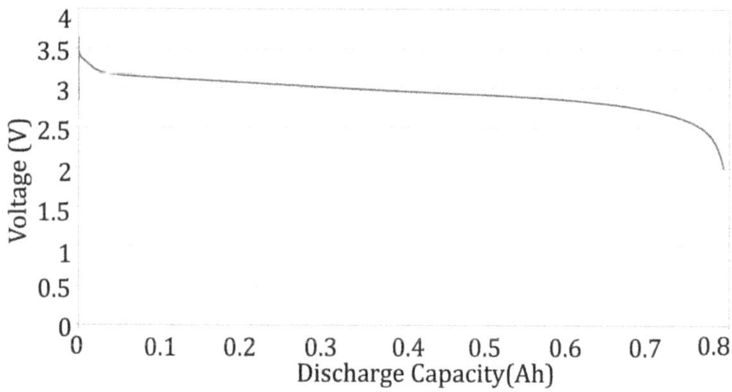

Figure 11.27 C/20 discharge curve at room temperature for a 0.8 Ah pouch cell containing $LiFePO_4$ *vs.* Si (145 Wh/kg).

11.4.4.2 High power: interest of nanotype materials

When it comes to high-power applications, nanomaterials offer the advantage of large surface area for rapid transfer of Li and small diffusion length inside the particle. As a matter of fact the Li insertion time is proportional to the square of the particle size, so decreasing it by an order of magnitude increases achievable rates by two orders of magnitude.[255] The negative electrode of choice for this application is lithium titanate $Li_4Ti_5O_{12}$ (LTO). Surface area of this material can be as high as 40 m^2/g, and the rates it can sustain both during charge and discharge are much larger than graphite. However it is higher potential *versus* Li (1.5 V) makes it only suitable for power applications. Nano $LiFePO_4$ (LFP) also has very good high-rate performance. Extensive studies have been made in our lab on the LFP-LTO couple for high-power applications. It has a constant potential of 1.9 V which makes it very stable. For instance self-discharge after 6 months has been measured below 5% at room temperature and was 15% at 40°C. Figure 11.28 shows the rate capability during fast charge up to 256 C. The performances are outstanding and make it an intermediate between super-capacitors and conventional Li-ion batteries. Energy densities of 50 Wh/kg can be achieved with this technology, far superior to super-capacitors (~5–10 Wh/kg). This fast charge capability is particularly interesting for hybrid vehicle application where regenerative braking needs high power on charge to recover as much energy as possible during those phases. Since LTO potential is much higher than Li, it does not suffer from the Li plating problem that can occur with graphite at low temperatures or high-charge rate. Another advantage of titanate is that it is a zero-strain material, so there is no volume expansion during Li intercalation. LFP undergoes negligible volume expansion so that LFP-LTO battery volume is constant, increasing the cycle life. More than 30,000 partial cycles simulating hybrid vehicle use have been performed on the technology representing 7 years of utilization.

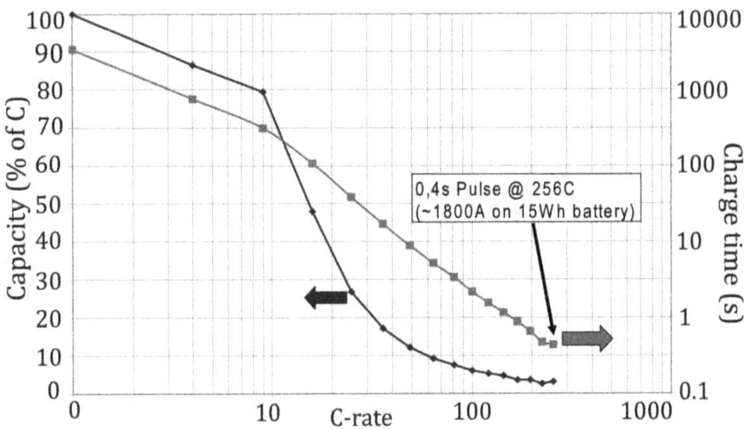

Figure 11.28 LFP-LTO cell performance on charge. Electrode loading = 1 mAh/cm^2.

Discharge power is also very high and can reach 4 kW/kg during 1 s pulses for 100 C rates (Fig. 11.29). This property can be extremely useful in hybrid vehicles when power surge are needed for short periods of time. A 10 kg battery pack with LFP-LTO cells could deliver 40 kW of power for 1 second, or still 30 kW during 5 seconds, enough to accelerate a car without problem.

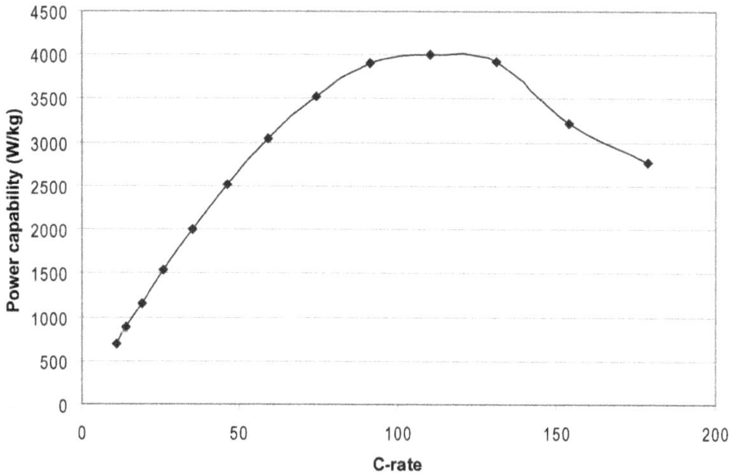

Figure 11.29 Power capability in W/kg of prototype for a 1 s pulse discharge at different rates.

11.5 Conclusion

In this article, we tried to give a broad overview of the potential interest of "nanos" for batteries:

- synthesis of nanoparticles for active material (e.g., LTO or TiO_2)
- electrode engineering at nanoscale (especially for Si–C composites)
- stabilizing the high-voltage material interface with electrolyte (coating or electrolyte additives)

One can see that nanoengineering is not only a reduction of the active material particle size. In fact, despite for power applications this size reduction is very interesting in terms of energy density; the key point with "nano" is more in electrode engineering and mastering of the material/electrode interface down to the nanoscale.

Acknowledgment

The authors want to thank all the CEA-LITEN-LCE team for their strong involvement in obtaining many of the results presented in this chapter.

References

1. A. Manthiram, J. B. Goodenough, *J. Solid State Chem.*, **71**, 349–360 (1987).
2. A. Manthiram, J. B. Goodenough, *J. Power Sources*, **26**, 403–408 (1989).
3. C. Delmas, F. Cherkaoui, A. Nadiri, P. Hagenmuller, *Mater. Res. Bull.*, **22**, 631–639 (1987).
4. C. Delmas, A. Nadiri, J. L. Soubeyroux, *Solid State Ionics*, **28–30**, 419–423 (1988).
5. A. Nadiri, C. Delmas, R. Salmon, P. Hagenmuller, *Rev. Chim. Miner.*, **21**, 537–544 (1984).
6. O. Tillement, J. Angenault, J. C. Couturier, M. Quarton, *Solid State Ionics*, **53–56**, 391–399 (1992).
7. O. Tillement, "Nasicons à valences mixtes, synthèse et caractérisation", *Thèse de Doctorat*, Paris VI (1992).
8. A. K. Padhi, *Mapping Redox Energies of Electrode Materials for Lithium Batteries*, Ph.D. Thesis, University of Texas, Texas (1997).
9. G. Cheruvally, *Lithium Iron Phosphate: A Promising Cathode-Active Material for Lithium Secondary Batteries*, Trans Tech Publications Inc., Switzerland (2008).
10. C. V. Ramana, A. Mauger, F. Gendron, C. M. Julien, K. Zaghib, *J. Power Sources*, **187**, 555–564 (2009).
11. S.-Y. Chung, J. T. Bloking, Y.-M. Chiang, *Nat. Mater.*, **1**, 123–128 (2002).
12. P. Gibot, M. Casa-Cabanas, L. Lafont, S. Levasseur, P. Carlach, S. Hamelet, J.-M. Tarascon, C. Masquelier, *Nat. Mater.*, **7**, 741 (2008).
13. P. Subramanya Herle, B. Ellis, N. Combs, L. F. Nazar, *Nat. Mater.*, **3**, 147–152 (2004).
14. D.-H. Kim, J. Kim, *Electrochem. Solid-State Lett.*, **9**, A439–A442 (2006).
15. N. Recham, L. Dupont, M. Courty, K. Djellab, D. Larcher, M. Armand, J.-M. Tarascon, *Chem. Mater.*, **21**, 1096–1107 (2009).
16. N. Ravet, J. B. Goodenough, S. Besner, M. Simoneau, P. Hovington, M. Armand, *Proceedings of the 196th Meeting of the Electrochemical Society, Hawai*, October 1999 (Abstract 127).
17. H. Gabrisch, J. D. Wilcox, M. M. Doeff, *Electrochem. Solid-State Lett.*, **9**, A360–A363 (2006).
18. J. Barker, M. Y. Saidi, J. L. Swoyer, *Electrochem. Solid-State Lett.*, **6**, A53–A55 (2003).

19. J. D. Wilcox, M. M. Doeff, M. Marcinek, R. Kostecki, *J. Electrochem. Soc.*, **154**, A289–A395 (2007).

20. N. Ravet, A. Abouimrane, M. Armand, *Nat. Mater.*, **2**, 702 (2003).

21. D. K. Kim, H. M. Park, S. J. Jung, Y. U. Jeong, J. H. Lee, J. J. Kim, *J. Power Sources*, **159**, 237–240 (2006).

22. S. Franger, C. Bourbon, F. LeCras, Boron-substituted lithium compounds, active electrode materials, batteries and electrochrome devices. *US patent No. 7529014* (2009).

23. J.Wolfenstine, J. Allen, *J. Power Sources*, **136**, 150–153 (2004).

24. G. Li, H. Azuma, M. Tohda, *Electrochem. Solid-State Lett.*, **5**, A135–A137 (2002).

25. A. Yamada, Y. Kudo, K.-Y. Liu, *J. Electrochem. Soc.*, **148**, A1153–A1158 (2001).

26. A. Yamada, Y. Kudo, K.-Y. Liu, *J. Electrochem. Soc.*, **148**, A747–A754 (2001).

27. A. Yamada, S.-C. Chung, *J. Electrochem. Soc.*, **148**, A960–A967 (2001).

28. C. Delacourt, P. Poizot, M. Morcrette, J.-M. Tarascon, C. Masquelier, *Chem. Mater*, **16**, 93–99 (2004).

29. M. Yonemura, A. Yamada, Y. Takei, N. Sonoyama, R. Kanno, *J. Electrochem. Soc.*, **151**, A1352–A1356 (2004).

30. C. Delacourt, L. Laffont, R. Bouchet, C. Wurm, J.-B. Leriche, M. Morcrette, J.-M. Tarascon, C. Masquelier, *J. Electrochem. Soc.*, **152**, A913–A921 (2005).

31. N.-H. Kwon, T. Drezen, I. Exnar, I. Teerlinck, M. Isono, M. Graetzel, *Electrochem. Solid-State Lett.*, **9**, A227–A280 (2006).

32. D. Wang, H. Buqa, M. Crouzet, G. Deghenhghi, T. Drezen, I. Exnar, N. H. Kwon, J. H. Miners, L. Poletto, M. Greatzel, *J. Power Sources*, **189**, 624–628 (2009).

33. S. Yang, P. Y. Zavalij, M. S. Whittingham, *Electrochem. Commun.*, **3**, 505–508 (2001).

34. S. Franger, F. Le Cras, C. Bourbon, H. Rouault, *Electrochem. Solid-State Lett.*, **5**, A231–A233 (2002).

35. N. N. Bramnik, K. Nikolowski, C. Baehtz, K. G. Bramnik, H. Ehrenberg, *Chem. Mater.*, **19**, 908–915 (2007).

36. P. Deniard, A. M. Dulac, X. Rocquefelte, V. Grigorova, O. Lebacq, A. Pasturel, S. Jobic, *J. Phys. Chem. Solids*, **65**, 229–233 (2004).

37. N. N. Bramnik, K. G. Bramnik, T. Buhrmester, C. Baehtz, H. Ehrenberg, H. Fuess, *J. Solid State Electrochem.*, **8**, 558–564 (2004).

38. M. E. Rabanal, M. C. Gutierrez, F. Garcia-Alvarado, E. C. Gonzalo, M. E. Arroyo-de Dompablo, *J. Power Sources*, **160**, 523–528 (2006).

39. J. Wolfenstine, J. Read, J. L. Allen, *J. Power Sources*, **163**, 1070–1073 (2007).

40. H. H. Li, J. Jin, J. P. Wei, Z. Zhou, J. Yan, *Electrochem. Commun.*, **11**, 95–98 (2009).

41. A. Nyten, A. Abouimrane, M. Armand, T. Gustafsson, J. O. Thomas, *Electrochem. Commun.*, **7**, 156–160 (2005).

42. M. E. Arroyo-de Dompablo, M. Armand, J. M. Tarascon, U. Amador, *Electrochem. Commun.*, **8**, 1292–1298 (2006).

43. R. Dominko, M. Bele, A. Kokalj, M. Gaberscek, *J. Power Sources*, **174**, 457–461 (2007).

44. A. Kokalj, R. Dominko, G. Mali, A. Meden, M. Gaberscek, J. Jamnik, *Chem. Mater.*, **19**, 3633–3640 (2007).

45. R. Dominko, *J. Power Sources*, **184**, 462–468 (2008).

46. A. R. West, F. P. Glasser, *J. Solid State Chem.*, **4**, 20–28 (1972).

47. C. Lyness, B. Delobel, A. R. Armstrong, P. G. Bruce, *Chem. Commun.*, (2007) 4890–4892.

48. A. S. Prakash, P. Rozier, L. Dupont, H. Vezin, F. Sauvage, J. M. Tarascon, *Chem. Mater.*, **18**, 407–412 (2006).

49. M. E. Arroyo y de Dompablo, U. Amador, J. M. Tarascon, *J. Power Sources*, **174**, 1251–1257 (2007).

50. G. G. Amatucci, N. Pereira, *J. Fluorine Chem.*, **128**, 243–262 (2007).

51. J. Barker, M. Y. Saidi, J. L. Swoyer, *J. Electrochem. Soc.*, **151**, A1670–A1677 (2004).

52. Y. Makimura, L. S. Cahill, Y. Iriyama, G. R. Goward, L. F. Nazar, *Chem. Mater.*, **20**, 4240–4248 (2008).

53. T. Jiang, G. Chen, A. Li, C. Wang, Y. Wei, *J. Alloys Compd.*, **478**, 604–607 (2009).

54. F. Sauvage, E. Quarez, J. M. Tarascon, E. Baudrin, *Solid State Sci.*, **8**, 1215–1221 (2006).

55. B. L. Ellis, W. R. M. Makahnouk, Y. Makimura, K. Toghill, L. F. Nazar, *Nat. Mater.*, **6**, 749–753 (2007).

56. O. V. Yakubovith, O. V. Karimova, O. K. Melnikov, *Acta Crystallogr.* C, **53**, 395–397 (1997).

57. M. Dutreilh, C. Chevalier, M. El-Ghozzi, D. Avignant, *J. Solid State Chem.*, **142**, 1–5 (1999).

58. S. Okada, M. Ueno, Y. Uebou, J. I. Yamaki, *J. Power Sources*, **146**, 565–569 (2005).

59. K. Mizushima, P. C. Jones, P. J. Wiseman, J. B. Goodenough, *Mater. Res. Bull.*, **15**, 783–789 (1980).

60. M. G. S. R. Thomas, W. I. F. David, J. B. Goodenough, P. Groves, *Mater. Res. Bull.*, **20**, 1137–1146 (1985).

61. J. Reed, G. Ceder, *Chem. Rev.*, **104**, 4513–4534 (2004).

62. T.-H. Cho, Y. Shiosaki, H. Noguchi, *J. Power Sources*, **159**, 1322 (2006).

63. J. P. B. Ammundsen, *Adv. Mater.*, **13**, 943 (2001).

64. E. Rossen, C. D. W. Jones, J. R. Dahn, *Solid State Ionics*, **57**, 311–318 (1992).

65. M. E. Spahr, P. Novak, B. Schynder, O. Haas, R. Nesper, *J. Electrochem. Soc.*, **145**, 1113 (1998).

66. Y. Makimura, T. Ohzuku, *J. Power Sources*, **119–121**, 156 (2003).

67. N. Yabuuchi, T. Ohzuku, *J. Power Sources*, **119–121**, 171 (2003).

68. K. M. Shaju, G. V. Subba Rao, B. V. R. Chowdari, *Electrochim. Acta*, **48**, 145 (2002).

69. S. Patoux, M. M. Doeff, *Electrochem. Commun.*, **6**, 767–772 (2004).

70. M. S. Islam, R. A. Davies, J. D. Gale, *Chem. Mater.*, **15**, 4280 (2003).

71. B. J. Hwang, Y. W. Tsai, D. Carlier, G. Ceder, *Chem. Mater.*, **15**, 3676 (2003).

72. J. Reed, G. Ceder, *Electrochem. Solid-State Lett.*, **5**, A145 (2002).

73. C. S. Johnson, J.-S. Kim, A. J. Kropf, A. J. Kahaian, J. T. Vaughey, L. M. L. Fransson, K. Edstrom, M. M. Thackeray, *Chem. Mater.*, **15**, 2313 (2003).

74. W.-S. Yoon, Y. Paik, X.-Q. Yang, M. Balasubramanian, J. McBreen, C. P. Grey, *Electrochem. Solid-State Lett.*, **5**, A263 (2002).

75. S. Venkatram, A. Manthiram, *Chem. Mater.*, **15**, 5003 (2003).

76. Y. Sun, C. Ouyang, Z. Wang, X. Huang, L. Chen, *J. Electrochem. Soc.*, **151**, A504 (2004).

77. I. Belharouak, W. Q. Lu, D. Vissers, K. Amine, *Electrochem. Commun.*, **8**, 329 (2006).

78. Y. Chen, G. X. Wang, K. Konstantinov, H. K. Liu, S. X. Dou, *J. Power Sources*, **119–121**, 184 (2003).

79. H. Ito, M. Shimakawa, T. Tanaka, T. Ohzuku, *5th Hawaii Battery Conference*, Waikaloa, HI (2003).

80. J. Wilcox, S. Patoux, M. M. Doeff, *J. Electrochem. Soc.*, **156**, A192–A198 (2009).
81. C. Li, H. P. Zhang, L. J. Fu, H. Liu, Y. P. Wu, E. Rahm, R. Holze, H. Q. Wu, *Electrochim. Acta*, **51**, 3872–3883 (2006).
82. Y. K. Sun, S. T. Myung, M. H. Kim, J. Prakash, K. Amine, *J. Am. Chem. Soc.*, **127**, 13411–13418 (2005).
83. J. Paulsen, J. S. Jeong, K. Y. Lee, *Electrochem. Solid-State Lett.*, **10**, A101–A105 (2007).
84. M. M. Thackeray, *Prog. Solid State Chem.*, **25**, 1 (1997).
85. R. J. Gummow, D. C. Liles, M. M. Thackeray, *Mater. Res. Bull.*, **28**, 1249 (1993).
86. I. Koetschau, M. N. Richard, J. R. Dahn, J. B. Soupart, J. C. Rousche, *J. Electrochem. Soc.*, **142**, 2906 (1995).
87. F. Capitaine, P. Gravereau, C. Delmas, *Solid State Ionics*, **89**, 499 (1996).
88. A. R. Armstrong, P. G. Bruce, *Nature*, **381**, 499 (1996).
89. J. Reed, G. Ceder, A. Van der Ven, *Electrochem. Solid-State Lett.*, **4**, A78 (2001).
90. J. Reed, *Ab Initio Study of Cathode Materials for Lithium Batteries*, Ph.D. Thesis, Massachusetts Institute of Technology, Cambridge, MA (2003).
91. B. Ammundsen, J. Paulsen, *Adv. Mater.*, **13**, 943 (2001).
92. M. H. Rossouw, M. M. Thackeray, *Mater. Res. Bull.*, **26**, 463–473 (1991).
93. D. Y. W. Yu, K. Yanagida, Y. Kato, H. Nakamura, *J. Electrochem. Soc.*, **156**, A417–A424 (2009).
94. K. Numata, C. Sakaki, S. Yamanaka, *Chem. Lett.*, **8**, 725–726 (1997).
95. B. Ammundsen, J. Paulsen, I. Davidson *et al.*, *J. Electrochem. Soc.*, **149**, A431–A436 (2002).
96. J. Jiang, K. W. Eberman, L. J. Krause, J. R. Dahn, *J. Electrochem. Soc.*, **152**, A1879–A1889 (2005).
97. Z. Lu, D. D. MacNeil, J. R. Dahn, *Electrochem. Solid-State Lett.*, **4**, A191–A194 (2001).
98. M. Tabuchi *et al.*, *J. Electrochem. Soc.*, **154**, A638–A648 (2007).
99. N. Tran, L. Croguennec, C. Labrugère, C. Jordy, P. Biensan, C. Delmas, *J. Electrochem. Soc.*, **153**, A261–A269 (2006).
100. N. Tran, L. Croguennec, M. Ménétrier, F. Weill, P. Biensan, C. Jordy, C. Delmas, *Chem. Mater.*, **20**, 4815–4825 (2008).

101. J. M. Kim, N. Kumagai, H. T. Chung, *Electrochem. Solid-State Lett.*, **9**, A494–A498 (2006).

102. J. M. Kim, N. Kumagai, T. H. Cho, *J. Electrochem. Soc.*, **155**, A82–A86 (2008).

103. Y. Wu, A. Manthiram, *Electrochem. Solid State Lett.*, **9**, A221–A224 (2006).

104. Y. Wu, A. Manthiram, *J. Power Sources*, **183**, 749–754 (2008).

105. J. S. Kim, C. S. Johson, M. M. Thackeray, *Electrochem. Commun.*, **4**, 205–209 (2002).

106. M. M. Thackeray, S.-H. Kang, C. S. Johnson, J. T. Vaughey, R. Benedek, S. A. Hackney, *J. Mater. Chem.*, **30**, 3112–3125 (2007).

107. S. H. Kang, M. M. Thackeray, *J. Electrochem. Soc.*, **155**, A269–A275 (2008).

108. S. H. Kang, K. Amine, *J. Power Sources*, **124**, 533–537 (2003).

109. J. H. Lim, H. Bang, Ki-Soo Lee, K. Amine, Y. K. Sun, *J. Power Sources*, **189**, 571–575 (2009).

110. J.-S. Kim, C. S. Johnson, M. M. Thackeray, *Electrochem. Commun.*, **4**, 205–209 (2002).

111. C. S. Johnson, *et al.*, *Electrochem. Commun.*, **9**, 787–795 (2007).

112. C. S. Johnson, N. Li, C. Lefief, M. M. Thackeray, *Electrochem. Commun.*, **9**, 787–795 (2007).

113. J. Gao, A. Manthiram, *J. Power Sources*, **191**, 644–647 (2009).

114. S. H. Kang, M. M. Thackeray, *Electrochem. Commun.*, **11**, 748–751 (2009).

115. M. M. Thackeray, W. I. F. David, P. G. Bruce, J. B. Goodenough, *Mater. Res. Bull.*, **18**, 461–472 (1983).

116. M. M. Thackeray, P. J. Johnson, L. A. DePicciotto, P. G. Bruce, J. B. Goodenough, *Mater. Res. Bull.*, **19**, 179–187 (1984).

117. R. J. Gummow, A. D. Kock, M. M. Thackeray, *Solid Sate Ionics*, **69**, 59 (1994).

118. G. Rousse, "Ordre de charges, Magnétisme et Electrochimie du spinelle $LiMn_2O_4$", *Thèse de Doctorat*, University Paris XI (2000).

119. J. M. Tarascon, E. Wang, F. K. Shokoohi, W. R. McKinnon, S. Colson, *J. Electrochem. Soc.*, **138**, 2859–2864 (1991).

120. K. Amine, H. Tukamoto, H. Yasuda, Y. Fujita, *J. Electrochem. Soc.*, **143**, 1607–1613 (1996).

121. K. Amine, H. Tukamoto, H. Yasuda, Y. Fujita, *J. Power Sources*, **68**, 604–608 (1997).

122. Q. Zhong, A. Bonakdarpour, M. Zhang, Y. Gao, J. R. Dahn, *J. Electrochem. Soc.*, **144**, 205–213 (1997).

123. S. Vivekanandhan, M. Venkateswarlu, N. Satyanarayana, *Mater. Lett.*, **58**, 1218–1222 (2004).

124. G. T. K. Fey, W. Li, J. R. Dahn, *J. Electrochem. Soc.*, **141**, 2279 (1994).

125. S. Patoux, L. Sannier, H. Lignier, Y. Reynier, C. Bourbon, S. Jouanneau, F. Le Cras, S. Martinet, *Electrochim. Acta*, **53**, 4137–4145 (2008).

126. S. Patoux, L. Daniel, C. Bourbon, H. Lignier, C. Pagano, F. Le Cras, S. Jouanneau, S. Martinet, *J. Power Sources*, **189**, 344–352 (2009).

127. D. Gryffroy, R. E. Vandenberghe, E. Legrand, *Mater. Sci. Forum*, **79–82**, 785–790 (1991).

128. K. Takahashi, M. Saitoh, M. Sano, M. Fujita, K. Kifune, *J. Electrochem. Soc.*, **151**, A173–A177 (2004).

129. J.-H. Kim, S.-T. Myung, C. S. Yoon, S. G. Kang, Y.-K. Sun, *Chem. Mater.*, **16**, 906–914 (2004).

130. M. Aklalouch, J. Manuel Amarilla, R. M. Rojas, I. Saadoune, J. M. Rojo, *J. Power Sources*, **185**, 501–511 (2008).

131. X. Zhang, J. K. Pugh, P. N. Ross, *J. Electrochem. Soc.*, **148**, E183–E188 (2001).

132. K. Kanamura, S. Toriyama, S. Shiraishi, M. Ohashi, Z. Takehara, *J. Electroanal. Chem.*, **419**, 77 (1996).

133. K. Kanamura, T. Umegaki, M. Ohashi, S. Toriyama, S. Shiraishi, Z. Takehara, *Electrochim. Acta*, **47**, 433 (2001).

134. D. Aurbach, B. Markovsky, A. Rodkin, E. Levi, Y. S. Cohen, H.-J. Kim, M. Schmidt, *Electrochim. Acta*, **47**, 4291 (2002).

135. A. Würsig, H. Buqa, M. Holzapfel, F. Krumeich, P. Novák, *Electrochem. Solid-State Lett.*, **8**, A34 (2005).

136. K. Abe, Y. Ushigoe, H. Yoshitake, M. Yoshio, *J. Power Sources*, **153**, 328 (2006).

137. S. S. Zhang, *J. Power Sources*, **162**, 1379 (2006).

138. H. Y. Xu, S. Xie, Q. Y. Wang, X. L. Yao, Q. S. Wang, C. H. Chen, *Electrochim. Acta*, **52**, 636 (2006).

139. K. Abe, T. Takaya, H. Yoshitake, Y. Ushigoe, M. Yoshio, H. Wang, *Electrochem. Solid-State Lett.*, **7**, A462 (2004).

140. A. Thenappan, D. J. Burton, *J. Fluorine Chem.*, **77**, 45 (1996).

141. J. I. Yamaki, T. Nakajima, H. Groult (Eds.), *Fluorinated Materials for Energy Conversion*, Elsevier, pp. 267 (2005).

142. M. Hasegawa, H. Ishii, Y. Cao, T. Fuchigami, *J. Electrochem. Soc.*, **153**, D162 (2006).

143. M. Hasegawa, T. Fuchigami, *Electrochim. Acta*, **49**, 3367 (2004).

144. M. Hasegawa, H. Ishii, T. Fuchigami, *Tetrahedron Lett.*, **43**, 1503 (2002).

145. M. Kobayashi, T. Inoguchi, T. Iida, T. Tanioka, H. Kumase, Y. Fukai, *J. Fluorine Chem.*, **120**, 105 (2003).

146. Q. Wang, P. Pechy, S. M. Zakeeruddin, I. Exnar, M. Gratzel, *J. Power Sources*, **146**, 813 (2005).

147. T. Nakajima, K. Dan, M. Koh, *J. Fluorine Chem.*, **87**, 221 (1998).

148. J. I. Yamaki, I. Yamazaki, M. Egashira, S. Okada, *J. Power Sources*, **102**, 288 (2001).

149. P. Yazami, R. Touzain, *J. Power Sources*, **9**(3–4) 365–371 (1983).

150. G. Li, H. Imoto, S. Yamada, T. Endo, H. Tanizaki. Material for negative electrode and nonaqueous-electrolyte battery incorporating the same. *US Patent No. 6300013* (2001).

151. Y. Nishi, *J. Power Sources*, **100**(1–2), 101–106 (2001).

152. U. Kasavajjula, C. Wang, A. J. Appleby, *J. Power Sources*, **163**(2), 1003–1039 (2007).

153. R. Alcántara, J. L. Tirado, J. C. Jumas, L. Monconduit, J. Olivier-Fourcade, *J. Power Sources*, **109**(2), 308–312 (2002).

154. H. Pfei-er, F. Tancret, M. P. Bichat, L. Monconduit, F. Favier, T. Brousse, *Electrochem. Commun.*, **6**(3), 263–267 (2004).

155. M. P. Bichat, T. Politova, J. L. Pascal, F. Favier, L. Monconduit, *J. Electrochem. Soc.*, **151**, A2074 (2004).

156. P. Poizot, S. Laruelle, S. Grugeon, L. Dupont, J. M. Tarascon, *Nature*, **407**, 496–499 (2000).

157. A. N. Dey, *J. Electrochem. Soc.*, **118**, 1547 (1971).

158. D. Fauteux, R. Koksbang, *J. Appl. Electrochem.*, **23**(1), 1–10 (1993).

159. R. A. Huggins, *J. Power Sources*, **81**, 13–19 (1999).

160. Y. Idota, T. Kubota, A. Matsufuji, Y. Maekawa, T. Miyasaka, *Science*, **276**(5317), 1395 (1997).

161. T. Brousse, R. Retoux, U. Herterich, D. M. Schleich, *J. Electrochem. Soc.*, **145**(1), 1–4 (1998).

162. J. Wolfenstine, J. L. Allen, J. Read, D. Foster, Army research lab Adelphi MD sensors and electron devices directorate. Chemistry and Structure of Sony's *Nexelion Li-ion Electrode Materials* (Progress report), 2006.

163. J. Hassoun, G. Mulas, S. Panero, B. Scrosati, *Electrochem. Commun.*, **9**(8), 2075–2081 (2007).

164. D. Larcher, S. Beattie, M. Morcrette, K. Edström, J. C. Jumas, J. M. Tarascon, *J. Mater. Chem.*, **17**(36), 3759–3772 (2007).

165. J. P. Maranchi, A. F. Hepp, P. N. Kumta, *Electrochem. Solid-State Lett.*, **6**, A198 (2003).

166. B. Gao, S. Sinha, L. Fleming, O. Zhou, *Adv. Mater.*, **13**(11), 816–819 (2001).

167. P. G. Bruce, *Solid State Ionics*, **179**, 752–760 (2008).

168. E. Ferg, R. J. Gummow, A. De Kock, M. M. Thackeray, *J. Electrochem. Soc.*, 141(11), L147–L150 (1994).

169. A. N. Jansen, A. J. Kahaian, K. D. Kepler, P. A. Nelson, K. Amine, D. W. Dees, D. R. Vissers, M. M. Thackeray, *J. Power Sources*, **81**, 902–905 (1999).

170. M. Dion, Y. Piffard, M. Tournoux, *J. Inorg. Nucl. Chem.*, **40** (1978) 917–918.

171. T. Ohzuku, A. Ueda, N. Yamamoto, *J. Electrochem. Soc.*, **142**, 1431–1435 (1995).

172. K. Nakahara, R. Nakajima, T. Matsushima, H. Majima, *J. Power Sources*, **117**, 131–136 (2003).

173. R. Marchand, L. Brohan, M. Tournoux, *Mater. Res. Bull.*, **15**, 1129–1133 (1980).

174. L. Brohan, R. Marchand, *Solid State Ionics*, **9 & 10**, 419–424 (1983).

175. M. Tournoux, R. Marchand, L. Brohan, *Prog. Solid State Chem.*, **17**, 33–52 (1986).

176. R. A. Sharma, R. N. Seefurth, *J. Electrochem. Soc.*, **123**, 1763 (1976).

177. T. B. Massalski, H. Okamoto, P. R. Subramanian, L. Kacprzak (eds.), *Binary Alloy Phase Diagrams*. ASM International, Ohio (1990).

178. D. Billaud, E. McRae, A. Hérold, *Mater. Res. Bull.*, **14**, 857 (1979).

179. J. Li, J. R. Dahn, *J. Electrochem. Soc.*, **154**, A156 (2007).

180. H. Li, X. Huang, L. Chen, G. Zhou, Z. Zhang, D. Yu, Y. Jun Mo, N. Pei, *Solid State Ionics*, **135**(1–4), 181–191 (2000).

181. P. Limthongkul, Y. I. Jang, N. J. Dudney, Y. M. Chiang, *Acta Mater.*, **51**(4), 1103–1113 (2003).

182. M. N. Obrovac, L. Christensen, *Electrochem. Solid-State Lett.*, **7**, A93 (2004).

183. M. N. Obrovac, L. J. Krause, *J. Electrochem. Soc.*, **154**, A103 (2007).

184. T. L. Kulova, Y. V. Pleskov, A. M. Skundin, E. I. Terukov, O. I. Konkov, *Russ. J. Electrochem.*, **42**(7), 708–714 (2006).

185. H. Jung, M. Park, Y. G. Yoon, G. B. Kim, S. K. Joo, *J. Power Sources*, **115**(2), 346–351 (2003).

186. J. Wolfenstine, *J. Power Sources*, **79**(1), 111–113 (1999).

187. W. R. Liu, Z. Z. Guo, W. S. Young, D. T. Shieh, H. C. Wu, M. H. Yang, N. L. Wu, *J. Power Sources*, **140**(1), 139–144 (2005).

188. I. Sandu, P. Moreau, D. Guyomard, T. Brousse, L. Roué, *Solid State Ionics*, **178**(21–22), 1297–1303 (2007).

189. M. Holzapfel, H. Buqa, F. Krumeich, P. Novak, F. M. Petrat, C. Veit, *Electrochem. Solid-State Lett.*, **8**, A516 (2005).

190. M. Holzapfel, H. Buqa, W. Scheifele, P. Novák, F. M. Petrat, *Chem. Commun.*, **12**, 1566–1568 (2005).

191. M. Holzapfel, H. Buqa, L. J. Hardwick, M. Hahn, A. Würsig, W. Scheifele, P. Novák, R. Kötz, C. Veit, F. M. Petrat, *Electrochim. Acta*, **52**(3), 973–978 (2006).

192. M. Alias, *Composés à base de silicium ultra-divisé pour électrodes négatives d'accumulateur lithium-ion*, Ph.D. Thesis, Université de Nantes, France (2008)

193. S. Bourderau, T. Brousse, D. M. Schleich, *J. Power Sources*, **81**, 233–236 (1999).

194. S. Ohara, J. Suzuki, K. Sekine, T. Takamura, *J. Power Sources*, **136**(2), 303–306 (2004).

195. J. Graetz, C. C. Ahn, R. Yazami, B. Fultz, *Electrochem. Solid-State Lett.*, **6**, A194 (2003).

196. J. Xie, G. S. Cao, X. B. Zhao, *Mater. Chem. Phys.*, **88**(2–3), 295–299 (2004).

197. I. C. Kim, D. Byun, J. K. E. E. Lee, *J. Electroceram.*, **17**(2–4), 661–665 (2006).

198. M. Noel, V. Suryanarayanan, *J. Power Sources*, **111**(2), 193–209 (2002).

199. I. S. Kim, P. N. Kumta, *J. Power Sources*, **136**(1), 145–149 (2004).

200. P. Zuo, Z. Wang, G. Yin, D. Jia, X. Cheng, C. Du, P. Shi, *J. Mater. Sci.*, **43**(9), 3149–3152 (2008).

201. C. S. Wang, G. T. Wu, X. B. Zhang, Z. F. Qi, W. Z. Li, *J. Electrochem. Soc.*, **145**(8) 2751–2758 (1998).
202. B. C. Kim, H. Uono, T. Sato, T. Fuse, T. Ishihara, M. Senna, *Solid State Ionics*, **172**(1–4), 33–37 (2004).
203. H. Y. Lee, S. M. Lee, *Electrochem. Commun.*, **6**(5), 465–469 (2004).
204. H. Uono, B. C. Kim, T. Fuse, M. Ue, J. Yamaki, *J. Electrochem. Soc.*, **153**, A1708 (2006).
205. X. Yang, Z. Wen, X. Xu, B. Lin, S. Huang, *J. Power Sources*, **164**(2), 880–884 (2007).
206. V. G. Khomenko, V. Z. Barsukov, *Electrochim. Acta*, **52**(8), 2829–2840 (2007).
207. N. Dimov, S. Kugino, M. Yoshio, *J. Power Sources*, **136**(1), 108–114 (2004).
208. G. X. Wang, J. Yao, H. K. Liu, *Electrochem. Solid-State Lett.*, **7**, A250 (2004).
209. N. Dimov, Y. Xia, M. Yoshio, *J. Power Sources*, **171**(2), 886–893 (2007).
210. L. Y. Beaulieu, K. W. Eberman, R. L. Turner, L. J. Krause, J. R. Dahn, *Electrochem. Solid-State Lett.*, **4**, A137 (2001).
211. C. K. Chan, H. Peng, G. Liu, K. McIlwrath, X. F. Zhang, R. A. Huggins, Y. Cui, *Nature*, **3**, 31–35 (2008).
212. B. Laïk, L. Eude, J. P. Pereira-Ramos, C. S. Cojocaru, D. Pribat, E. Rouvière, *Electrochim. Acta*, **53**(17), 5528–5532 (2008).
213. T. Kim, S. Park, S. M. Oh, *J. Electrochem. Soc.*, **154**, A1112 (2007).
214. M. Morita, T. Ohmi, E. Hasegawa, M. Kawakami, M. Ohwada, *J. Appl. Phys.*, **68**, 1272 (1990).
215. Y. M. Lee, J. Y. Lee, H. T. Shim, J. K. Lee, J. K. Park, *J. Electrochem. Soc.*, **154**, A515 (2007).
216. A. M. Andersson, K. Edström, *J. Electrochem. Soc.*, **148**, A1100 (2001).
217. H. Buqa, M. Holzapfel, F. Krumeich, C. Veit, P. Novák, *J. Power Sources*, **161**(1), 617–622 (2006).
218. W. R. Liu, M. H. Yang, H. C. Wu, S. M. Chiao, N. L. Wu, *Electrochem. Solid-State Lett.*, **8**, A100 (2005).
219. B. Lestriez, S. Bahri, I. Sandu, L. Roué, D. Guyomard, *Electrochem. Commun.*, **9**(12), 2801–2806 (2007).
220. N. S. Hochgatterer, M. R. Schweiger, S. Koller, P. R. Raimann, T. Wöhrle, C. Wurm, M. Winter, *Electrochem. Solid-State Lett.*, **11**, A76 (2008).

221. J. Li, R. B. Lewis, J. R. Dahn, *Electrochem. Solid-State Lett.*, **10**, A17 (2007).

222. R. K. B. Gover, J. R. Tolchard, H. Tukamoto, T. Murai, J. T. S. Irvine, *J. Electrochem. Soc.*, **146**, 4348–4353 (1999).

223. J. Kim, J. Cho, *J. Electrochem. Soc.*, **154**(6), A542–A546 (2007).

224. J. Kim, J. Cho, *Electrochem. Solid State Lett.*, **10**(3), A81–A84 (2007).

225. G. G. Amatucci, F. Badway, A. Du Pasquier, T. Zheng, *J. Electrochem. Soc.*, **148**, A930–A939 (2001).

226. T. Brousse, R. Marchand, P-L. Taberna, P. Simon, *J. Power* Sources, **158**, 571–577 (2006).

227. S. Jouanneau, F. LeCras, C. Bourbon, H. Lignier, Commissariat à l' Energie Atomique, *French Patent Application No.* FR 2 874 603; WO 2006/027449.

228. R. Marchand, L. Brohan, M. Tournoux, *Mater. Res. Bull.*, **15**, 1129–1133 (1980).

229. Amit Singhal, Ganesh Skandan. Nanostructured $Li_4Ti_5O_{12}$ powders and method of making the same. *US Patent No. US6827921B1* (2002).

230. Jinyoung Kim, Jaephil Cho, *Electrochem. Solid-State Lett.*, **10**(3), A81–A84 (2007).

231. A. R. Armstrong, G. Armstrong, J. Canales, P. G. Bruce, *J. Power Sources*, **146**(1–2), 501–506 (2005).

232. A. R. Armstrong, G. Armstrong, J. Canales, R. Garcia, P. G. Bruce, *Adv. Mater.*, **17**(7), 862–865 (2005).

233. B. Zachau-Christiansen, K. West, T. Jacobsen, S. Atlung, *Solid State Ionics*, **28–30**, 1176–1182 (1988).

234. G. Nuspl, K. Yoshizawa, T. Yamabe, *J. Mater. Chem.*, **7**, 2529–2536 (1997).

235. H. Kawamura, Y. Muranishi, T. Miura, T. Kishi, *Denki Kagaku*, **59**, 766–772 (1991).

236. B. Zachau-Christiansen, K. West, T. Jacobsen, S. Skaarup, *Solid State Ionics*, **53–56**, 364–369 (1992).

237. A. R. Armstrong, G. Armstrong, J. Canales, P. G. Bruce, *Angew. Chem., Int. Ed.*, **43**, 2286–2288 (2004).

238. M. Zukalovă, M. Kalbăč, L. Kavan, I. Exnar, M. Graetzel, *Chem. Mater.*, **17**, 1248–1255 (2005).

239. L. J. Fu, H. Liu, H. P. Zhang, C. Li, T. Zhang, Y. P. Wu, H. Q. Wu., *J. Power Sources*, **159**, 219–222 (2006).

240. Hui Qiao, Yawen Wang, Lifen Xiao, Lizhi Zhang, *Electrochem. Commun.*, **10**, 1280–1283 (2008).
241. Same as Ref. 250
242. Liang Cheng, Hai-Jing Liu, Jing-Jun Zhang, Huan-Ming Xiong, Yong-Yao Xia., *J. Electrochem. Soc.*, **153**(8), A1472–A1477 (2006).
243. Kye Sang Yoo, Tai Gyu Lee, Jinsoo Kim., *Micropor. Mesopor. Mat.*, **84**, 211–217 (2005).
244. G. Armstrong, A. R. Armstrong, J. Canales, P. G. Bruce, *Chem. Commun.*, **19**,2454–2456 (2005).
245. B. Kang, G. Ceder, *Nature*, **458**, 190–193 (2009).
246. M. Gaberscek, *J. Power Sources*, **189**, 22–27 (2009).
247. W. Porcher *et al.*, *J. Electrochem. Soc.*, **156**(3), A133 (2009).
248. H. Yang *et al.*, *J. Power Sources*, **161**, 573–579 (2006).
249. M. Kerlau *et al.*, *J. Electrochem. Soc.*, **153**(9), A1644 (2006).
250. Y. M. Kang *et al.*, *Electrochem. Commun.*, **9**, 959 (2007).
251. S. D. Beattie *et al.*, *J. Electrochem. Soc.*, **155**(2), A158 (2008).
252. X. Li *et al.*, *Carbon*, **44**, 1298 (2006).
253. F. Joho *et al.*, *J. Power Sources*, **78** (2001).
254. M. S. Wu *et al.*, *J. Mater. Sci.*, **42**, 259 (2007).
255. J. Graetz *et al.*, *J. Electrochem. Soc.*, **151**(5), A698 (2004).

Index

www.ingramcontent.com/pod-product-compliance
Ingram Content Group UK Ltd.
Pitfield, Milton Keynes, MK11 3LW, UK
UKHW021502240926
475702UK00035B/119
* 9 7 8 9 8 1 4 3 1 6 4 0 8 *